NEUROSCIENCE DATABASES
A Practical Guide

NEUROSCIENCE DATABASES
A Practical Guide

edited by

Rolf Kötter
Heinrich Heine University Düsseldorf,
Düsseldorf, Germany

KLUWER ACADEMIC PUBLISHERS
Boston / Dordrecht / London

Distributors for North, Central and South America:
Kluwer Academic Publishers
101 Philip Drive
Assinippi Park
Norwell, Massachusetts 02061 USA
Telephone (781) 871-6600
Fax (781) 681-9045
E-Mail: kluwer@wkap.com

Distributors for all other countries:
Kluwer Academic Publishers Group
Post Office Box 322
3300 AH Dordrecht, THE NETHERLANDS
Telephone 31 786 576 000
Fax 31 786 576 474
E-Mail: services@wkap.nl

Electronic Services < http://www.wkap.nl>

Library of Congress Cataloging-in-Publication Data

A C.I.P. Catalogue record for this book is available
from the Library of Congress.

NEUROSCIENCE DATABASES A Practical Guide / edited by Rolf Kötter
ISBN 1-4020-7165-5

Printed on acid-free paper.

Printed in the United States of America.

Contents

Contributors

Jamila Ahdidan
Max Planck Institute of Experimental Endocrinology, Feodor-Lynen-Str. 7, D-30165 Hannover, Germany.

Zeynep F. Altun
Albert Einstein College of Medicine, Department of Neuroscience, 1410 Pelham Parkway South, R#614 Bronx, NY 10461-1101, USA.

Michael A. Arbib
Dept. of Neurobiology, University of Southern California, Los Angeles, CA 90088, USA.

Giorgio A. Ascoli
Computational Neuroanatomy, Krasnow Institute for Advanced Study and Psychology, George Mason University, Fairfax, VA, USA.

Javed Aslam
Computer Science Department, Dartmouth College, 6162 Moore Hall, Hanover, NH 03755, USA.

Kim. T. Blackwell
Krasnow Institute of Advanced Studies and School of Computational Studies, George Mason University, Fairfax, VA 22030, USA.

Mihail Bota
Dept. of Neurobiology, University of Southern California, Los Angeles, CA 90088, USA.

Thomas Boulin
Columbia University, Department of Biochemistry and Molecular Biophysics, 630 West 168th Street, New York, NY 1003, USA.

Douglas M. Bowden
Department of Psychiatry and Behavioral Sciences and Washington National Primate Research Center, University of Washington, Seattle, WA, USA.

Ahmet Bozkurt
Computational | Systems | Neuroscience Group, C. & O. Vogt Brain Research Institute, Heinrich Heine University, D-40225 Düsseldorf, Germany.

Gully A.P.C Burns
USC, Hedco Neurosci Bldng, 3614 Watt Way, Los Angeles, CA 90089, USA.

Robert C. Cannon
Theoretical Biology, Born-Bunge Foundation, University of Antwerp, Wilrijk, Belgium.

Nansheng Chen
Cold Spring Harbor Laboratory, 1 Bungtown Road, Cold Spring Harbor, NY 11724, USA.

Chiquito J. Crasto
Yale Center for Medical Informatics, Yale University School of Medicine, New Haven, CT, USA.

Andrew P. Davison
Department of Neurobiology, Yale University School of Medicine, New Haven, CT, USA.

Michael S. Gazzaniga
The fMRI Data Center, Center for Cognitive Neuroscience, Dartmouth College, 6162 Moore Hall, Hanover, NH 03755, USA.

Jeffrey S. Grethe
The fMRI Data Center, Dartmouth College, 6162 Moore Hall, Hanover, NH 03755, USA.

Jonas D.-Johnsen
Computational | Systems | Neuroscience Group, C. & O. Vogt Brain Research Institute, Heinrich Heine University, D-40225 Düsseldorf, Germany.

Mark Dubach
Department of Psychiatry and Behavioral Sciences and Washington National Primate Research Center, University of Washington, Seattle, WA, USA.

Gregor Eichele
Max Planck Institute of Experimental Endocrinology, Feodor-Lynen-Str. 7, D-30165 Hannover, Germany.

Mark H. Ellisman
National Center for Microscopy and Imaging Research and Department of Neurosciences, and San Diego Supercomputer Center, University of California, San Diego, San Diego CA, USA.

Rainer Förster
Institute for Theoretical Biology, Humboldt-Universität zu Berlin, Invalidenstraße 43, D-10115 Berlin, Germany.

Karl J. Friston
Wellcome Dept. of Imaging Neuroscience, UCL, UK.

Andreas Geissler
Computational | Systems | Neuroscience Group, C. & O. Vogt Brain Research Institute, Heinrich Heine University, D-40225 Düsseldorf, Germany.

Ina Gerken
Computational | Systems | Neuroscience Group, C. & O. Vogt Brain Research Institute, Heinrich Heine University, D-40225 Düsseldorf, Germany.

Seth G.N. Grant
Centre for Neuroscience, University of Edinburgh, Edinburgh, EH8 9JZ, UK.

Jo Grothe
Department of Psychiatry III, University of Ulm, D-89075 Ulm, Germany.

Amarnath Gupta
San Diego Supercomputer Center, University of California, San Diego, San Diego CA, USA.

Jeanette Hellgren Kotelski
Royal Institute of Technology and Karolinska Institute, S100-44 Stockholm, Sweden.

Andreas V.M. Herz
Institute for Theoretical Biology, Humboldt-Universität zu Berlin, Invalidenstraße 43, D-10115 Berlin, Germany.

Claus C. Hilgetag
International University Bremen, Campus Ring 1, D-28759 Bremen, Germany.

Michael L. Hines
Department of Neurobiology, Yale University School of Medicine, New Haven, CT, USA.

Holger Husi
Centre for Neuroscience, University of Edinburgh, Edinburgh, EH8 9JZ, UK.

Lars Kamper
Computational | Systems | Neuroscience Group, C. & O. Vogt Brain Research Institute, Heinrich Heine University, D-40225 Düsseldorf, Germany.

Rolf Kötter
Computational | Systems | Neuroscience Group, C. & O. Vogt Brain Research Institute, and Institute of Anatomy II, Heinrich Heine University, D-40225 Düsseldorf, Germany.

Raymond Y.N. Lee
Howard Hughes Medical Institute, Division of Biology, California Institute of Technology, Pasadena, CA 91125, USA.

Nian Liu
Yale Center for Medical Informatics, Yale University School of Medicine, New Haven, CT, USA.

Bertram Ludäscher
San Diego Supercomputer Center, University of California, San Diego, San Diego CA, USA.

Wolfgang Maass
Institute for Theoretical Computer Science, Technische Universität Graz, Graz, Austria.

Jürgen Maier
Computational | Systems | Neuroscience Group, C. & O. Vogt Brain Research Institute, Heinrich Heine University, D-40225 Düsseldorf, Germany.

Luis Marenco
Center for Medical Informatics, Yale University School of Medicine, New Haven, CT, USA.

Henry Markram
Brain Mind Institute, Ecole Polytechnique Federale de Lausanne, Lausanne, Switzerland

Maryann E. Martone
National Center for Microscopy and Imaging Research and Department of Neurosciences, University of California, San Diego, San Diego CA, USA.

Michele Migliore
Department of Neurobiology, Yale University School of Medicine, New Haven, CT, USA, and Institute of Advanced Diagnostic Methodologies, National Research Council, Palermo, Italy.

Mark Montague
The fMRI Data Center, Computer Science Department, Dartmouth College, 6162 Moore Hall, Hanover, NH 03755, USA.

Thomas M. Morse
Department of Neurobiology, Yale University School of Medicine, New Haven, CT, USA.

Thomas Natschläger
Institute for Theoretical Computer Science, Technische Universität Graz, Graz, Austria.

Mark A. O'Neill
Department of Psychology, University of Newcastle upon Tyne, Newcastle upon Tyne, NE2 4HH, UK.

Stefano Panzeri
Department of Psychology, University of Newcastle upon Tyne, The Henry Wellcome Building for Neuroecology, Medical School, Framlington Place, Newcastle upon Tyne, NE2 4HH, UK.

Rasmus S. Petersen
Cognitive Neuroscience Sector, International School for Advanced Studies, Via Beirut 2/4, 34014 Trieste, Italy.

Gianni Pola
Department of Psychology, University of Newcastle upon Tyne, The Henry Wellcome Building for Neuroecology, Medical School, Framlington Place, Newcastle upon Tyne, NE2 4HH, UK.

Raphael Ritz
Institute for Theoretical Biology, Humboldt-Universität zu Berlin, Invalidenstraße 43, D-10115 Berlin, Germany.

Daniel Rockmore
Computer Science Department, Dartmouth College, 6162 Moore Hall, Hanover, NH 03755, USA.

Daniela Rus
Computer Science Department, Dartmouth College, 6162 Moore Hall, Hanover, NH 03755, USA.

Konrad Rybacki
Computational | Systems | Neuroscience Group, C. & O. Vogt Brain Research Institute, Heinrich Heine University, D-40225 Düsseldorf, Germany.

Volker Schmitt
Department of Neural Information Processing, University of Ulm, D-89069 Ulm, Germany.

Simon R. Schultz
Howard Hughes Medical Institute and Center for Neural Science, New York University, 4 Washington Place, New York, NY 10003, USA.

Gordon M. Shepherd
Section for Neurobiology, Yale University School of Medicine, New Haven, CT, USA.

Geoffrey Simonds
The fMRI Data Center, Dartmouth College, 6162 Moore Hall, Hanover, NH 03755, USA.

Friedrich T. Sommer
Department of Neural Information Processing, University of Ulm, D-89069 Ulm, Germany.

Olaf Sporns
Department of Psychology, Indiana University, Bloomington, IN 47405, USA.

Lincoln D. Stein
Cold Spring Harbor Laboratory, 1 Bungtown Road, Cold Spring Harbor, NY 11724, USA.

Klaas E. Stephan
C. & O. Vogt Brain Research Institute, Heinrich Heine University, D-40225 Düsseldorf, and IME, Research Centre Jülich, D-52425 Jülich, Germany.

Paul W. Sternberg
Howard Hughes Medical Institute, Division of Biology, California, Institute of Technology, Pasadena, CA 91125, USA.

Dennis A. Turner
Neurosurgery and Neurobiology, Duke University Medical Center and Durham VAMC, Durham, NC, USA.

Bennet Vance
The fMRI Data Center, Dartmouth College, 6162 Moore Hall, Hanover, NH 03755, USA.

John Darrell Van Horn
The fMRI Data Center, Center for Cognitive Neuroscience, Dartmouth College, 6162 Moore Hall, Hanover, NH 03755, USA.

Axel Visel
Max Planck Institute of Experimental Endocrinology, Feodor-Lynen-Str. 7, D-30165 Hannover, Germany.

Andrzej Wichert
Department of Psychiatry III, University of Ulm, D-89075 Ulm, Germany.

Jeffrey B. Woodward
The fMRI Data Center, Dartmouth College, 6162 Moore Hall, Hanover, NH 03755, USA.

Ilya Zaslavsky
San Diego Supercomputer Center, University of California, San Diego, San Diego CA, USA.

Acknowledgements

I thank Birgit Kowohl for her rigorous organization and her incessant efforts to improve the formatting of the book chapters. The contributors showed admirable enthusiasm and compliance going through repeated revisions to improve contents and format of the chapters within a very tight time schedule. They also made many valuable comments and suggestions after cross-reading the chapters of other contributors. C. Opfermann-Rüngeler composed the cover image. I am grateful to the members of the Computational | Systems | Neuroscience group at the C. & O. Vogt Brain Research Institute for their unanimous support; and to the institute director, Professor K. Zilles, as well as Professor H.-G. Hartwig, director of the Institute of Anatomy II, Heinrich Heine University Düsseldorf, for leaving me the freedom to tackle a book project between research work and teaching duties. Joanne Tracy and Dianne Wuori of Kluwer Academic Publishers gave energetic support to see this book taking shape.

Preface

Neuroscience databases are a young and dynamic field with many developments still under way. Nevertheless, the databasing approach is likely to make a major impact on the way neuroscience is being conducted and perceived, similar to the developments that we have seen in the field of genome databases and bioinformatics. Databasing already gives a new flavour to the term neuroinformatics emphasizing high-throughput technologies for data generation, systematic large-scale data collation and presentation, and the development of computational tools that allow us to extract characteristic features and relationships among ever-growing amounts of data.

The current speed of development of neuroscience databases has the consequence that any textbook aiming to explain how to build, use or appraise neuroscience databases runs the danger of being outdated by the time of its publication. Thus, producing a detailed reference book is not warranted at this stage. What this book attempts is to give a starter on current concepts, implementations and limitations, couched as a "practical guide" to many of the most advanced and ambitious neuroscience database projects that currently exist. The authors were asked to point to web sites where additional information will be available and updated in the future. Where web sites did not exist the authors set them up on this occasion and made software available for download. In this sense creating the book already fostered the growth of the field. Clearly, the book is not meant for standalone reading but is best used in conjunction with the database or software described, which will then most likely provide more and enhanced features than were available at the time of writing.

The book includes not only databases in a narrow sense but extends to software tools that are used to display, analyze or mine the contents of such databases. As far as concepts, applications, prospects and limitations are discussed the contents of this book is inseparable from the field of neuroinformatics / computational neuroscience and lays a basis for the analysis of neuronal systems and the computational modeling of their mechanisms.

The chapters are organized in a loose sequence leading from the classical C. elegans data via molecules and their interactions to single cells and microcircuits from the point of morphology, modeling and information theory. Further contributions address connectivity data and methods for network analysis, functional brain images and their statistical evaluation, atlas-based brain information systems and, last not least, a meta-database of neuroinformatics activities. This organization contrasts divergent approaches to similar problems (for example: chapters 7 and 8 on neuron and microcircuit data for modeling) and brings together data and analysis tools (for example: chapters 11 and 12 on connectivity data and their analysis). A strict partition of the book into parts focusing on molecular, cellular, connectivity and imaging issues did not seem warranted there are many strong cross-links between seemingly unconnected topics (for example: chapters 2 and 11 for the display of density and distribution of label) or the complementary information provided when tackling the same problems from different ends (for example: chapters 11 and 18 for the identification of brain regions).

I hope that this book not only serves the reader but also fosters the field of neuroscience databases, which is on the way to provide essential data for the analysis of complex brain systems at all levels of description. It will become clear from this collection of contributions that much work is still to be done in this direction: There are many important considerations concerning the acquisition, representation, retrieval, display, exchange, evaluation, security etc. of neuroscience data and these will need to be addressed in detail. It is well to keep in mind, however, that databasing is not an end in itself but an important means with the aim to generate insights into the organization and workings of the brain.

Further reading:

Chicurel, M. (2000) Databasing the brain. *Nature* 406, 822-825.

Theme issue "Neuroscience databases: tools for exploring brain structure-function relationships" (2001) *Phil Trans R Soc Lond B* 356(1412), 1107-1339.

http://www.hirnforschung.net/cneuro

Rolf Kötter

Chapter 1

Neuroinformatics for *C. elegans*: Relating Mind and Body in Wormbase

Nansheng Chen[1], Raymond Y.N. Lee[2], Zeynep F. Altun[3], Thomas Boulin[4], Paul W. Sternberg[2] and Lincoln D. Stein[1]

[1]Cold Spring Harbor Laboratory, 1 Bungtown Road, Cold Spring Harbor, NY 11724, USA. [2]Howard Hughes Medical Institute, Division of Biology, California, Institute of Technology, Pasadena, California 91125 USA. [3]Albert Einstein College of Medicine Department of Neuroscience 1410 Pelham Parkway South, R#614 Bronx, NY 10461-1101 USA. [4]Columbia University, Department of Biochemistry and Molecular Biophysics, 630 West 168th Street, New York, NY 10032.

Correspondence to: chenn@cshl.org

Abstract: In this chapter, we have described how neuroscience data for the nematode *C. elegans* are deposited to and accessed from database. Research on neuroscience of *C. elegans* has attracted tremendous attention primarily because of its simple nervous system with 302 neurons. Data generated by comprehensive reconstruction and other efforts were initially deposited to and retrieved from ACeDB. More recently, Wormbase was created, which allows users to access data remotely and easily.

Key words: *C. elegans*, ACeDB, Wormbase, Database, Neuroscience, Wiring, Gene Ontology

1. *C. ELEGANS* NERVOUS SYSTEM: STRUCTURE AND FUNCTION

1.1 Historical Review

The worm, *Caenorhabditis elegans*, together with the bacterium *Escherichia coli*, the yeast *Saccharomyces cerevisiae*, the fruit fly *Drosophila melanogaster*, the plant, *Arabidopsis thaliana*, and the mouse *Mus musculus*, are the most prominent of the model organisms (Alberts, 1997). Although *C. elegans* is widely used in all aspects of biological studies, the initial choice of *C. elegans* as an experimental animal was focused on studying the nervous system. Sydney Brenner, who first recommended *C. elegans* as a model animal, wrote in 1974: "...one concerned with the question of the genetic specification of nervous systems and the other with the way nervous systems work to produce behaviour. Both require that we must have some way of analysing a nervous system...Some eight years ago, when I embarked on this problem, I decided that what was needed was an experimental organism which was suitable for genetics study and in which one could determine the complete structure of the nervous system. *Drosophila*, with about 10^5 neurons, is much too large, and looking for a simpler organism, my choice eventually settled on the small nematode, *Caenorhabditis elegans...*" (Cited in Brenner, 1997).

A large volume of data generated by neuroanatomy reconstruction projects motivated the creation of a database in 1987, when Richard Durbin, Jean Thierry-Mieg and colleagues established a database called ACeDB, (Durbin, 1987). In this chapter, we will review how neuroscience data are deposited in ACeDB and how it is used by the *C. elegans* research community, either directly or indirectly through the WormBase web site (http://www.wormbase.org).

1.2 Early Reconstructions of *C. elegans* nervous system

Studies on the *C. elegans* nervous system were launched in the mid-1960s beginning with a series of electronic microscopic reconstructions of the nervous system. Brenner and others started with a small part of the nervous system, the retrovesicular ganglion (RVG) at the anterior end of the ventral nerve cord, based on the idea that it might be possible to grasp some general principles (Riddle *et al.,* 1997). More reconstruction work was done in the 1970s (sensory ultrastructure of the wild type animal, Ward *et al.,* 1975, Ware *et al.,* 1975; pharynx, Albertson and Thomson, 1976; ventral nerve cord, White *et al.,* 1976; sensory ultrastructure of mutants defective in chemotaxis, Lewis and Hodgkin, 1977, tail ganglia, Hall, 1977, mail tail, Sulston *et al.,* 1980). In 1986, White et al. reported the reconstruction of the 381-cell male nervous system at the EM level (White *et al.,* 1986), which set up a milestone in the study of *C. elegans* nervous system and elegantly realized Brenner's proposal that its complete circuitry could be determined by EM serial sections.

1.3 White Project: "The Mind of a Worm"

White, Southgate, Thomson and Brenner (1986) pushed the reconstruction effort a step further than earlier reconstruction endeavours.

They aimed to reconstruct the nervous system of the whole animal (except the pharynx, since it had been reconstructed earlier by Albertson and Thomson, 1976). This laborious project was motivated by the same ideas as stated earlier in 1974 by Sydney Brenner.

The project, which was published in the monograph "***The Mind of a Worm***" in 1986 (White et al., 1986), presented a large volume of information about the nervous system of *C. elegans*, including among others detailed morphology of individual neurons, neuronal connectivity, and circuitry. Neurons were categorized into classes according to their morphologies and connectivity.

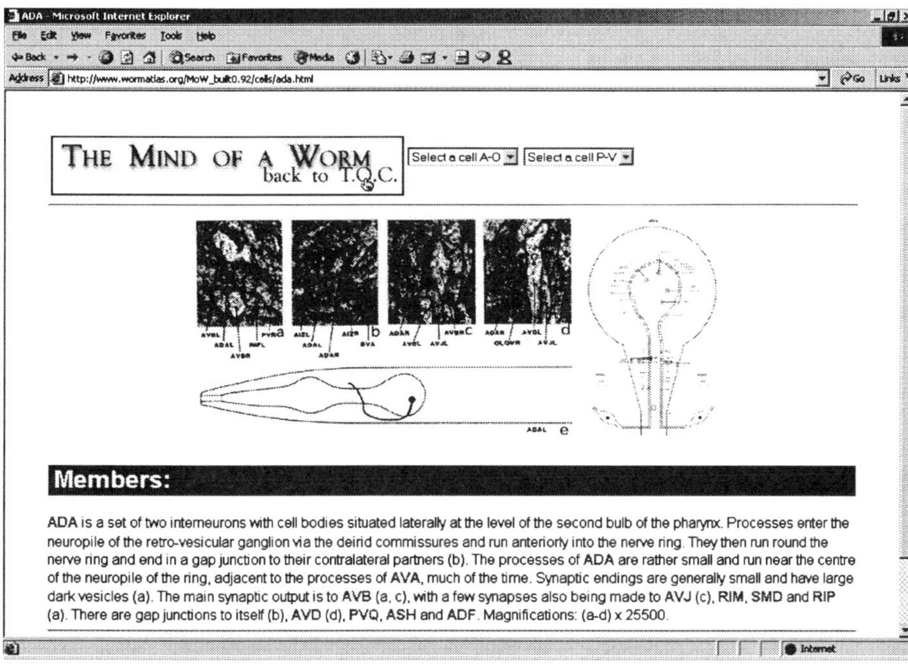

Figure 1. A Sample WormAtlas page showing general description of neurons ADAL, ADAR.

The reconstructions were done using data from five overlapping series, which were designated as N2T (hermaphrodite), N2U (hermaphrodite), JSH (L4 larva, for nerve ring and anterior ventral cord), N2Y (adult male, posterior body), and JSE (hermaphrodite). Figure 1 shows representative serial electron micrographs from N2U (http://www.wormatlas.org/MoW_built0.92/cells/ada.html, for ADAL and ADAR. See section 2.4.1 "**Accessing WormBase: general description of a selected neuron**" for details). Reconstructions were done mostly by hand from a total of about

8000 such prints, although some use was made of a computer-aided reconstruction system (White *et al.,* 1986) in later phases of the project. It should be noted that due to technical restrictions, none of the five animals was reconstructed completely (White *et al.,* 1986).

1.4 Nervous System

Although the nervous system is the most complex organ in *C. elegans*, the hermaphrodite individual has an invariant number of only 302 neurons and 56 glial and associated cells (in males the nervous system is more complex, with 381 neurons and 92 glial and supporting cells) (White *et al.,* 1986; Chalfie and White, 1988). The difference in the number of neurons in hermaphrodites and males can be explained by sex-specific behaviors such as egg laying in hermaphrodites and the more complex mating behavior of males. Neurons with similar morphologies and connectivity have been grouped together into classes, and for hermaphrodite there are 118 such classes (White *et al.,* 1986).

The *C. elegans* nervous system can be broken down into two almost independent systems: pharyngeal nervous system and the general body nervous system, which are joined through a pair of neurons, the RIP cells. The pharynx contains 20 cells; the remaining neurons are distributed throughout the body of the animal. Most of the neurons seem to be organized as ganglia (Table 1, White *et al.,* 1986).

1.5 Structure of neurons

Neurons in *C. elegans* have strikingly simple morphologies with few branches, in contrast to mammalian brain neurons, which usually possess tremendous number of axonal and dendrite processes (also, functionally, they lack action potentials since there are no voltage-dependent sodium channels in *C. elegans,* Goodman *et al.,* 1998). Nevertheless, each of the 302 neurons in adult hermaphrodite has a unique combination of properties, such as morphology, connectivity and position. One caution about the classification of 302 neurons into 118 classes is that they are grouped according to morphological, connectivity, and position information. Therefore it would be a potential mistake to conflate the classification with functional role.

Many neurons in *C. elegans* are monopolar with a single unbranched process; with a small number of neurons have more than two processes. Morphology and other properties of neurons tend to be rather conservative, with little variations among different individuals animals.

Each adult *C. elegans* hermaphrodite has about 5000 chemical synapses, 2000 neuromuscular junctions, and 600 gap junctions (White *et al.,* 1986).

1.6 Wiring pattern and circuitry

The 302 neuron nervous system of *C. elegans* is usually broken down into smaller circuits according to what is known about their connectivity and function where there is some relevant experimental data (White *et al.,* 1986).

Table 1. Ganglia in *C. elegans*

Name	Location	Details
Pharyngeal nervous system	Pharynx	20 neurons
Circumpharyngeal nerve ring	-	Major areas of neuropil
Ventral nerve cord	-	Open and continuous with preanal ganglion
Dorsal nerve cord	-	-
Anterior ganglion	Anterior to the nerve ring	Structurally limited
Small dorsal ganglion	Posterior to the nerve ring	-
Lateral ganglion	Posterior to the nerve ring	There are two such ganglia, open and continuous with lumbar ganglion
Ventral ganglion	Posterior to the nerve ring	Structurally limited
Retrovesicular ganglion	-	Open and continuous with ventral cord
Preanal ganglion	Tail	-
Lumbar ganglion	Tail	There are two such ganglia
Dorso-rectal ganglion	Tail	Structurally limited
Posterior lateral ganglion	Posterior body	There are two such ganglia

1.7 Non-anatomical Neuroscience Data

A detailed review of neuroscience data other than neuroanatomy is not intended in this chapter. For readers who are interested in these topics, we recommend to review the book *C. elegans* (II) published by the Cold Spring Harbor Laboratory Press in 1997 as well as more recent data published about the subject in various papers and books. Chapters related to *C. elegans* Nervous System include Chapter 20 on Patterning the nervous system (Ruvkun); Chapter 22 on Synaptic Transmission (Rand and Nonet); Chapter 23 on Mechanotransduction (Driscoll and Kaplan); Chapter 24 on Feeding and Defecation (Avery and Thomas); Chapter 25 on Chemotaxis and Thermotaxis (Bargmann and Mori); Chapter 27 on Neural Plasticity (Jorgensen and Rankin).

2. NEUROINFORMATICS FOR *C. ELEGANS*: FROM ACEDB TO WORMBASE

2.1 Neuroanatomy and ACeDB

As part of his Ph.D. thesis, Richard Durbin (1987) established a database holding "wiring" data for neurons in *C. elegans*. It was immediately welcomed by the *C. elegans* community because it made the large amount of reconstruction data "live" and searchable. The database was called ACeDB for "A *C. elegans* DataBase" (Durbin and Thierry-Mieg, 1991; Stein and Thierry-Mieg, 1998). It must be stressed that Durbin limited his effort only to the nerve ring, which contains altogether 3452 chemical synapses and 754 gap junctions between the 183 neurons (for JSH series reconstruction). Data for other neurons are currently being curated and deposited into ACeDB.

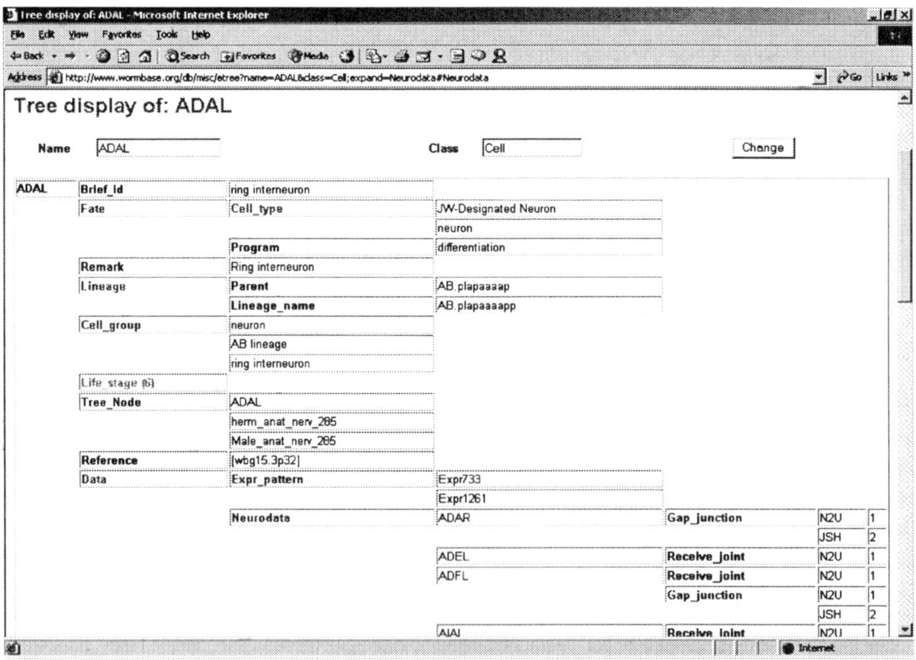

Figure 2. Representation of neuron ADAL in ACeDB.

In ACeDB, data are organized into "models" (classes) that represent specific biological objects. For example, a *Cell model* contains information about which group a cell belongs to, the developmental fate of the cell, and the published expression patterns of genes in the cell. If the cell happens to be a neuron, a section of the model provides its associated neurobiology data.

In ACeDB, each of the 183 cell objects carries a "Neurodata" tag, which has a data format illustrated in Figure 2. This illustrates how an AceDB object is a hierarchical set of tags and values information about a neuron stored in AceDB. The cell object lists the types (Receive, Receive_joint, Send, Send_joint, and Gap_junction, for details, see Durbin, 1998) and numbers of connections between neuron ADAL and all other neurons in the nerve ring.

One confusing situation that confronts novice ACeDB users is that the obvious way to retrieve the "magic" 302 hermaphrodite neurons by searching for all cells that have the "Neurodata" tag will not work. This is because neurons from the male, and glial cells also share the Neurodata tag. The correct way to find the classic set of 302 neurons is to search for cells with the tag "Cell_type" equal to "JW-Designated Neuron" (see next section for details).

2.2 Talking to ACeDB about neurons

Neuron data for each neuron deposited in ACeDB can be retrieved from the database either locally or remotely with the help of a program written in Perl programming language (Wall *et al.*, 2000) called *ace.pl* written by Lincoln Stein. The program *ace.pl* is included in the AcePerl package freely available on the CPAN (http://www.cpan.org) website or Lincoln Stein's home page (http://stein.cshl.org). To run the program, either Internet access is required or a local AceDB is installed. The program needs information about server port number. For example, if you want to talk to the wormbase using ace.pl, you need to explicitly state ace.pl –host www.wormbase.org -port 2007. If you talk to your locally installed AceDB, you can run ace.pl first and try the following examples.

To get all neurons with Neurodata tag:

```
query find cell Neurodata; list
```

To get neuron named ADAL:

```
find cell ADAL; show
```

To get all classes of neurons:

```
query find cell Neurodata; follow cell_group
```

To get the "magic" 302 neurons:

```
query find cell Neurodata AND cell_type = "JW-Designated
Neuron"
```

2.3 Talking to AceDB about others

The goal of WormBase is to provide its users the best navigability among all scientific data about *C. elegans* and other nematodes. Thus, in addition to cells, WormBase offers data from the perspective of sequences, genes, and the organism (behavior), and links them together. We will illustrate this point with the following specific examples.

2.3.1 From Sequence to Gene

A user who studies GABAergic neural transmission is interested in learning what is known about *C. elegans* homologs of mammalian vesicular GABA transporter (GenBank accession number: NP_542119). One performs a blastp search at http://www.wormbase.org/db/searches/blast with the sequence in FASTA format retrieved from Genbank. The search result indicates that the top hit is T20G5.6. Clicking on it brings us to a new page, which shows that the corresponding gene is *unc-47*.

2.3.2 From Gene to Behavior

Following the link of *unc-47*, there are descriptions of mutant phenotype: "very poor backing, good slow forward movement, slight shrinker (contracts both dorsally and ventrally when prodded), small." Thus, *unc-47* is involved in particular aspects of nematode rhythmic locomotion.

2.3.3 From Gene to Cell

Also on the gene page, there are descriptions of *unc-47* expression in "all GABAergic neurons, which include the 19 D neurons in the ventral cord, the four RMEs, AVL, and RIS in the head, and DVB in the tail". These particular cells have also been independently classified as GABAergic neurons based on immunohistochemistry studies. From here, one can learn about the connectivity and other aspects of these GABAergic cells by way of the neuron page.

2.4 The WormBase project: putting ACeDB online to allow web-based access

Although AceDB provides an intuitive object-object oriented view of neurobiological data, rapid response times, and allows user to access data in AceDB, in practice this requires users' familiarity with data models and the AQL (AceDB query language) syntax.

WormBase (http://www.wormbase.org) builds upon ACeDB with a user-friendly interface. It also expands the range of subjects (Stein *et al.,* 2001, Figure 3). Web Pages designed for neuroscience data (http://www.wormbase.org/db/searches/neuron_display, Figure 4) were recently implemented. While this section will introduce the neuroscience pages, for a thorough reference guide readers are encouraged to read the "User's Guide for WormBase" at http://vermicelli.caltech.edu/~wen/userguide/index.html.

The goal of WormBase is to capture, curate and distribute information about *C. elegans* biology through collaborative efforts (Stein *et al.,* 2001). It originated from ACeDB (http://www.acedb.org). There is a dramatic shift in focus from AceDB to WormBase. ACeDB database was used in the course of the *C. elegans* sequencing project to coordinate the sequencing effort and to integrate the worm sequence with the genetic and physical maps, while WormBase is heavily committed to the curation and interpretation of the *C. elegans* literature, and has moved from a genome-centric perspective to one that more evenly balances the worm genome with other aspects of its biology (Stein *et al.,* 2001).

2.4.1 Accessing WormBase: general description of a selected neuron

The complete reconstruction of *C. elegans* nervous system (White *et al.,* 1986) provided information on the exact position of the cell body, axodendritic morphology and synaptic connectivity of all hermaphrodite neurons (except 20 pharyngeal neurons).

Recently this work has been made available online by the Wormatlas consortium (http://www.wormatlas.org). This adaptation preserves the organization of the original publication. It is split into 5 sections (introduction, materials and methods, general description, the nervous system, conclusions) and three appendices (individual cell pages, discussion on the connectivity data, remaining ambiguities) that can be accessed via the main table of contents (T.O.C.). Throughout these pages, all graphic material is made available in two resolutions: screen resolution and print resolution. Print resolution images are opened into separate browser windows by clicking into the screen picture.

Individual pages for each cell type have a consistent organization exemplified by the ADA neurons (ADAL and ADAR) class in Figure 1. Each page starts by a banner containing a link back to the table of contents and two popup menus that allow the user to jump directly from one cell page to another. Below the banner, small electron micrographs show details of synapses and neuron processes (N.B. synapses are made en-passant in *C. elegans*, not by terminal synaptic differentiations). Below these micrographs, a schematic drawing shows the position of the cell body and axodendritic morphology of the neuron(s) in the class. On the right side of the page, a diagram indicates the connectivity of the neuron(s) with other neurons. On the connectivity diagrams, arrows pointing to the process indicate presynaptic partners, arrows originating from the process indicate postsynaptic partners and arrows ending by a vertical bar indicate gap junctions.

The paragraph at the bottom of the page gives general information about the neuron class. The name of each neuron on top of the paragraph is HTML-linked to relevant wormatlas neuron pages, which in turn, have optional links to wormbase neuron pages. Furthermore, the neuron names mentioned within the text is HTML-linked to the cell query pages in the wormbase. Using these features the reader is able to see all the information pertaining to a specific neuron of interest all at once in both databases.

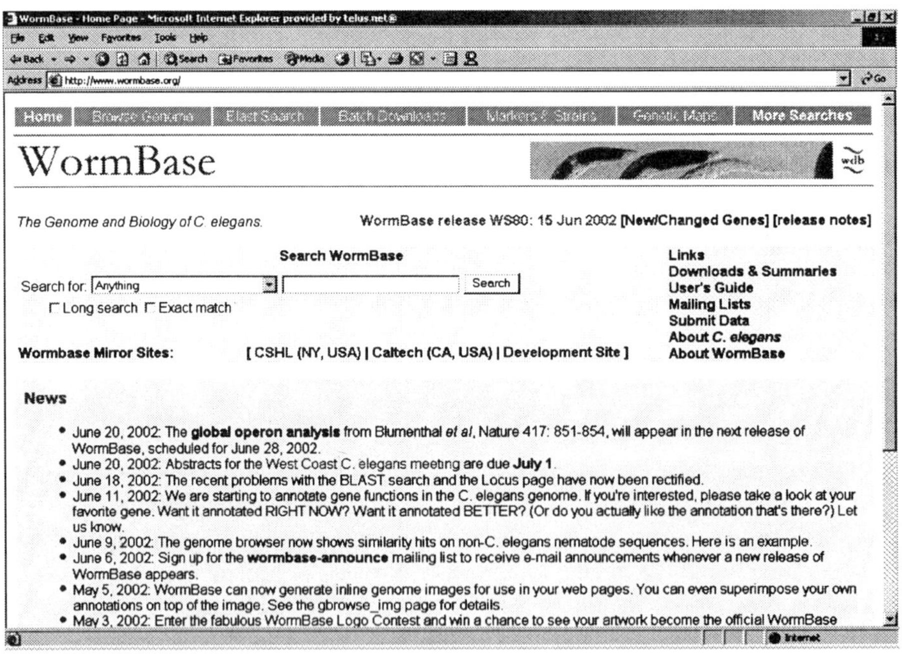

Figure 3. Home page of the wormbase (http://www.wormbase.org)

2.4.2 Accessing WormBase: Cell summary page for a selected neuron

The simplest way to get to the cell summary page for a neuron is via the Cell Search page, reachable from the navigation bar at the top of each WormBase page. From the Cell Search page, enter the name of the neuron you are interested in (if you do not know its name, you have other options to find the neuron, see Neuron Search page for details) and click on "Show Cell" button. A pedigree tree shows up with name of the selected neuron highlighted in yellow. Clicking on this link will retrieve the Cell Summary page for the neuron of interest.

Figures 5 and 6 show the Cell Summary page for the neuron ADAL. In this page, users can find detailed information about ADAL. For example, OVERVIEW shows the neuron class based on White classification (White *et al.,* 1986), which, in this case is "Ring interneuron". GROUP denotes cell group such as "part of the neuron cell group", and FATE is "terminally differentiated". Other fields show the developmental lineage of the cell and the list of genes that are known to be expressed in this cell.

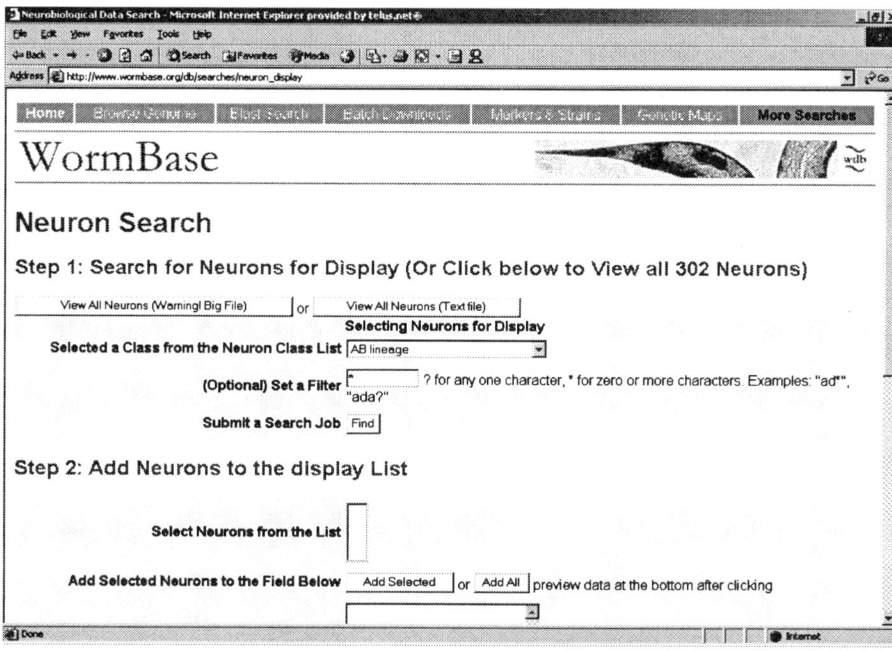

Figure 4. Neuron Search page

At the bottom of each Cell Summary page for a neuron is the SYNAPSE DATA table (Figure 6), which displays wiring data from the two individuals JSH and N2U described in White et al. (1986). The first column of the table lists all neurons that have synaptic or gap junction contacts with the neuron of interest. The numbers in each cell of the table represent the number of the specific types of connections. Due to the difficulties associated with the processes of microscopic reconstruction, ambiguities and mistakes could have occurred at the time these reconstructions were done. This fact potentially limits the confidence in determining the actual number and/or type of connections. Numbers in bold indicate that the number of the specified connection is sure, while numbers in grey and italic indicates that the number is uncertain (corresponding to the "send_joint" and "receive_joint" fields described earlier).

At the top of this page, in the yellow navigation bar, is a link to the "The Mind of a Worm" page displaying the detailed description of the neuron shown in the previous section.

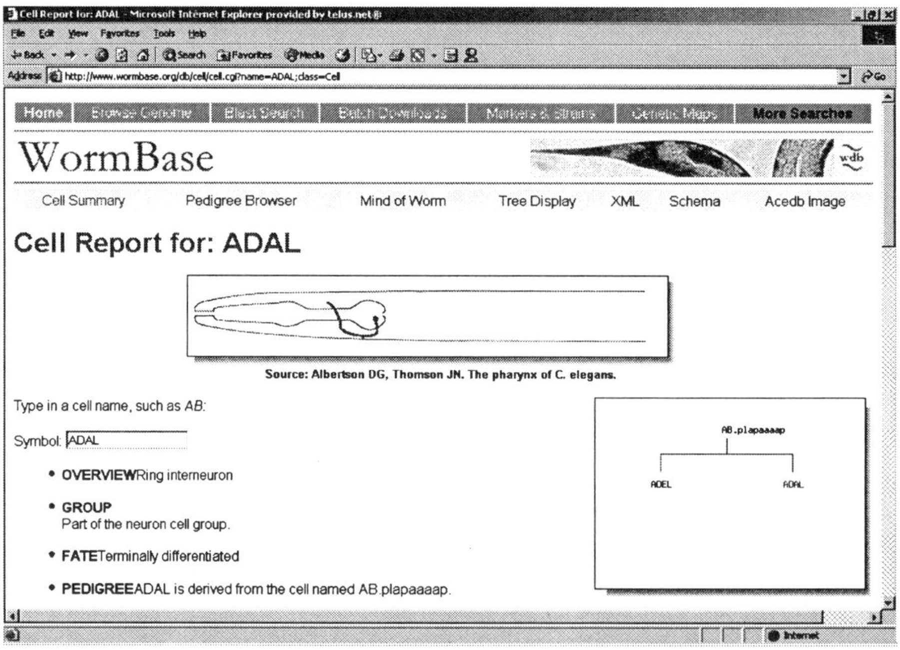

Figure 5. Cell summary page for neuron ADAL

2.4.3 Accessing WormBase: The Neuron search page

Neuron Search page was launched earlier this year (2002, Figure 4). This page gives users multiple ways to search the underlying ACeDB database for neuroanatomy information. For example, suppose you are interested in neurons in *C. elegans* that have an unusually large number of interacting neurons. The easiest way is to click on the "View All Neurons" button to request a page containing data on all of the 302 neurons in hermaphrodite *C. elegans* (Because of its size, this page may take up to a couple of minutes to retrieve.) It contains a brief summary of each neuron, including its name, number of neurons connected, total number of connections (synaptic inputs, synaptic outputs and gap junctions), a link to cell summary page, and a link to a graph page (introduced later). Browsing through the table, we can find that neuron AIBL interacts with 47 other neurons.

From the Neuron Search Page, user can access Cell Summary page easily if the user knows the names of neurons. The user can simply type the names (separated by spaces) in the "Selected Neurons for Display" field in the "Step 2: Add Neurons to the display List" section and click "View Summary Data" button in the "Step 3: Display Data for Selected Neurons" section. A new menu pops up at the bottom of the page asking the user to select properties to be displayed. Click on "Go" button to the right after selection and a table will be displayed with a title "Search Result". The user, therefore, can go to Cell Summary page for each neuron by following the links (in the Name column).

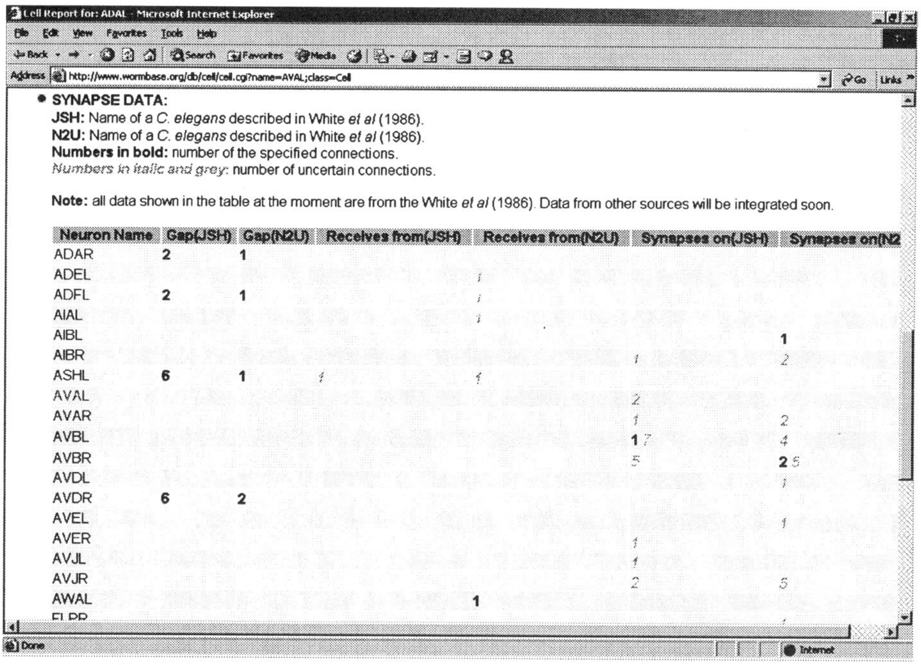

• **SYNAPSE DATA:**
JSH: Name of a *C. elegans* described in White *et al* (1986).
N2U: Name of a *C. elegans* described in White *et al* (1986).
Numbers in bold: number of the specified connections.
Numbers in italic and grey: number of uncertain connections.

Note: all data shown in the table at the moment are from the White *et al* (1986). Data from other sources will be integrated soon.

Neuron Name	Gap(JSH)	Gap(N2U)	Receives from(JSH)	Receives from(N2U)	Synapses on(JSH)	Synapses on(N2
ADAR	**2**	**1**				
ADEL				*1*		
ADFL	**2**	**1**		*1*		
AIAL				*1*		
AIBL						**1**
AIBR					*1*	*2*
ASHL	**6**	**1**	*1*	*1*		
AVAL					*2*	
AVAR					*1*	*2*
AVBL					**1** *7*	*4*
AVBR					*5*	**2** *5*
AVDL						*1*
AVDR	**6**	**2**				
AVEL						*1*
AVER					*1*	
AVJL					*1*	
AVJR					*2*	*5*
AWAL				**1**		
FLPR						

Figure 6. Cell summary page for neuron ADAL (continued)

"I want to get the Cell Summary pages for some neurons. But I don't know their names. All I know is the class they belong to. What should I do?" If you have such a question, the Neuron Search page has a solution for you. Say, if you want to find some neurons in class "amphid neurons", you first select "amphid neurons" from the class list menu labelled with "Select a

Class from the Neuron Class List" in "Step 1: Search for Neurons for Display" section. Clicking on "Find" button and all neurons in the class "amphid neurons" show up in the "Select Neurons from the List" list. At this step, you have the option to select any neurons (or select all neurons by clicking on "Add All") from this list and add to the field below it for further processing. You can repeat this process many times until you get all of the neurons you are interested in and each time a summary table will show up, displaying summary data for all selected neurons.

You can restrict your search by adding a filter (a mask) in the "Set a Filter" field. A default "*" in the field indicates that you don't filter anything away. "ad*" in the field indicates that you are only interested in neurons whose names start with "ad", for example ADAL and ADAR.

Once names of neurons are selected, we can move to the next step to view the properties we want to view. Two choices are supplied, to view "Summary Data" or to view "Detailed Wiring Data".

2.4.4 Accessing WormBase: neuron-wiring data

Wiring data for each neuron can be viewed in a table or graphically (Figure7). Once you have collected all neurons of interests into the "Selected Neuron for Display" field in "Step 2: Add Neurons to the Display List" section, you are ready to view the wiring data. Click on the "View Wiring Data" in "Step 3: Display Data for Selected Neurons" section, a new table will pop up as the bottom of the page with a title "For Detailed Wiring Data, Check Radio Buttons and Click Go Button". This new table allows you to select kind of wiring data you are interested in. Click on the "Go" button and a new page with wiring data will show up.

We have designed a separate page to display neurons graphically. The page can be accessed through Cell Summary page, Neuron Summary page and Wiring data table. Figure 7 displays wiring data from cell RIPR. Neuron RIPR sits at the top of each graph, while all of the neurons with connections to this neuron line up at the bottom of the graph. Wiring data obtained from specimens JSH and N2U are displayed separately. In these graphics, specific shapes and colours are used to represent different types of neurons and the number of connections between neurons. A legend table at the lower part of the page can be consulted for details for colour and shape coding. Arrow direction indicates directions of synaptic projections, and gap junctions are represented with thick dots at both ends of the connecting line.

3. REMAINING PROBLEMS AND FUTURE DIRECTIONS

AceDB and Wormbase projects have received extensive attention because on the one hand, they have been extremely successful in experimenting how to deposit and present neuroscience and other biological data; on the other hand, they have made *C. elegans* data easily retrievable. The most important problem associated with *C. elegans* neuroscience data is that the database is far from up-to-date. We propose three lines of efforts should be put on the neuroscience database in order to better server the *C. elegans* and other communities.

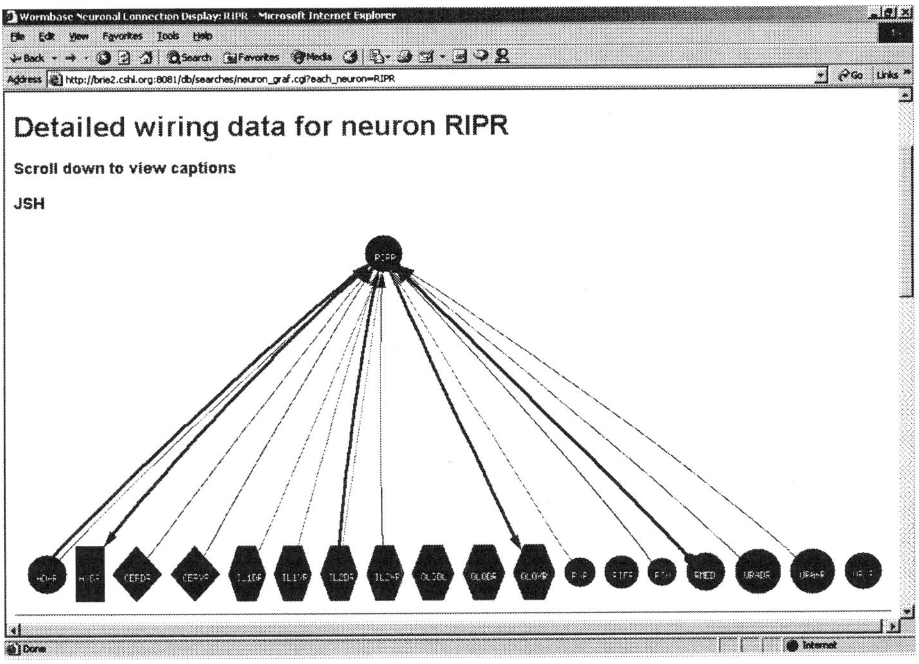

Figure 7. Graphical representation of neuronal connections.

3.1 Data updating and validation

As mentioned earlier, neuroscience data in the current Wormbase were mostly deposited by Durbin in 1987, which means that a substantial amount of data from the White project are still missing from the database, not mentioning the data generated since then. We need therefore to deposit the wiring data from the "Mind of a worm" project, as well as from sources other than the "Mind of a worm" project. We also need addition of data from earlier developmental stages and from male-specific neurons.

3.2 More features for the Neuron Search Page

More effort should be put to create more features in the Neuron Search page to allow users easily access neuronal circuits information, interaction between classes of neurons, etc.

3.3 Integration of GeneOntology

Although WormBase strives to provide its users an integrated view of scientific research results, it is not yet particularly user-friendly, especially to researchers not familiar with the worm as an experimental system. To work towards the integration of knowledge learned from nematodes with that from other model organisms, WormBase has begun to implement Gene Ontology (http://www.geneontology.org); assigning GO controlled vocabulary to gene products based on what is known about them. Furthermore, as part of the GO consortium, WormBase actively participates in the continued effort to build better biological ontologies.

SELECTED REFERENCES

Albert P.S. and Riddle D.L., 1983. Developmental alterations in sensory neuroanatomy of the *Caenorhabditis elegans* dauer larva. *J. Comp. Neurol.* 219, 461-481.

Alberts, B., 1997. Preface. In *C. elegans* II (editors: D.L. Riddle, T. Blumenthal, B.J. Meyer and J.R. Priess). pp XI-XIV.

Albertson D.G. and Thomson J.N., 1976. The pharynx of *Caenorhabditis elegans* . *Philos. Trans. R. Soc. Lond. B Biol. Sci.* 275, 299-325.

Brenner, S., 1997 Preface. In *C. elegans* II (editors: D.L. Riddle, T. Blumenthal, B.J. Meyer and J.R. Priess). pp XV-XVII.

Chalfie M. and White J. 1988 In The Nematode *Caenorhabditis elegans*. pp 337-391.

Durbin, R.M., 1987. Studies on the development and organization of the nervous system of Caenorhabditis elegans, Ph.D. thesis, King's College, Cambridge.

Durbin, R.M. and Thierry-Mieg J., 1991. A *C. elegans* database. Documentation, code and data available from anonymous FTP servers at lirmm.lirmm.fr, cele.mrc-lmb.cam.ac.uk and ncbi.nlm.nih.gov.

Goodman M.B., Hall D.H., Avery L., and Lockery S.R., 1998. Active Currents Regulate Sensitivity and Dynamic Range in C. elegans Neurons, *Neuron* 20, 763-772.

Hall D.H., 1977. The posterior nervous system of *Caenorhabditis elegans*, Ph.D. thesis, California Institute of Technology.

Lewis J.A. and Hodgkin J.A., 1997. Specific neuroanatomical changes in chemosensory mutants of the nematode *Caenorhabditis elegans. J. Comp. Neurol.* 172, 489-510.

Mori, I. (1999). Genetics of chemotaxis and thermotaxis in the nematode caenorhabditis elegans. *Annu. Rev. Genet.* 33, 399-422

Rand and Nonet, 1997. Neurotransmitter assignments for specific neurons. In *C. elegans* II (editors: D.L. Riddle, T. Blumenthal, B.J. Meyer and J.R. Priess). pp 1049-1052.

Riddle, DL, Blumenthal, T, Meyer, BJ, and Priess JR, 1997. Introduction to *C. elegans*. In *C. elegans* II (editors: D.L. Riddle, T. Blumenthal, B.J. Meyer and J.R. Priess). pp 1-22.

Stein L.D., Sternberg, P., Durbin, R., Thierry-Mieg J, and Spieth, J., 2001. WormBase: network access to the genome and biology of *Caenorhabditis elegans. Nucl. Acids. Res.* 29, 82-86.

Stein L. D. and Thierry-Mieg J., 1998. Scriptable access to the *C. elegans* genome sequence and other ACEDB databases. *Genome Research.* 8(12), 1308-1315.

Stein L.D., 1999. Internet access to the *C. elegans* genome [Internet] *Trends in Genetics*, 15, 10, 425-427.

Sternberg, P. W., 2001. Working in the post-genomic *C. elegans* world. *Cell* 105, 173-176.

Sulston J.E. and White J.G., 1980. Regulation and cell autonomy during postembryonic development of *Caenorhabditis elegans. Dev. Biol.* 78, 577-597.

Wall, L., Christiansen, T., Orwant, J., 2000. Programming Perl, 3rd edition. O'Reilly Press.

Ward S., Thomson N., White J.G., and Brenner S., 1975. Electron microscopical reconstruction of the anterior sensory anatomy of the nematode *Caenorhabditis elegans* . *J. Comp. Neurol.* 160, 313-337.

Ware R.W., Clark D., Crossland K., and Russell R.L., 1975. The nerve ring of the nematode *Caenorhabditis elegans:* Sensory input and motor output. *J. Comp. Neurol.* 162, 71-110.

White J.G., Southgate E., Thomson J.N., and Brenner S., 1976. The structure of the ventral nerve cord of *Caenorhabditis elegans. Philos. Trans. R. Soc. Lond. B Biol. Sci.* 275, 327-348.

White J.G., Southgate E., Thomson J.N., and Brenner S., 1986. The structure of the nervous system of the nematode *C. elegans. Philos. Trans. R. Soc. Lond. B Biol. Sci.* 314, 1-340.

Chapter 2

A Gene Expression Map of the Mouse Brain

Genepaint.org - A Database of Gene Expression Patterns

Axel Visel, Jamila Ahdidan, and Gregor Eichele
Max Planck Institute of Experimental Endocrinology, Feodor-Lynen-Str. 7, D-30165 Hannover, Germany
Correspondence to: Gregor.Eichele@mpihan.mpg.de

Abstract: Tens of thousands of novel genes have been identified by systematic DNA sequencing of a variety of genomes. Such gene sequences create a tremendous opportunity for addressing gene function by a variety of experimental approaches. To be attractive for genome-wide analysis, experimental strategies need not only to be efficient (high-throughput) but results also have to be systematized and made accessible in a way that they can be promptly leveraged by the scientific community. Having a genome-wide atlas of gene expression patterns with cellular resolution would be a rich resource to tackle a broad spectrum of biological and medical questions. The techniques to analyze gene expression have long been invented: mRNA *in situ*-hybridization (ISH) and antibody-based immunohistochemistry (IC). Traditionally, both ISH and IC are carried out on a gene-by-gene basis. Here we describe instrumentation (termed GenePaint) that permits automation of many aspects of ISH and image data acquisition. The throughput that is achieved renders investigations at a chromosome- or even genome-wide scale highly feasible. To effectively disseminate gene expression data, we have developed a database (www.genepaint.org). Systematic annotation of gene expression data in reference to a standardized catalogue of anatomical structures makes it possible to compare expression patterns between specimens. For viewing expression patterns, GenePaint.org provides a "virtual microscope" which enables zooming and translocation of expression patterns. GenePaint is currently applied to large-scale studies of gene expression in the CNS of the mouse.

Key words: functional genomics, high-throughput technology, *in situ*-hybridization, brain, CNS, embryo, web database, virtual microscopy

1. INTRODUCTION

1.1 Challenges Arising from Structural Genomics

The complete genomes of humans and of a number of non-mammalian species have now been sequenced (Adams et al., 2000; *C.elegans* Sequencing Consortium 1998; Lander et al., 2001; Venter et al., 2001). Sequence determination of the mouse genome is on-going, and extensive sequence data for other mammals are being produced, spurred by the demand for such data by researchers from various fields.

Genomic sequences represent very accurate data that can be contended in minor details only. By contrast, annotation of sequence is far more difficult especially in higher eukaryotes where exons are buried in noncoding DNA, some of which regulates gene expression, but that for the most part has an elusive function. Because current gene prediction algorithms fail to unambiguously predict exons, the exact number of human genes is still under debate. The existence of mRNA splicing variants, of pseudogenes and of low abundance transcripts further complicate the annotation task. Thus a first major challenge of the post-genomics era is the comprehensive identification of all expressed genes. A second task is to determine the "function" of gene products. Genetics, biochemistry, physiology, pharmacology, medicine, to name just a few, will greatly benefit from such efforts. Based on sequence homology with well-characterised proteins, structure and biochemical function of unknown proteins can be predicted to some extent, but elucidation of their *in vivo* function will require experimental approaches. A full functional characterization of a protein would include primary sequence, three-dimensional structure, regulation of expression, tissue and subcellular localizations, consequence of misexpression or loss of expression, functionally important interactivity with other proteins, the protein's role in signalling networks, and type and regulation of enzymatic activity if present.

1.2 The Need for Comprehensive Cell-Based Gene Expression Analysis

It has long been appreciated that determining when and where a gene is expressed often provides clues to its function. Genes encoding receptors are expressed in cells that can respond to ligands, genes coding for metabolic enzymes are expressed in tissues active in energy turnover, and genes for contractile proteins are e.g. found in muscles. Knowing the expression pattern of a gene often suggests a focus of subsequent experimental work, but it goes without saying one cannot assume if a gene is expressed in a certain cell that it has an essential function in this cell. Expression may reflect the nature of the regulatory region and expression hence be tolerated but not required by the cell in question. Such apparently adventitious expression, however, would still be informative if it can be used as a read-out for underlying physiological and pathophysiological states. There are numerous other areas for which knowledge of gene expression is useful. Sets of molecular markers are used in developmental biology to characterise experimentally or genetically altered animal embryos. In the case of drug development knowing gene expression patterns is critical since drug side-effects could be mediated by the presence of the target outside the tissue where the drug is intended to be efficacious. Cancers of diverse types can be characterized by gene expression profiles (Nielsen et al., 2002; van 't Veer et al., 2002).

When it comes to simultaneous monitoring of the expression of large numbers of genes under different developmental, genetic, pharmacological, or otherwise defined conditions, massive-parallel array hybridization approaches provide impressive amounts of semiquantitative or quantitative expression data. However, a limitation of these methods is their lack of cellular resolution (fluorescence-activated cell-sorting or laser-assisted microdissection can in part overcome this limitation). By contrast, *in situ*-hybridization (ISH) and immunohistochemistry reveal gene expression at cellular or even subcellular levels. This is a major advantage in the case of the brain, in which gene expression and other biochemical processes need to be tied to intricate structures composed of many cell types and to neuronal circuits. Single cell resolution-based expression analyses are performed manually one gene/protein at a time, which is an anachronism in view of the large number of genes and proteins to be examined. Thus there is an urgent need to develop technology to substantially accelerate histological procedures. This has prompted us to build (1) equipment that automates and hence accelerates *in situ*-hybridization and immunohistochemisty and (2) adapt microscopy to make it suitable for automated image data acquisition (Herzig et al., 2001). Collectively, the technology is referred to as "GenePaint". Of particular relevance to this chapter, we have established a web database termed "GenePaint.org" that contains *in situ*-hybridization images and their associated meta data (information about probe sequence, specimen, annotation of expression level and pattern etc.).

2. HIGH THROUGHPUT *IN SITU*-HYBRIDIZATION

Although the focus of this chapter is on the GenePaint.org database and issues directly related to it, we will briefly review the technology by which the data found in GenePaint.org are generated. We limit our discussion to

Table 1. Developmental stages of the mouse examined by GenePaint-based ISH

Stage	Type of Specimen	Plane of Sectioning	Reference Atlas
Embryonic E10.5	whole embryo fixed, gelatin or paraffin embedding	sagittal	Kaufman (1992)
Embryonic E14.5	whole embryo fresh frozen, OCT embedding	sagittal	Kaufman (1992), Schambra et al. (1992)
Embryonic E15.5	head of embryo fresh frozen, OCT embedding	sagittal	Kaufman (1992), Schambra et al. (1992)
Postnatal P7	brain fresh frozen, OCT embedding	sagittal	Valverde (1998)
Postnatal P56	brain fresh frozen, OCT embedding	sagittal	Paxinos & Franklin (2001)

mouse tissues and to gene expression analysis by ISH on slide-mounted sections.

2.1 Tissue and Riboprobe Preparation

All expression data are based on sectioned material (Table 1). Except for embryonic stage E10.5, all tissues are prepared and directly embedded in Tissue-Tek medium (Sakura Finetek) and immediately frozen. E10.5 embryos are fixed in paraformaldehyde and embedded in either gelatin or in paraffin. Sections are cut in a cryostat (20 μm thickness) or, in the case of paraffin-embedded E10.5 specimens, in a standard microtome (7 μm thickness). Special equipment is used to orient specimens during the embedding and sectioning process so as to produce well-oriented sections. This is critical for two reasons. First, it permits comparison of data to reference sections found in published atlases (Table 1) and second one can readily compare between the sections stored in GenePaint.org.

2.2 Automated *in situ*-Hybridization

Gene expression is detected using digoxygenin-labeled antisense and sense riboprobes (non-radioactive) generated by *in vitro*-transcription from DNA templates using T3-, T7- or Sp6-bacteriophage RNA polymerases. An important feature of the detection procedure is tyramide-biotin signal amplification without which digoxygenin-based ISH is not nearly as sensitive as radioactive ISH (Adams 1992; Kerstens et al., 1995). Expression signal is thus detected as follows. A digoxygenin-tagged RNA is hybridized to cellular mRNA in the section. Peroxidase-coupled anti-digoxygenin antibody is added, followed by addition of biotinylated tyramide. The peroxidase moiety of the anti-digoxygenin antibody activates the biotin-tyramide conjugate resulting in covalent attachment of biotin to proteins in the vicinity of the peroxidase enzyme. Biotin is subsequently detected by an alkaline phosphatase-based BCIP/NPT color reaction.

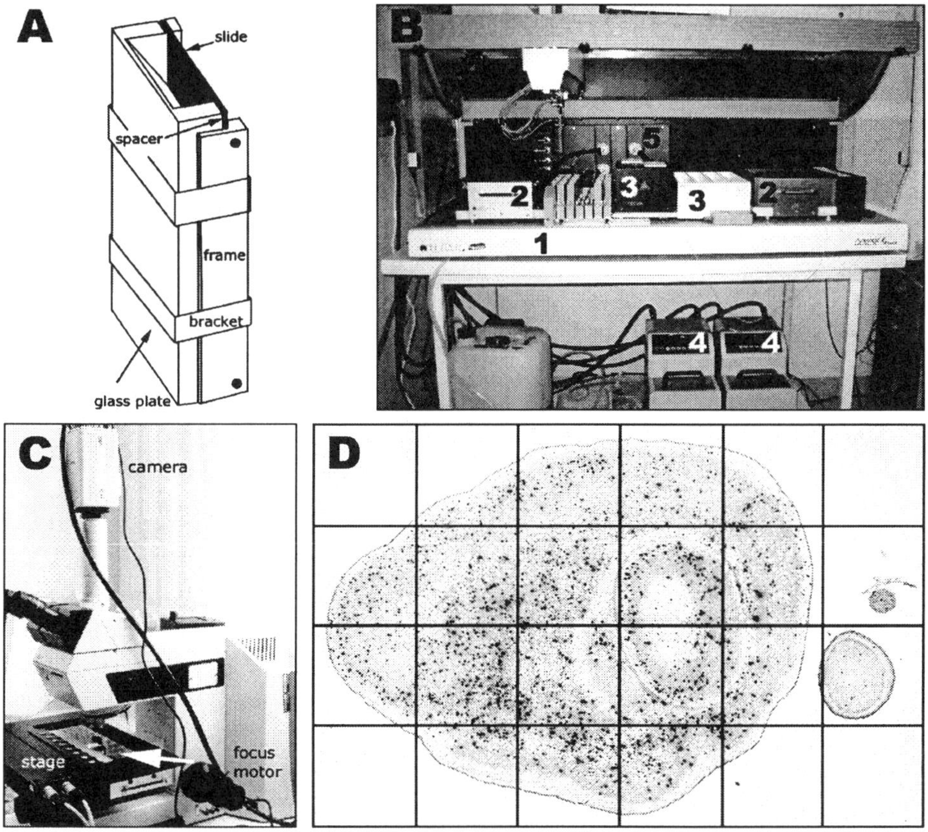

Figure 1. Hardware used for high-throughput *in situ*-hybridization. A - Design of the flow-through chambers that are the heart of the technology. A glass slide carrying tissue sections is separated by two 80-μm thick spacers from a 5-mm thick glass plate into which a depression has been cut. By assembling these items into a frame, a flow-through chamber is created with a well on the top. Liquid is delivered into the well, enters in and flows through the chamber, but capillary forces retain the liquid at the level defined by the bottom of the well where the two glass plates are just 80 μm apart. B - Hybridization is performed on a Genesis pipetting robot platform (1) carrying up to four chamber racks (2) that house up to 48 hybridization chambers each. Receptacles containing reagents (3), circulatory water baths for temperature control (4), and check valves that control water flow (5) are indicated. Operation of the system is controlled by a PC (not shown). C - A Leica scanning microscope carrying eight parallel slides on the motorized stage (arrow). D - High resolution images of sections are assembled in a mosaic pattern as indicated by the grid from single exposures captured by the CCD camera.

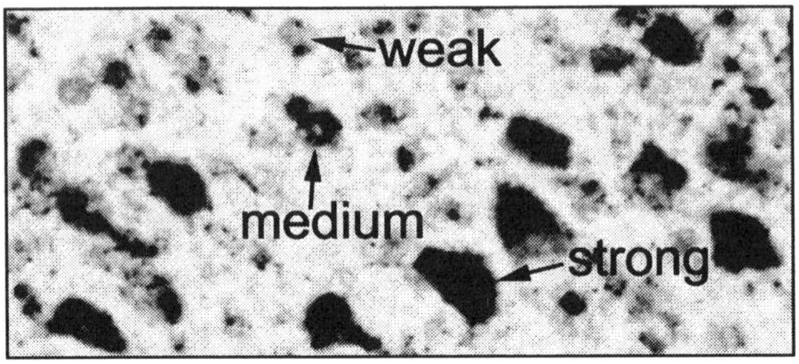

Figure 2. Annotation of intensity of expression exemplified with *trkB* (a high affinity receptor for nerve growth factor) expression in the CNS of a postnatal day 7 mouse, using three levels of scoring.

Slides carrying sections are assembled into a flow-through hybridization chamber (see Fig. 1A). Up to 192 hybridization chambers can be placed into 4 chamber racks that are positioned on the platform of a Tecan Genesis liquid handling robot (see Fig. 1B for details). A PC-controlled protocol has been developed enabling the Genesis liquid handling platform to analyze 192 slides (768 brain sections) in less than 24 hours. Manual steps required are preparation of solutions, the assembly of hybridization chambers and placement of chambers and solutions onto the platform. After completion of the color reaction, hybridization chambers are dismantled and coverslips are applied using an automated coverslipper.

Expression signal becomes apparent as small grains of brown to purple precipitate. Level of gene expression is estimated by viewing the size of the precipitate (Fig. 2). A few distinct grains of precipitate indicate "weak expression", if precipitate fills a considerable proportion of the cytoplasm this is scored as "medium expression", and if precipitate completely fills the cytoplasm of a cell "strong expression" is scored. This scale was established by comparing data generated by GenePaint with published data obtained with radioactive probes (an extensive validation of GenePaint is currently in progress in our laboratory).

2.3 Microscopy and Image Acquisition

Slides are digitally photographed using the microscope depicted in Fig. 1C. They are placed on an 8-position motorized stage. The stage translocates each section in a step-by-step fashion and in a meander pattern in front of the microscope objective. An attached CCD video camera (572 x 572 pixels) captures at each step multiple images to compute the best focal plane followed by a single picture taken at that plane. The imaging frequency is approximately one image every two seconds. After all single images for a section have been recorded (around fifty), they are automatically pasted together to form a nearly seamless mosaic of the type shown in Fig. 1D. With this image acquisition strategy, the resolution of images is not limited by the chip resolution of the camera, but only by the power and numerical aperture of the objective used. We found it optimal to use either 5x or 10x objectives with numerical apertures of 0.15 and 0.40. This translates into an image resolution of 3.4 μm/pixel and 1.7 μm/pixel. A sagittal section of an adult mouse brain amounts to approximately 4500 x 3500 pixels after cropping when a 5x objective is used. The image size of the corresponding bitmap file is 47 MB (4500 x 3500 x 3 bytes). Doubling the microscope magnification quadruples the image storage size and image acquisition time. Thus collecting an adult brain section with a 10x instead of 5x objective will produce a bitmap file of 188 MB. Bitmap mosaic images are compressed without data loss using the Lempel Ziv Welch (LZW) method and stored in TIFF (Tagged Image File Format) files. This results in files in the range of 15 to 30 MB for a brain section scanned with a 5x objective.

2.4 Annotation of Expression Levels and Patterns

Availability of a stereotyped annotation of gene expression patterns is crucial for the success of an expression database. GenePaint.org extends beyond merely displaying images of gene expression patterns and provides interactivity and searchable information on patterns and levels of expression.

Annotation of gene expression patterns is best discussed by means of an example for adult mouse brain. Fifteen structures (olfactory bulb, cortex, amygdala, striatum, ventral portion of striatum, septum, globus pallidus, basal forebrain, thalamus, hypothalamus, hippocampus, midbrain, pons, medulla and cerebellum) are annotated. In the case of the adult mouse brain, there is reasonable agreement in the neuroscience literature regarding the boundaries of each of these structures.

GenePaint analysis involves approximately forty sagittal sections beginning at the plane containing the lateral-most portion of the cerebellum and ending slightly beyond the midline. Among these forty sections, ten standard sections are selected corresponding to sections # 102, 105, 108, 111, 114, 117, 120, 123, 126, and 129 in the mouse brain atlas of Paxinos and Franklin (2001). Following selection of standard levels, the structures to be annotated are delineated and the level of expression is determined. The symbols for expression level are "strong", "medium" and "weak" expression (see above). In addition, there can be explicitly *no detectable expression* in a structure. A structure can be absent (e.g. there is no amygdala near the midline) resulting in a blank space in the annotation tables. Simultaneously, the pattern of expression is determined in each of the structures that are present in a given section. Patterns can be *ubiquitous* (i.e. the gene is expressed in every cell within the structure), *regional* (in some parts of the structure, but not in others) or *scattered* (in a subset of cells that occur dispersed throughout the structure). Although an expression pattern is a three-dimensional tissue property, we found it more practical to annotate the pattern on a section-by-section basis.

Within a structure there can be more than one expression level or pattern. In this case, one selects the visually most prominent pattern. For example, if a gene is weakly and ubiquitously expressed in thalamus but there is in addition also strong expression in one of the thalamic nuclei, then "regional" and "strong" are selected. The result of the annotation process for a given gene and stage is entered into a table showing level and pattern of expression for each of the ten standard sections. These annotations, plus the associated sections and if available the non-annotated interspersed sections, are imported into GenePaint.org.

The catalogue of structures to be annotated is still evolving. The example discussed above is based on an ongoing project supported by the National Institute of Neurological Disorders and Stroke (USA) and represents a consensus reached by the contributing investigators (Drs. N. Heintz, E. Hatten, both Rockefeller University; Dr. A. Joyner (NYU); Dr. G. Leblanc (NINDS); and one of us, G.E.). A comprehensive, tree-based anatomical dictionary for the mouse is being developed by the MRC Human Genetics Unit in Edinburgh (UK) and the Department of Anatomy, University of Edinburgh (http://genex.hgu.mrc.ac.uk/). These important efforts have the benefit of integrating both anatomy and development. This sophisticated dictionary contains hundreds of tissues; we are currently designing another dictionary, hierarchically organised, focussing on the nervous system, spanning embryonic as well as adult, but simpler, so expression can be annotated in a reasonable amount of time. Eventually, image processing software will be developed to identify anatomical structures, to automatically detect expression and identify expression patterns. Current annotation of expression patterns is done manually by skilled neuroanatomists, thus limiting how much can be annotated.

3. THE GENEPAINT.ORG DATABASE

3.1 Structure of the Database

The vast majority of currently existing public domain biological databases are equipped with a web-based user interface. The major advantage of using HTML-encoded pages for input of queries and output of search results and data is the wide distribution of web browser programs (Explorer, Netscape etc.) and their availability for virtually all existing hardware platforms and operating systems.

GenePaint.org consists of two parts that are closely interlinked and accessible through the web. The first part is primarily reserved for and accessible by laboratories that contribute experimental and annotation data. Furthermore, this part contains tools for specimen and laboratory management, administration of image files and associated experimental and annotation data. A detailed discussion of these functions is beyond the scope of this article. Here we will focus on the second part, which is accessible to the public. It allows searches for images of gene expression patterns on standardized sections, for annotation of levels and patterns of expression in reference to a standard anatomy dictionary, for sequence of the template used for probe synthesis, and for selected data of the specimen used and experimental conditions employed.

Data sets in Genepaint.org are identified by a unique reference number (GenePaint set ID) and consist of n individual sections from the same developmental stage and hybridized with the same riboprobe. In addition, each data set contains data about the riboprobe, the tissue type and annotation of the expression pattern. More than one data set will exist for a gene if it has been tested at different developmental stages or has been tested on different kinds of specimens, e.g. animals with a different genetic background.

GenePaint.org consists of an IBM DB2 database system and an Apache web server running under the AIX operating system. The Magic development environment (Magic Software Enterprises, Irvine, CA) is used for programming and maintenance of the database and associated programs. Server-client communication is managed by the Magic application server.

3.2 Zoomserver Technology Creates a Virtual Microscope

The primary mosaic image files that are produced by the digital camera are in bitmap format. They are assembled and converted to TIFF, which is a standard format for high-resolution graphics files, but at the same time it has the drawback that the files are relatively large. The image file of a section is generally about 15 to 30 MB in size, which is acceptable for long-term storage but is too large for instantaneous viewing over the internet. Lossy compression formats such as JPEG (Joint Photographic Experts Group) can be used to address bandwidth issues, but are not an ideal solution as data quality is noticably reduced. Moreover, a JPEG image usually creates as large of a memory footprint as a TIFF when viewed, which at these sizes would diminish performance on many computers currently in use.

While the original TIFF data are archived for reference purposes, the GenePaint.org server providing the web-based public user interface uses two different image files for each tissue section. One is a 70 kB low resolution JPEG image used as a thumbnail. Whenever a more detailed view is required, the server uses a file that has been created in Flash Pix (FPX) format. FPX is a multi-resolution image format that has been developed for viewing high-resolution images whenever performance is limited by factors such as maximum network bandwidth or memory constraints of the client computer. FPX files are usually about thirty percent larger than the TIFF files from which they were created, which means that in our case typical sections require 20 to 40 MB of space on the server. For conversion of images to FPX format and dynamic creation of JPEG files (see below), we use the Zoom Image Server software package (Roxio, Santa Clara, CA).

When users request an image, which is displayed within an HTML page, an image of the desired size and resolution is dynamically created in JPEG format and transmitted (see also Fig. 5). This means that the resolution of the picture (i.e., the number of micrometers represented by one pixel) is reduced to fit the resolution of the client's screen. If users wish to zoom into a subregion and thus approach the resolution of the primary image data, they can do so by pointing to and clicking into the screen image using the mouse. The requested position and zooming factor are transferred to the server that returns the appropriate JPEG image to the user's computer. In addition to zooming in and out, the zoom server allows panning, vertically and horizontally, within the image while maintaining the current zoom factor. In other words, sections can be handled on the computer screen reminiscent of a microscope in which sections can be viewed at different magnifications and be translocated by stage movement.

4. USING GENEPAINT.ORG

In this section we will introduce the most important features of GenePaint.org and explain how they can be used. Queries are launched either from the *entry page* (www.genepaint.org) or from a more detailed *query page*. This *query page* contains additional search features, e.g. restricting search to specimens of certain stages, sequence homology searches for templates used for probe production, and searches for user-defined expression patterns. The same page serves to display lists of *search results* next to the search criteria used. Sets of images of expression patterns can be browsed either in a *thumbnail viewer* or, along with the annotation of the expression pattern, in the *image and annotation viewer*. For each section, a *zoom viewer* window is available to view sections at high magnification. Annotations of expression patterns of several genes to anatomical structures can be viewed and compared side by side on the *expression pattern overview* page. In the following section we discuss the salient features of each of the pages and explain how they are connected.[1]

4.1 Entry Page

The *entry page* (Fig. 3A) at www.genepaint.org contains a project description, a newsletter and on the left side an entry field for simple searches. Gene names and symbols (including common synonyms), GenBank accession numbers and GenePaint set IDs can be directly entered into and searched from this field (section 4.2.1). The results of this simple search are displayed on the right hand side of the *query and results page* (Fig. 3B, see below).

4.2 Query and Results Page

More sophisticated queries, including those asking for expression in a given set of anatomical structures, are launched from the detailed query form that is retrieved by clicking the "advanced search" button on the *entry page.* The bipartite *query and results page* (Fig. 3B) contains a detailed interface for searching GenePaint.org as well as a list of hits that are retrieved by the search. Searches can encompass gene names, gene symbols, GenBank accession numbers, and GenePaint set IDs. Furthermore one can search for expression in a structure or combinations thereof. Finally, a BLAST routine permits entry of DNA sequence that is blasted against all sequences used as templates in GenePaint. If multiple search criteria are defined, only data sets that satisfy all of them are displayed in the list of hits.

4.2.1 Searches for General Identifiers

Terms that can be typed into the entry field (Fig. 3B), are gene name (e.g. *paired box-6*), GenBank accession number (e.g. X63963), gene symbol (e.g. *Pax-6*) or synonyms (e.g. *small eye* or *sey* for *Pax-6*), and GenePaint set ID. Restrictions of developmental stage, species, strain/genotype can be applied by selection in the appropriate combo box. The entry field accepts multiple search terms connected by Boolean operators (AND, OR etc.).

[1] It is expected that most, if not all of these features will be fully operable by late 2002.

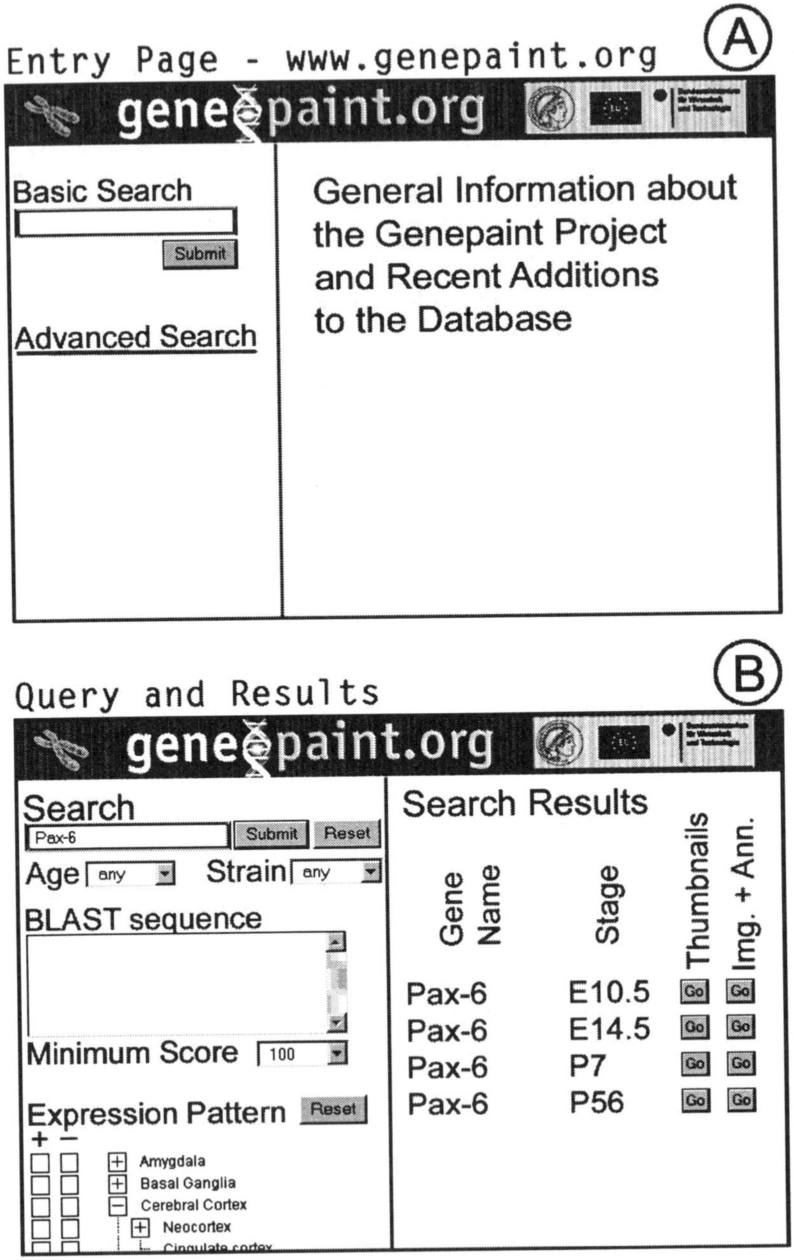

Figure 3. Scheme showing the main features of the *entry page* (A) and the *query and results page* (B). See text for details.

4.2.2 Searches for Homologous Sequences

Identification of data sets within molecular biology databases just based on gene names or gene descriptions can be obstructed by inconsistent gene names, synonyms, hyphenation, use of underscore etc. Another problem is multiple GenBank accession numbers for one and the same gene. To circumvent most of these problems, GenePaint.org features a search for sequence homologies against all DNA templates used for riboprobe synthesis for the GenePaint project. This search is based on the BLAST algorithm (Altschul et al., 1990) and can also be used to search for expression patterns of genes that are not identical but homologous to the query sequence. The BLAST field (Fig. 3B, center left) permits entry of a BLAST minimum score. E.g. entry of 100 would retrieve all image data sets that represent genes that score 100 or higher with respect to the query sequence.

4.2.3 Searches for Expression in Specific Structures

Searches can be driven by user-defined expression patterns, because expression is annotated at least in a rudimentary form (section 2.4). In order to provide a user-friendly interface to initiate such searches, the *query and results page* contains a hierarchical list of anatomical structures that are in the annotation dictionary (see Fig. 3B, bottom left). The structure(s) in which expression is to be searched are checked. Check marks can be placed into a positive or negative (not expressed) box. The anatomical dictionary we developed contains more terms than are currently annotated but only those structures for which at least one set of annotated data exist can be checked.

4.2.4 Search Results

All hits fulfilling the user-defined search criteria are listed in the results table. Each row represents one GenePaint set and lists e.g. GenePaint set ID, gene name, accession number, age and type of specimen. Data sets can be opened either in the *thumbnail viewer* (section 4.3 and Fig. 4A) or in the *image and annotation viewer* (section 4.3 and Fig. 4B) by clicking the appropriate button. Alternatively, users can select all or some of the hits and compare the annotation of the expression patterns in the *expression pattern overview* (section 4.6). Because the search results are displayed on the same page as the original query, the user can conveniently modify or refine the search parameters if there are too many or too few hits.

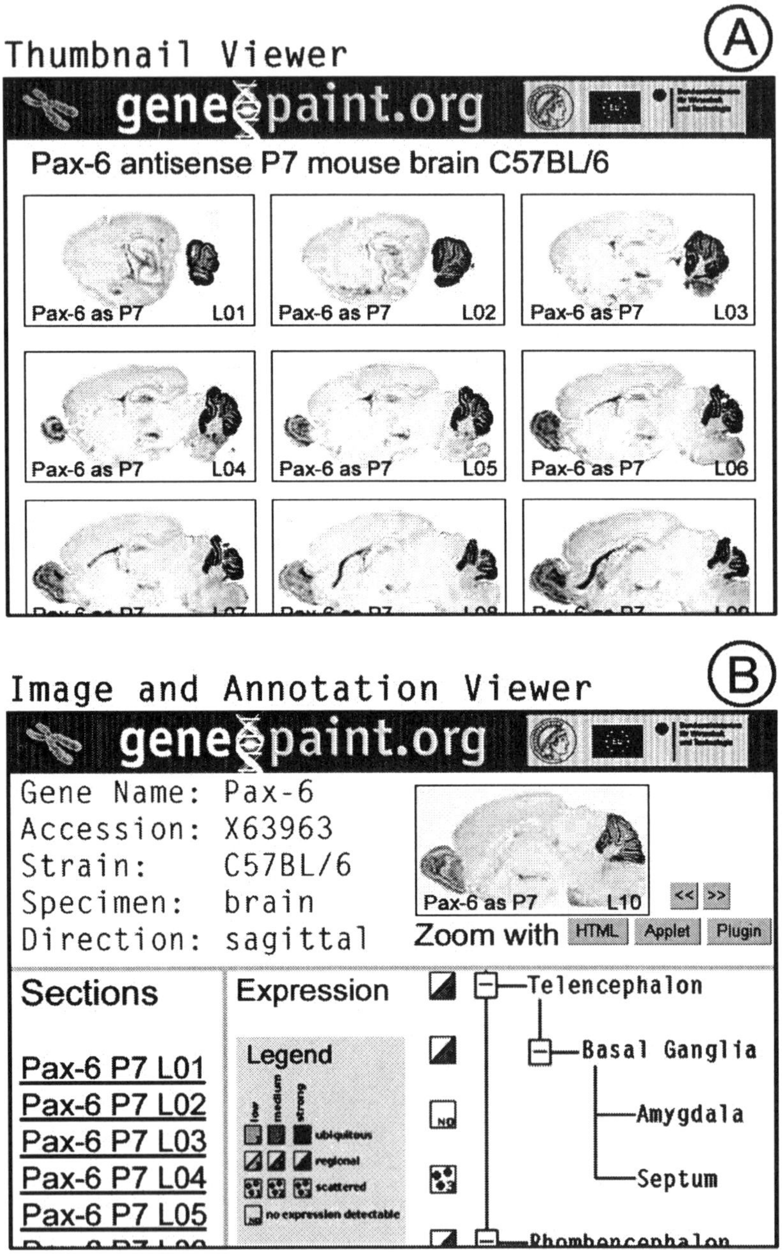

Figure 4. Scheme showing the main features of the *thumbnail viewer* (A) and the *image and annotation viewer* (B). See text for details.

4.3 Thumbnail Viewer

The *thumbnail viewer* shows small JPEG representations of all images from one GenePaint set on a single page (Fig. 4A). While size and resolution of these images are limited to facilitate loading, they provide an overview of the data in order to identify sections with structures of interest. Each thumbnail is linked to the corresponding image in the *image and annotation viewer*, where in turn zoom and annotation functions are available.

4.4 Image and Annotation Viewer

This page shows expression patterns in form of images of sections and as a table (Fig. 4B). All available images are tabulated in the left bottom frame, sorted by the position of each section within the GenePaint set. Clicking a particular line in this list will produce the appropriate thumbnail in the mini-viewer window at the top of the page. Instead of directly selecting single sections from the list, it is also possible to browse through the thumbnails using forward and back buttons. For viewing high resolution images, one of the three zoom viewer options (HTML, applet, or plugin) is selected, which then opens a new window containing the *zoom viewer* (section 4.5).

The right bottom frame contains a hierarchical tree of the anatomical structures (see section 2.4). Square icons indicate the level and pattern of expression. Thus for each GenePaint set there is such an annotated tree and images to be viewed using the zoom server.

4.5 Zoom Viewer

Three different *zoom viewers* are available: (1) a universal viewer that is strictly HTML-based, (2) a Java applet, and (3) a browser plugin that can be downloaded free of charge and is available for Windows and MacIntosh operating systems. The universal viewer runs with any web browser without the need to download and run an applet or to install a plugin. It is therefore best suited for use with slow computers, low bandwidth internet connections or with uncommon browser software or operating systems.

The applet and and plugin versions are more restrictive regarding the user's hard- and software, but in return they provide a more convenient interface for zooming and sliding the image. While a complete new JPEG image has to be loaded each time a user zooms or scrolls the image in the universal viewer, the applet and plugin versions feature soft scrolling with dynamic loading and displaying of high resolution image data. In addition, a movable "magnifying-glass" window with an adjustable zoom factor is available in the applet version (Fig. 5).

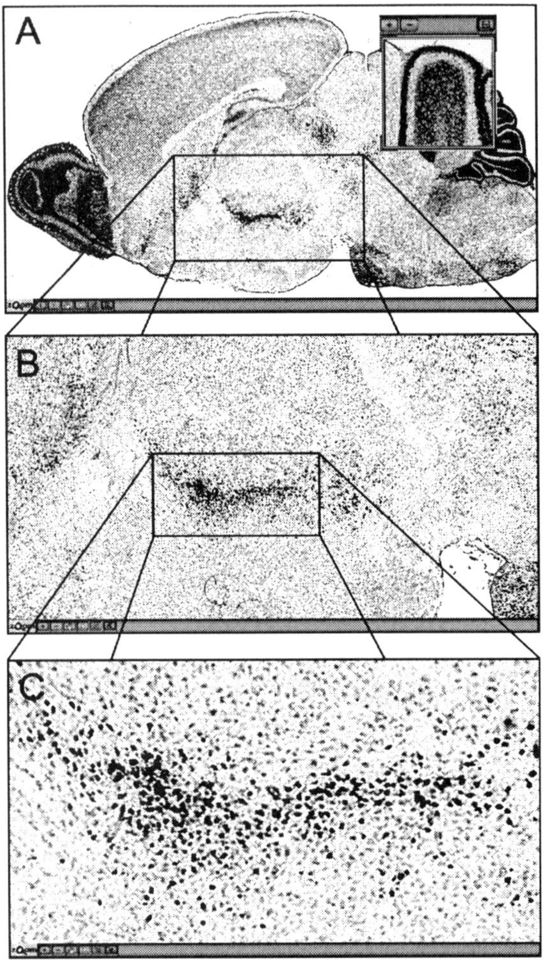

Figure 5. Functionality of the zoom viewer, demonstrated on a sagittal cryosection of a postnatal day 7 mouse brain, C57BL/6 strain, hybridized with an antisense riboprobe for the transcription factor *paired-box 6* (*pax-6*). A - Overview at low magnification. B - Zooming into the septal and ventral thalamus region using the Java applet viewer. C - Expression in the ventral thalamus at maximum zoom factor. Single *pax-6*-positive (black) and negative (opaque) neurons can be distinguished. Inset in A: The magnifying glass tool is activated and slid to the cerebellum to examine it at higher resolution.

4.6 Expression Pattern Overview

The list of search results may be very extensive and it would take a considerable amount of effort to inspect many sets of images using either the *thumbnail* or the *image and annotation viewer*. Therefore, we have created the *expression pattern overview page*, which is a tabulated side-by-side comparison of expression levels and patterns (as defined in section 2.4) of several GenePaint sets. In this table, each row represents one GenePaint set, while each column represents one distinct anatomical structure. Each row contains buttons that directly open the respective GenePaint set either in the *thumbnail viewer* or in the *image and annotation viewer*.

ACKNOWLEDGMENTS

We would like to thank James Carson (Baylor College of Medicine) for helpful comments on the manuscript. The development and use of GenePaint is funded by the National Institute of Neurological Disorders and Stroke (NO1-NS-0-2332), the German Ministry of Research (01 KW9965), the European Union (QLG2-1999-00793), the Merck Genome Research Institute (152), and the Burroughs Wellcome Fund (1001299). A.V. is supported by a Boehringer Ingelheim Fonds scholarship.

REFERENCES

Adams, J. (1992) Biotin amplification of biotin and horseradish peroxidase signals in histochemical stains. *J Histochem Cytochem*, 40, 1457-1463.

Adams, M., Celniker, S., Holt, R., Evans, C., Gocayne, J., Amanatides, P., et al. (2000) The genome sequence of Drosophila melanogaster. *Science*, 287, 2185-2195.

Altschul, S., Gish, W., Miller, W., Myers, E., Lipman, D. (1990) Basic local alignment search tool. *J Mol Biol*, 215, 403-410.

C.elegans Sequencing Consortium (1998) Genome sequence of the nematode C. elegans: a platform for investigating biology. *Science*, 282, 2012-2018.

Herzig, U., Cadenas, C., Sieckmann, F., Sierralta, W., Thaller, C., Visel, A., Eichele, G. (2001) Development of high-throughput tools to unravel the complexity of gene expression patterns in the mammalian brain. *Novartis Found Symp*, 239, 129-146.

Kaufman, M. H. (1992) *The Atlas of Mouse Development* London: Academic Press

Kerstens, H., Poddighe, P., Hanselaar, A. (1995) A novel in situ hybridization signal amplification method based on the deposition of biotinylated tyramine. *J Histochem Cytochem*, 43, 347-352.

Lander, E., Linton, L., Birren, B., Nusbaum, C., Zody, M., Baldwin, J., et al. (2001) Initial sequencing and analysis of the human genome. *Nature*, 409, 860-921.

Nielsen, T., West, R., Linn, S., Alter, O., Knowling, M., O'Connell, J., et al. (2002) Molecular characterisation of soft tissue tumours: a gene expression study. *Lancet*, 359, 1301-1307.

Paxinos, G., Franklin, K. (2001) *The Mouse Brain in Stereotaxic Coordinates*. 2nd ed. San Diego: Academic Press.

Schambra, U., Lauder, J., Silver, J., eds (1992) *Atlas of the Prenatal Mouse Brain*. San Diego: Academic Press.

Valverde, F. (1998) *Golgi Atlas of the Postnatal Mouse Brain* Wien: Springer-Verlag

van 't Veer, L., Dai, H., van de Vijver, M., He, Y., Hart, A., Mao, M., et al. (2002) Gene expression profiling predicts clinical outcome of breast cancer. *Nature*, 415, 530-536.

Venter, J., Adams, M., Myers, E., Li, P., Mural, R., Sutton, G., et al. (2001) The sequence of the human genome. *Science*, 291, 1304-1351.

Chapter 3

Databases for the Functional Analyses of Olfactory Receptors

Chiquito J. Crasto[1], Nian Liu[1] and Gordon M. Shepherd[2]
[1]Yale Center for Medical Informatics and [2]Section for Neurobiology, Yale University School of Medicine, New Haven, Connecticut, USA
Correspondence to: gordon.shepherd@yale.edu

Abstract: The olfactory receptors constitute over 1000 genes, making it the largest family in the genome. The completion of the human and mouse genomes presents a tremendous challenge to understanding the odor ligand affinities of these receptors and their relation to odor processing in the olfactory system. We describe briefly the practical use of a set of databases in the SenseLab project (http://senselab.med.yale.edu/) that is devoted to this problem. Olfactory Receptor Database (ORDB) archives almost 2000 sequences, and provides tools for their analysis by type. OdorDB stores data about the structures of odor molecules and their affinities for different olfactory receptors, as established by studies in expression systems. Odor simulation gives rise to odor maps in the olfactory bulb which is believed to be a basis for odor discrimination. OdorMapDB provides a site for integration of data from odor mapping studies by different methods. These databases thus enable students and researchers to integrate molecular analysis of olfactory receptors with the functional organization of the olfactory system to understand the neural basis of the sense of smell.

Key words: olfactory receptors, odorant molecules, odor maps, databases, molecular models and nomenclature

1. INTRODUCTION

Olfactory receptors belong to the superfamily of GTP-binding protein coupled receptors (GPCRs) that are characterized by seven transmembrane helical regions (Buck and Axel, 1991). Olfactory or olfactory like receptor genes are expressed in high numbers in the olfactory epithelium (Parmentier et al., 1992; Dryler, 1996; Weth, Nadler and Korsching, 1996). Interestingly, these genes are found beyond the olfactory system, often expressed in tissues as far removed as the testes (Goto et al., 2001) and the heart (Hillier et al., 1996)—the different locations point to the varied functionality of the proteins encoded by these genes.

A publicly funded consortium of researchers (Lander et al., 2001) and, in the private sector, researchers at Celera Genomics (Venter et al., 2001) independently published a first draft of the human genome. Recently, Celera Genomics published the first draft of the mouse genome. This seminal work provides a new base for research on olfactory receptors and their genes.

Two groups separately mined olfactory receptors from the human genome in 2001 from the public human genome sequence (Glusman et al. 2001; Zozulya et al., 2001), and two groups separately mined olfactory receptors from the mouse genome provided by Celera in 2002 (Zhang and Firestein, 2002; Young et al., 2002). Sources of genomic information such as GENBANK at the National Center for Biotechnology Information (http://www.ncbi.nih.gov/Genbank/) are storehouses of gene and protein sequences for olfactory receptors for different species deposited by researchers. It is timely therefore, to synthesize the newly mined information with that already available to provide new insights into the evolution of olfactory receptors and new directions for research on the molecular and neural basis of the sense of smell.

For many years it has been estimated that mammals have some 1000 olfactory genes, making it the largest family in the genome (Mombaerts, 1999a; 1999b; Pilpel, Sosinsky and Lancet, 1998). This family is estimated to constitute 3% of the genes that comprise the human genome. There is a progression from mostly functional genes in fish and rodents to a high proportion of non-functional genes (pseudogenes) in humans (Roquier et al, 2000; Glusman et al. 2001; Zozulya et al., 2001; Zhang and Firestein, 2002). Glusman et al. (2001) and Zozulya et al. (2001) who mined these sequences from the public draft of the human genome have identified respectively, approximately, 360 and 347 olfactory receptor genes of the greater than 900 sequences projected as ORs. The remaining genes remain non-identified or have been identified as non-functional. Glusman et al. (2001) have performed additional pseudogene analysis and identified two thirds of the human olfactory receptor subgenome as pseudogenes. In the mouse, 1296 OR genes have been identified of which some 220 are pseudogenes.

The implication is that as species require less olfactory acuity, molecular disruptions accumulate and erode the functionality of olfactory receptor genes. Recent psychophysical studies indicate however, that humans have a sense of smell that equals or in some cases exceeds that of rats and dogs (Lehmer, Gluck and Laska, 1999; Laska and Hubener, 2001). Understanding how the olfactory receptors mediate and constrain the sense of smell is one of the main aims of current research in our laboratory and many others.

2. OLFACTORY RECEPTOR DATABASE (ORDB)

ORDB (http://senselab.med.yale.edu/senselab/ORDB) at the Yale University School of Medicine is a central repository for olfactory receptors. It is part of a suite of databases of the SenseLab project dedicated to archiving neuronal membrane properties in order to support research on gene expression patterns and the integrative properties of neurons and neuronal dendrites (http://senselab.med.yale.edu/). Three of the six databases in SenseLab are related to olfactory receptors and the sense of olfaction: ORDB, OdorDB and OdorMapDB. ORDB is a member of the GPCR databases that includes other similar databases: GPCRDB (http://www.gpcr.org/) maintained at the Center for Molecular and Biomolecular Informatics (CMBI) in the Netherlands, and GRAP (http://tinygrap.uit.no/GRAP) based at the University of Tromsø in Norway is the GPCR mutant database. With the publications of the human and mouse olfactory receptors, additional databases have been developed and maintained; namely, HORDE, a storehouse of human olfactory receptors housed at the Weizmann Institute of Science in Israel (http://bioinformatics.weizmann.ac.il/HORDE/); and, mouse olfactory receptors independently, experimentally cloned and mined from the Celera Mouse Genome are housed at the Fred Hutchinson Cancer Research Center (http://www.fhcrc.org/labs/trask/OR) (Young et al. 2002); the latter also contains links to human olfactory receptor orthologs in HORDE.

The ORDB has an extensive user list which includes laboratories involved in olfactory receptor or GPCR research. The database is overseen by an Advisory Committee, reporting to the Director of the National Institute on Deafness and Other Communicative Disorders, which reviews the functioning of the database and makes recommendations as to its future directions. Regular newsletters apprise the Advisory Committee and User Group of recent changes and innovations to the database.

To date, ORDB contains more than 2000 chemosensory receptor gene sequences representing the work of more than one hundred laboratories worldwide for 35 species and expressed in more than thirty different organs and tissues. ORDB is expected to triple in size after all the recently mined human and mouse olfactory receptors are deposited. Besides olfactory receptors, ORDB also houses *C. elegans* chemosensory receptors (CeCR), vomeronasal receptors (VNR), insect olfactory receptors (IOR), fungal pheromone receptors (FPR) and taste papilla receptors (TPR)—the last, with olfactory receptors, drawing particular attention in the field of chemosensory receptors. Vomeronasal or pheromone receptors expressed in the vomeronasal organ are of interest to perfumes manufacturers. Each of the sequences of all the above types are described by: the source species, chromosomal location, GENBANK and MEDLINE links, the laboratory where the sequences were cloned and identified, the source of the sequences—usually GENBANK, the chromosomal location, the organism strain, gene length—full length or partial, the ligand or molecule of olfaction that is expressed as binding to the receptor, the nucleotide and protein sequences, the type of sequences—genomic, cDNA or pseudogenic, and a molecular model, if any exists. These properties or attributes serve as descriptors for each olfactory receptor.

2.1 Structure of the SenseLab Databases

Each of the databases in SenseLab is constructed using a unique database structure called EAV/CR (Entity Attribute Value with Classes and Relationships) (Nadkarni et. al. 1999). This schema is an extension of the often-used EAV schema (Chen et al., 2000; Nadkarni, Brandt and Marenco, 2000) and is used to store heterogeneous data. Rather than incorporate all the data in one table, different tables are used to store different types of data.

Figures 1 and 2 illustrate the working of the EAV/CR schema for olfactory receptors. The example is for I7, one of the first olfactory receptors cloned. All the sequences in ORDB belong to the **Class** Olfactory Receptor. ORL11 is an **entity** (ORDB name given to the rat I7 olfactory receptor) related to the class of olfactory receptors. Each entity is stored as objects identified by a unique identification number. The entities are described by **attributes**. Each attribute is qualified by **values**. The overall database schema then **relates** the classes, entities, attributes and values.

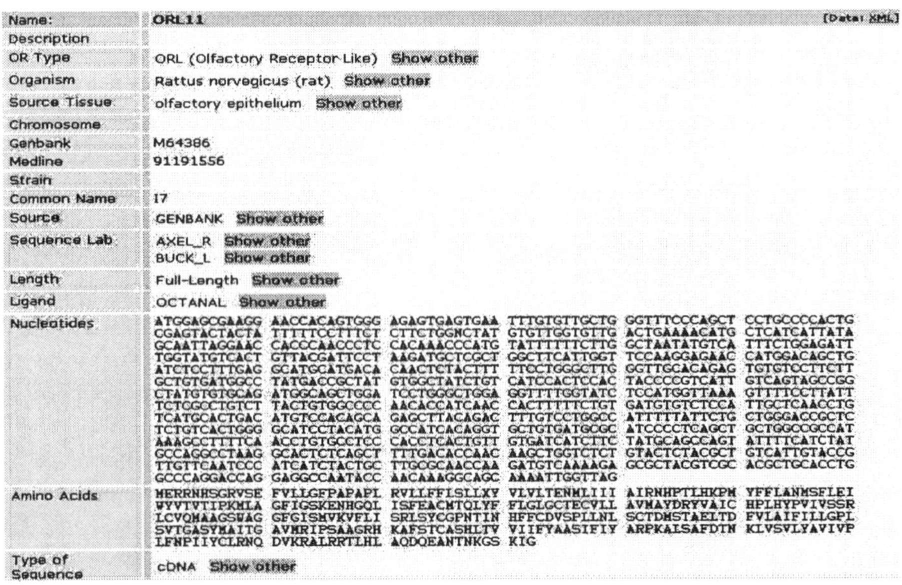

Name:	ORL11 [Data: XML]
Description	
OR Type	ORL (Olfactory Receptor Like) Show other
Organism	Rattus norvegicus (rat) Show other
Source Tissue	olfactory epithelium Show other
Chromosome	
Genbank	M64386
Medline	91191556
Strain	
Common Name	I7
Source	GENBANK Show other
Sequence Lab	AXEL_R Show other BUCK_L Show other
Length	Full-Length Show other
Ligand	OCTANAL Show other
Nucleotides	ATGGAGCGAAGG AACCACAGTGGG AGAGTGAGTGAA TTTGTGTTGCTG GGTTTCCCAGCT CCTGCCCCACTG CGAGTACTACTA TTTTTCCTTTCT CTTCTGGNCTAT GTGTTGGTGTTG ACTGAAAACATG CTCATCATTATA GCAATTAGGAAC CACCCAACCCTC CACAAACCCATG TATTTTTTCTTG GCTAATATGTCA TTTCTGGAGATT TGGTATGTCACT GTTACGATTCCT AAGATGCTCGCT GGCTTCATTGGT TCCAAGGAGAAC CATGGACAGCTG ATCTCCTTTGAG GCATGCATGACA CAACTCTACTTT TTCCTGGGCTTG GGTTGCACAGAG TGTGTCCTTCTT GCTGTGATGGCC TATGACCGCTAT GTGGCTATCTGT CATCCACTCCAC TACCCCGTCATT GTCAGTAGCCGG CTATGTGTGCAG ATGGCAGCTGGA TCCTGGGCTGGA GGTTTTGGTATC TCCATGGTTAAA GTTTTCCTTATT TCTCGGCTGTCT TACTGTGGCCCC AACACCATCAAC CACTTTTTCTGT GATGTGTCTCCA TTGCTCAACCTG TCATGCACTGAC ATGTCCACAGCA GAGCTTACAGAC TTTGTCCTGGCC ATTTTTATTCTG CTGGGACCGCTC TCTGTCACTGGG GCATCCTACATG GCCATCACAGGT GCTGTGATGCGC ATCCCCTCAGCT GCTGGCCGCCAT AAAGCCTTTTCA ACCTGTGCCTCC CACCTCACTGTT GTGATCATCTTC TATGCAGCCAGT ATTTTCATCTAT GCCAGGCCTAAG GCACTCTCAGCT TTTGACACCAAC AAGCTGGTCTCT GTACTCTACGCT GTCATTGTACCG TTGTTCAATCCC ATCATCTACTGC TTGCGCAACCAA GATGTCAAAAGA GCGCTACGTCGC ACGCTGCACCTG GCCCAGGACCAG GAGGCCAATACC AACAAAGGCAGC AAAATTGGTTAG
Amino Acids	MERRNHSGRVSE FVLLGFPAPAPL RVLLFFLSLLXY VLVLTENMLIII AIRNHPTLHKPM YFFLANMSFLEI WYVTVTIPKMLA GFIGSKENHGQL ISFEACMTQLYF FLGLGCTECVLL AVMAYDRYVAIC HPLHYPVIVSSR LCVQMAAGSWAG GFGISMVKVFLI SRLSYCGPNTIN HFFCDVSPLLNL SCTDMSTAELTD FVLAIFILLGPL SVTGASYMAITG AVMRIPSAAGRH KAFSTCASHLTV VIIFYAASIFIY ARPKALSAFDTN KLVSVLYAVIVP LFNPIIYCLRNQ DVKRALRRTLHL AQDQEANTNKGS KIG
Type of Sequence	cDNA Show other

Figure 1. Web page for ORL11 the gene entry for rat I7 as it appears in the ORDB showing Objects, Attributes and Values of different data-types illustrating the EAV/CR schema

For instance, the entities in the class olfactory receptors are described by attributes: organism, common name, chromosomal location, DNA and protein sequence, and sources of sequences. Each of these attributes is associated with a value that can either be short strings of text, long strings or memos and integers. Values that are likely to recur in the database are themselves stored as objects (blue links in figure 1).

At the right hand corner of figure 1 is a link called "XML." Clicking this link results in a page for ORL11 (Figure 2). XML (eXtensible Markup Language) is now a commonly used method of restructuring data for data transfer, data mining and web presentation (Means, 2001). XML tags starting with "a" followed by the unique number denote attributes. "37" is the attribute identifier for "Sequence Lab" or the laboratory where the sequence was identified or cloned. The values for this olfactory receptor entry attribute "a37" are "o848" and "o855." Note that the names of the researchers involved are not identified as text.


```
- <data>
  - <d4>
    - <c22>
      - <o559 umls_cid="" version="5" version_date="7/17/2001 10:53:03 AM">
          <name>ORL11</name>
          <a80 object_name="ORL (Olfactory Receptor Like)">o1364</a80>
          <a30 serial_no="" object_name="Rattus norvegicus (rat)">o804</a30>
          <a31 object_name="olfactory epithelium">o821</a31>
          <a45>M64386</a45>
          <a44>91191556</a44>
          <a35>I7</a35>
          <a36 object_name="GENBANK">o835</a36>
          <a37 serial_no="" object_name="AXEL_R">o848</a37>
          <a37 serial_no="" object_name="BUCK_L">o855</a37>
          <a39 object_name="Full-Length">o1362</a39>
          <a82 serial_no="" object_name="OCTANAL">o124</a82>

           <a40>ATGGAGCGAAGGAACCACAGTGGGAGAGTGAGTGAATTTGTGTTGCTGGGTTT

           <a41>MERRNHSGRVSEFVLLGFPAPAPLRVLLFFLSLLXYVLVLTENMLIIIAIRNHPTLHK
          <a42 object_name="cDNA">o842</a42>
          <a33>M64386</a33>
          <a46>OLF7_RAT.SW.html</a46>
          <a99>rat_I7.mpg</a99>
        </o559>
      </c22>
    </d4>
  </data>
```


Figure 2. XML file of the ORL11 entry or rat I7 olfactory receptor gene sequence. The attribute and object numbers are as stored in ORDB. The nucleotide and protein sequences have been truncated.

For the olfactory receptor ORL11, the information for each data type is stored in different tables each connected by the unique object identifier- "559" for ORL11. The ingenuity of the EAV/CR schema is that the entity

ORL11 comes under the data type "object" and is stored in the same table as say, the Sequence Laboratory object. Common web based scripts for each data type can be invoked if information is desired for any "object." The page denoted in Figure 1 can be obtained by typing the link http://senselab.med.yale.edu/senselab/ORDBe/eavData.asp?o=559; in the above URL, if the number 559 is replaced by 848, the same script will present information about Professor Richard Axel at Columbia University for which the attributes are Name, Affiliation, Phone number and Email address (Figure 3). "AXEL_R" being the value for attribute Sequence Lab for ORL11 stored as an object.

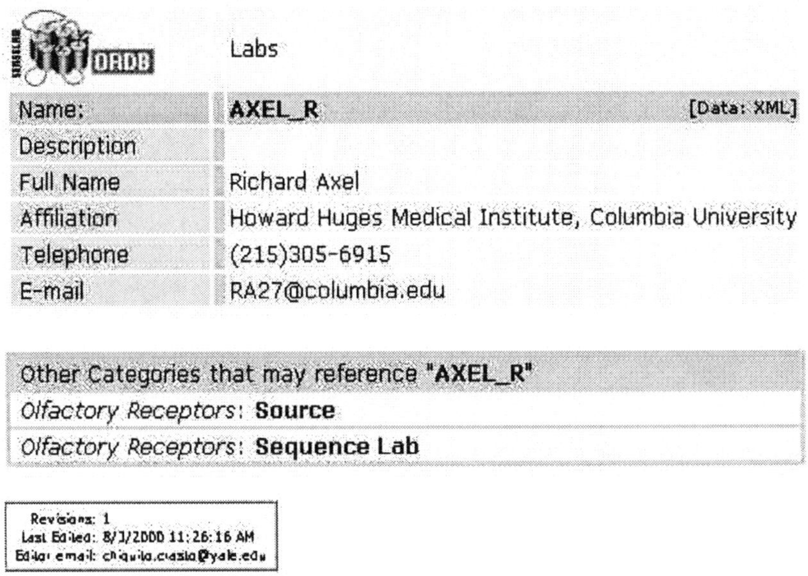

Figure 3. An Illustration of generic script that presents object descriptions in ORDB. The same script presents Object "AXEL_R" as ORL11 in Figure 1 to ORDB users

ORDB enables users to access the information by means of attribute and object searches. It also contains a local copy of the National Center for Biotechnology Information (NCBI) BLAST program (Altschul et al., 1990; 1997) that enables users to identify sequence similarities in their target sequences. The generalized schema enables users to use ORDB as their private database by maintaining their own sequences (prior to publication). Privately stored sequences are only available to the users who deposit them.

2.2 Populating the databases

Populating the OR database rapidly and efficiently so as to make information available is one of its curators' primary responsibilities. To this end, an automatic population method has been developed that mines information from sources of gene (GENBANK) and protein sequences (SWISSPROT) or user-supplied sequences (Crasto et al., 2002). These sequences are then deposited one at a time or *en masse* into ORDB. The automatic population program AUTO/POP is web based and sequences can be deposited remotely (with user privileges). AUTO/POP is available as a black box with a simple interface that does not require the user to know the inner workings of the algorithm. The ease of using the AUTO/POP interface as a means of communication between ORDB and GENBANK once again makes use of an XML file. AUTO/POP converts information mined from GENBANK into an XML file similar to the file in Figure 2 after integrating the mined data with ORDB architecture. The same program that extracts relevant information, presents it as an XML file and uses information created in the XML format to deposit it into the database. The AUTO/POP program is available at http://senselab.med.yale.edu/senselab/autoput/autopop.pl.

Recently two additional web pages have been added to ORDB. These represent sequence analyses of human and mouse olfactory receptors. The olfactory receptor gene sequences mined from the mouse and human genomes by different groups were analysed for sequence identity and similarities. The sequences were processed for complete and partial matches, and functionality. A color code identifies sequences as full-length vs. partial and functional vs. pseudogene. The name given to the sequence by each of the sources where the sequences were mined is included in the web page. Each sequence then links back to the database of the source laboratories. If such a database is not available, the sequence name links back to the GENBANK entry. The human and mouse OR sequence analyses results are at:
http://senselab.med.yale.edu/senselab/ORDB/files/humanorseqanal.html and
http://senselab.med.yale.edu/senselab/ORDB/files/mouseorseqanal.html.

2.3 Nomenclature for a large gene family

Providing a unified nomenclature for olfactory receptors has been under consideration by researchers in the community. Often, sequence identity has been a starting point. The sequence identity between any two randomly selected olfactory receptor genes is generally 40% or higher. The lowest reported percentage identity is 20%. Glusman et al. (2001), who reported average protein sequence identities among olfactory receptors in agreement with those reported by Zozulya et al. (2001), also calculated sequence relationships with 55 non-olfactory GPCR sequences. The average percentage identity of 27% for olfactory:non-olfactory pairs was less than that for olfactory:olfactory pairs. For the human olfactory receptors: Glusman et al. (2001) use a classification based on families and sub families.

The HORDE classification hOR10H11 means that this receptor belongs to family 10 and is the eleventh gene of sub family H. That olfactory receptors sequences with a percentage sequence identity of 40% or greater constitute a family, and sequences with sequence identity of 60% or greater a subfamily. Zozulya et al. (2001) at Senomyx Inc., on the other hand, use a different scheme. For example, hOR14.02.03 where the gene is identified as localized on chromosome 14, and is the third member of family 2. In this scheme, the sequence identity required for an olfactory receptor to qualify as a member of a family was 43%. For the mouse olfactory receptors housed at the Fred Hutchinson Cancer Research Center the nomenclature used is given to sequences by researchers at Celera. Some researchers believe that chromosomal localization is not the best form of classification because of possible misplacement during data mining or genome sequencing which may become evident when future experimental work is carried out to confirm sequencing results (Barbara Trask and Janet Young, Fred Hutchinson Cancer Research Centre, Personal Communications). ORDB nomenclature for olfactory receptors consists of sequential numbering, based on the availability of the sequence to the database, with the prefix "ORL" (Olfactory Receptor Like). This numbering scheme does not presuppose chronology or priority as sequences from different sources become available at different times.

2.4 Olfactory receptor models

Olfactory receptors, like all membrane proteins, are especially difficult to purify and crystallize. In order to elucidate the mechanism of olfactory receptor–odor interaction, theoretical, computationally derived models are used. These models often use the principles of homology modelling using the x-ray crystallography derived structure of the protein rhodopsin and bacteriorhodopsin. The first such modelling and docking study was completed by Michael Singer who studied the interaction of rat I7 (ORL11 in ORDB) with octanal (Singer, 2000). Singer's studies suggest that octanal binds with I7 in a pocket approximately 10Å from the extracellular surface formed by transmembrane (TM) domains 3–7. More recently, Floriano et al. (Floriano, Vaidehi and Goddard, 2001) modelled the mouse receptor OR S25 (ORL466 in ORDB) and predicted that TMs 3, 5, and 6 have residues directly involved in the binding pocket for hexanol and heptanol, with TM4 playing an indirect role. The GPCRDB (http://www.gpcr.org/7tm/models) website maintains models (including olfactory receptors) contributed by several researchers. Several programs are available to study docking in proteins: DOCK housed at University of California at San Francisco (http://www.cmpharm.ucsf.edu/kuntz/dock.html) (Kuntz et al., 1992), AUTODOCK (http://www.scripps.edu/pub/olson-web/doc/autodock) at the

Scripps Research Institute (Morris et al., 1998) and at the University of Illinois at Chicago—CASTp (http://cast.engr.uic.edu/cast/) and can be used to study docking of molecules of olfaction with olfactory receptors (Lie, Edelsbrunner and Woodward, 1998). A recently added feature to ORDB is an attribute called molecular models. Visual graphical displays (movie and picture files) of computationally derived models of the seven helical TM domains, with or without the interacting ligands, can be stored in ORDB. A sample molecular model of I7-octanal created by Singer (Singer, 2000) http://senselab.med.yale.edu/senselab/ORDB/molmodels/rat_I7.mpg can be viewed by ORDB users.

3. ODOR DATABASE (ODORDB)

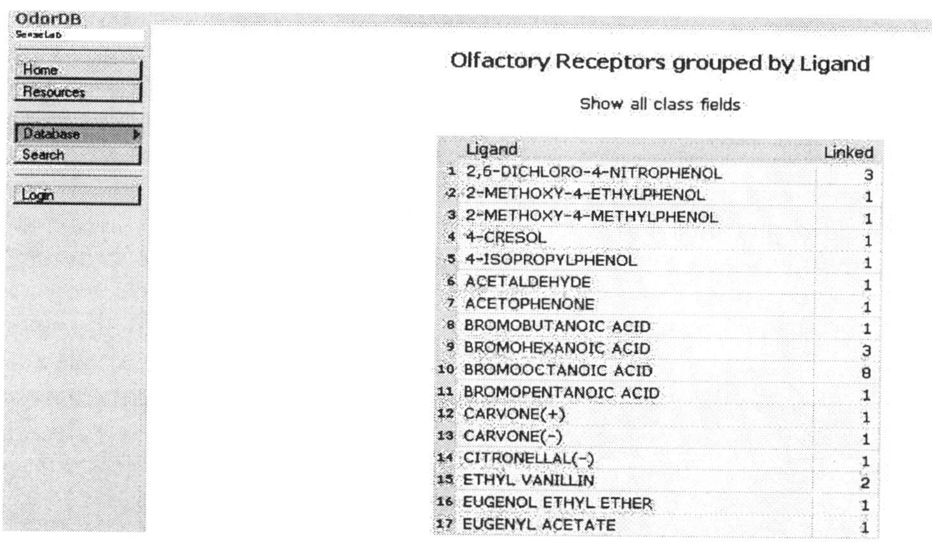

Figure 4. List of odors in OdorDB

OdorDB (http://senselab.med.yale.edu/senselab/OdorDB), a companion database to ORDB serves as a repository for organic molecules of smell. OdorDB is not an arbitrary list of organic compounds. Only those odor molecules that are identified as binding to olfactory receptors are stored in OdorDB.

The obvious connection then is between the list of ligands in ORDB and the odors in OdorDB. Each odor in OdorDB links back to its olfactory receptor in ORDB. Figures 4 and 5 provide examples of the odor (ligand) list in ORDB and OdorDB.

In each database the list of odors is extracted from the same location in an Oracle Database using Active Server Pages (ASP) that present information to the user on the Internet while retaining the unique features of each database. The database also includes features that allow database administrator and users to add (or delete) information and make modifications to the database architecture using internet tools—a distinct advantage for remote database management and curation.

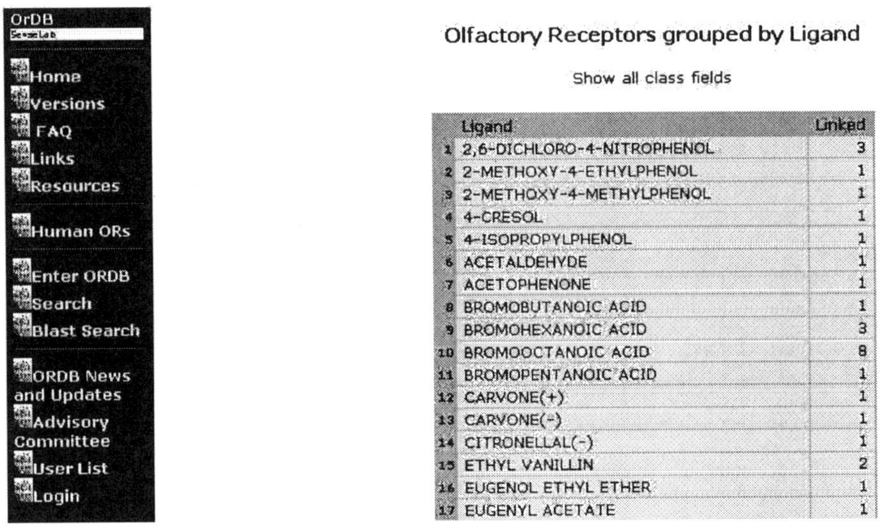

Figure 5. List of ligands in ORDB

4. ODOR MAP DATABASE (ODORMAPDB)

The most recent of the six SenseLab databases is Odor Map Database (http://senselab.med.yale.edu/senselab/OdorMapDB). It is aimed at supporting the experimental analysis of the molecular and functional organization of the main olfactory bulb and to further our understandings of the basis for the perception of smell. Odor maps represent spatial activity patterns in the olfactory bulb glomerular layer elicited by stimulation with different odors (Xu et al., 2000a). Glomeruli are structural and functional modules formed by the axon terminals of projecting olfactory receptor cells, the apical dendrites of mitral cells and processes of the surrounding periglomerular neurons. In essence, odor maps are unique fingerprints of specific odor molecules in a determined concentration on the glomerular layer.

Odor maps can be obtained using a range of different activity imaging methods. The first method used 2-deoxyglucose (2-DG) to study the activity patterns in the olfactory bulb glomerular layer elicited by exposure to different odors such as camphor and amyl acetate (Stewart et al., 1979). The discovery of distinguishing but overlapping densities in radio-labeling in the bulb in response to these two different odors indicated that neural space is used in this system to encode odor identity as well as odor concentration.

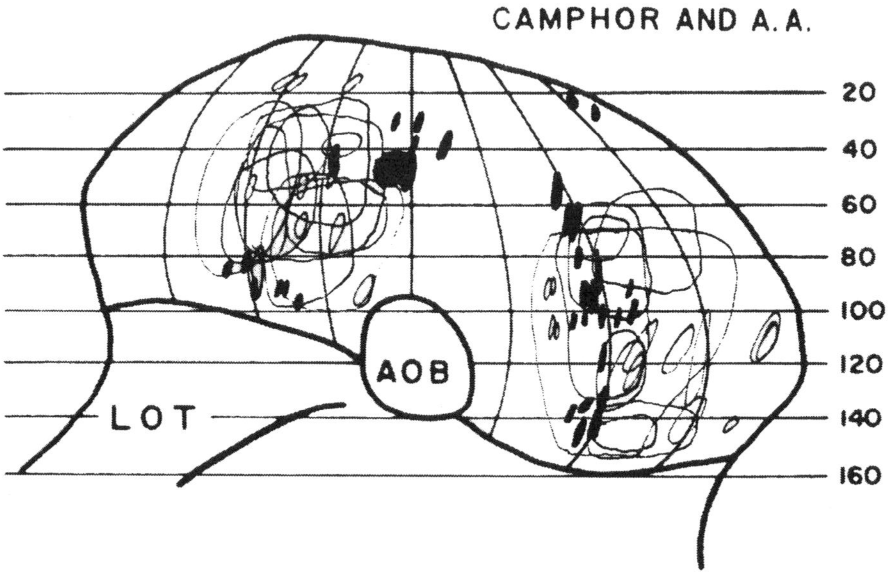

Figure 6. Odor maps for odorants camphor and amyl acetate (AA) super-imposed on the mapping coordinates. The activity patterns in rat elicited by stimulation of camphor (filled circles) and AA (open circles) are partially overlapped but distinguishable. LOT, lateral olfactory tract; AOB, accessory olfactory bulb. Adapted from Stewart et al., 1979.

The 2-DG technique has been further exploited by Leon and co-workers at the University of California in Irvine (Johnson and Leon, 2000), who have posted many of their odor maps at their website (http://leonlab.bio.uci.edu). Many other methods have been introduced to visualize and analyse the odor activity maps. Each has its advantages and disadvantages (summarized in Xu et al, 2000a), which will require careful correlations between the maps obtained with the different methods and with the maps of the olfactory receptor neuron projections to the glomeruli. Among the most recent methods is high-resolution fMRI (Xu et al., 2000b).

The archived data in OdorMapDB are from original studies using 2-DG and current studies using fMRI in mouse and rat. OdorMapDB is integrated with the two other olfactory databases in SenseLab: ORDB and OdorDB. The linkages provide more detail about the odor molecule or the receptor related to a particular odor map, and vice versa. As a general rule for

SenseLab databases, OdorMapDB also contains links that help users' access external related resources, a first step in forming a community of odour mappers.

REFERENCES

Altschul S.F., Gish W., Miller W., Myers E.W., Lipman D.J. (1990) Basic local alignment search tool. *J Mol Biol* 215, 403-10

Altschul S.F., Madden T.L., Schaeffer A.A., Zhang J., Zhang Z., Miller W., Lipman D.J. (1997) Gapped BLAST and PSI-BLAST: a new generation of protein database search programs. *Nucleic Acids Res* 25, 3389-402

Buck L., Axel R. (1991) A novel multigene family may encode odorant receptors: a molecular basis for odor recognition. *Cell* 65, 175-81

Chen R.S., Nadkarni P., Marenco L., Levin F., Erdos J., Miller P.L. (2000) Exploring performance issues for a clinical database organized using an entity-attribute-value representation. *J Am Med Inform Assoc* 7, 475-87

Crasto C., Marenco L., Miller P., Shepherd G. (2002) Olfactory Receptor Database: a metadata-driven automated population from sources of gene and protein sequences. *Nucl Acids Res* 30, 354-60

Floriano W.B., Vaidehi N., Goddard W.A. 3rd, Singer M.S., Shepherd G.M. (2000) Molecular mechanisms underlying differential odor responses of a mouse olfactory receptor. *Proc Natl Acad Sci USA* 97, 10712-6

Freitag J., Ludwig G., Andreini P., Roessler P., Breer H. (1998) Olfactory receptors in aquatic and terrestrial vertebrates. *J Comp Physiol* 183, 635-50

Glusman G., Yanai I., Rubin I., Lancet D. (2001) The Complete Human Olfactory Subgenome. *Genome Res* 11, 685-702

Goto T., Salpekar A., Monk M. (2001) Expression of a testis-specific member of the olfactory receptor gene family in human primordial germ cells. *Mol Hum Reprod* 7, 553-8

Hillier L., Lennon G., Becker M., Bonaldo M.F., Chiapelli B., Chissoe S., Dietrich N., DuBuque T., Favello A., Gish W., Hawkins M., Hultman M., Kucaba T., Lacy M., Le M., Le N., Mardis E., Moore B., Morris M., Parsons J., Prange C. et al. (1996) Generation and analysis of 280,000 human expressed sequence tags. *Genome Res* 6, 807-28

Lander et al. International Human Genome Sequencing Consortium. (2001) Initial sequencing and analysis of the human genome. *Nature* 409, 860-921

Laska M., Hubener F. (2001) Olfactory discrimination ability for homologous series of aliphatic ketones and acetic esters. *Behav Brain Res* 119, 193-201

Lehmer J.P., Gluck J., Laska M. (1999) Odor identification, consistency of label use, olfactory threshold and their relationships to odor memory over human lifespan. *Chem Senses* 24, 337-46

Liang J., Edelsbrunner H., Woodward C. (1998) Anatomy of Protein Pockets and Cavities: Measurement of Binding Site Geometry and Implications for Ligand Design. *Prot Sci* 7, 1884-1897

Johnson B.A., Leon M. (2000) Modular representations of odorants in the glomerular layer of the rat olfactory bulb and the effects of stimulus concentration. *J Comp Neurol* 422, 496-509

Kuntz, I.D. (1992) Structure-based strategies for drug design and discovery. *Science* 257, 1078-82

Means, Harold E. (2001) *XML in a Nutshell: A Desktop Quick Reference*. Cambridge, MA: O'Reilly.
Mombaerts P. (1999a) Seven-transmembrane proteins as odorant and chemosensory receptors. *Science* 286, 707-11
Mombaerts P. (1999b) Molecular biology of odorant recpetors in vertebrates. *Annu Rev Neurosci* 9, 315-20
Morris G. M., Goodsell D.S., Halliday R.S., Huey R., Hart, W.E., Belew R.K., Olson, A.J. (1998) Automated Docking Using a Lamarckian Genetic Algorithm and Empirical Binding Free Energy Function. *J Computational Chemistry* 19, 1639-62
Nadkarni P.M., Marenco L., Chen R., Skoufos E., Shepherd G., Miller P. (1999) Organization of heterogeneous scientific data using the EAV/CR representation. *J Am Med Inform Assoc* 6, 478-93
Nadkarni P.M., Brandt C.M., Marenco L. (2000) WebEAV: automatic metadata-driven generation of web interfaces to entity-attribute-value databases. *J Am Med Inform Assoc* 7, 343-56
Parmentier M., Libert F., Schurmans S., Schiffmann S., Lefort A., Eggerickx D., Ledent C., Mollereau C., Gerard C., Perret J. (1992) Expression of members of the putative olfactory receptor gene family in mammalian germ cells. *Nature* 355, 453-455
Pilpel Y., Sosinsky A., Lancet D. (1998) Molecular biology of olfactory receptors. *Essays Biochem* 33, 93-104
Roquier S., Taviaux S., Trask B., Brand-Arpon V., van den Engh G., Demaille J., Giorgi D. (1998) Distribution of olfactory receptor genes in the human genome. *Nature Genet* 18, 243-50
Roquier S., Blancher A., Dominique G. (2000) The olfactory receptor gene repertoire in primates and mouse: Evidence for reduction of the functional fraction in primates. *Proc Nat Acad Sci USA* 97, 2870-74.
Singer M.S. (2000) Analysis of the Molecular Basis for Octanal Interactions in the expressed Rat I7 Olfactory Receptor. *Chem Senses* 25, 155-65
Springer M.S, Burk A., Kavanagh J.R., Waddell V.G., Stanhope M.J. (1997) The interphotoreceptor retinoid binding protein gene in therian mammals: Implications for higher-level relationships and evidence for loss of function in the marsupial mole. *Proc Nat Acad Sci USA* 94, 13754-9
Stewart W.B., Kauer J.S., Shepherd G.M. (1979) Functional organization of rat olfactory bulb analysed by the 2-deoxyglucose method. *J Comp Neurol* 185, 715-34
Venter J.C., Adams M.D., Myers E.W., Li P.W., Mural R.J., Sutton G.G., Simth H.O., Yandell M., Evans C.A., Holt R.A. et al. (2001) The Sequence of the Human Genome. *Science* 291, 1304-51
Weth F., Nadler W., Korsching S. (1996) Nested expression domains for odorant receptors in zebra fish olfactory epithelium. *Proc Natl Acad Sci USA* 93, 13321-6
Xu F., Greer C.A., Shepherd G.M. (2000a) Odor maps in the olfactory bulb. *J Comp Neurol* 422, 489-95
Xu F., Kida I., Hyder F., Shulman R.G. (2000b) Assessment and discrimination of odor stimuli in rat olfactory bulb by dynamic functional MRI. *Proc Natl Acad Sci USA* 97, 10601-6
Young J.M., Friedman C., Williams E.M., Ross J.A., Tonnes-Priddy L., Trask B.J. (2002) Different evolutionary processes shaped the mouse and human olfactory receptor gene families. *Hum Mol Genet* 11, 535-46
Zhang X., Firestein S. (2002) The olfactory receptor gene super family of the mouse. *Nat Neurosci* 5, 124-33

Zozulya S., Echeverri F., Nguyen T. (2001) The human olfactory receptor repertoire. *Genome Biology* 2, 1-12

Chapter 4

Construction of a Protein-Protein Interaction Database (PPID) for Synaptic Biology

Holger Husi and Seth G.N. Grant
Centre for Neuroscience, University of Edinburgh, Edinburgh, EH8 9JZ, United Kingdom
Correspondence to: Seth.Grant@ed.ac.uk

Abstract: Large molecular datasets from genomic and proteomic studies present logistical problems of analysis, integration and ultimately usefulness, in part because of a shortage of suitable tools for their organization and maintenance. This problem was encountered when we began to investigate the molecular network of synapse proteins controlling neuronal plasticity and learning (Husi et al., 2000). Here, the N-methyl-D-aspartate receptor was part of a multiprotein complex (NRC) comprising 77 proteins, which themselves are part of the post synaptic density (PSD) that comprises over 700 proteins. We describe progress toward the construction of a Protein-Protein Interaction Database (PPID) comprising the NRC and PSD proteins and proteins known to interact with them. This database should serve several purposes including clustering and integration of datasets to facilitate the display and comprehensive analysis of biological systems including the synapse.

Key words: NMDA receptor, protein interaction, mammalian, synapse, cluster analysis, genome, proteome.

1. INTRODUCTION

The physiology of synaptic function has been studied for almost 100 years, yet it is only in the last decade that a significant amount of molecular information has been discovered. The synapse is generally considered as the fundamental unit of cellular anatomy in the nervous system for the processing and transmission of information between nerve cells. The simplest aspect of synapse function is its ability to transmit a signal between two connected nerve cells – synaptic transmission. The strength of synaptic transmission can be modified, a property known as synaptic plasticity. This is usually observed as a change in the efficiency of synaptic transmission in response to patterns of neuronal activity. It is this synaptic plasticity of synaptic transmission that has become the most widely studied cellular model for higher cognitive processes, especially learning and memory.

Substantial progress toward understanding the first molecular aspects of synaptic transmission and synaptic plasticity was made during the 1980's, using a combination of biochemical, pharmacological and physiological methods. Molecular cloning techniques, and more recently proteomic methods, have increased the number of proteins known at the synapse by over 10 fold to ~1000 proteins. Indeed, the rate of discovery of molecules has far outstripped the rate at which the function of these molecules can be understood. This explosion of molecular information is well illustrated by the studies of the N-methyl-D-aspartate receptor (NMDAR) a subtype of glutamate receptor, which is found at most excitatory synapses of the mammalian central nervous system.

The NMDAR forms an ion channel that is connected to a larger complex (NMDA Receptor Complex, NRC) of intracellular proteins (Figure 1). There have been ~100 proteins identified within these complexes, of which 1/3 are known to be important in synaptic plasticity. Moreover, there is genetic evidence suggesting the NMDAR signals to multiple downstream pathways mediated by distinct pathways in the NRC. Beyond these general conclusions the striking molecular complexity of the NRC almost certainly suggests a rich array of subtle functions that need a directed bioinformatic approach to guide further experimental study.

2. LARGE SCALE MOLECULAR NETWORKS FOR PROTEIN INTERACTIONS

Proteomic methods based on mass spectrometry and biochemical isolation of complexes generate lists of interacting proteins without specific information on their binary protein-protein interactions. In contrast, yeast 2-hybrid screens identify binary interactions without defining the supramolecular composition of complexes. The fact that proteins that interact are often involved with the same cellular function (often referred to as 'guilt by association') allows a predictive value in having a description of the protein interaction network (Schwikowski et al., 2000; Grant and Blackstock, 2001; Tucker et al., 2001).

Here we will use the term network to indicate sets or lists of proteins, which can interact through specific binary interactions. It is clear that there is not a simple correlation between a (theoretical) network of proteins and

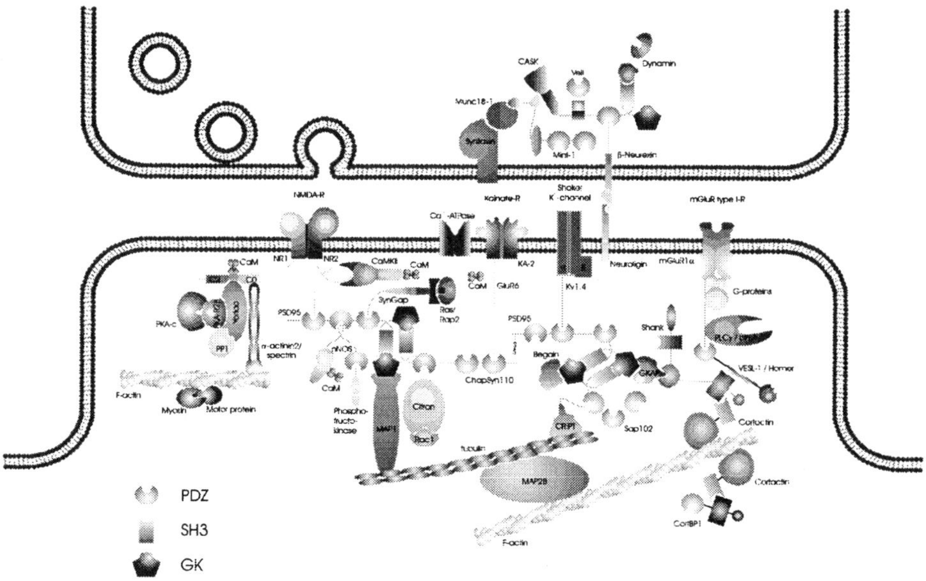

Figure 1. Schematic assembly of proteins found in the NMDA receptor complex (NRC).

Molecules found in the analysis of the NRC are depicted and interactions based on published profiles are indicated. Some proteins, which are composed of known domains, are shown in a modular fashion, e.g. PSD-95 with three PDZ-domains, an SH3 domain and a guanylate kinase (GK) domain. Each of these are Protein-Protein Interaction domains. Abbreviations used: CaM, calmodulin; CaMKII, calcium calmodulin kinase II; MAP, microtubule associated protein; mGluR, metabotropic glutamate receptor; nNOS, neuronal nitric oxide synthase; NR1, NMDA receptor subunit 1; PLA, phospholipase A; PLC, phosospholipase C; PKA, protein kinase A; PP1, protein phosphatase 1; PSD-95, post-synaptic density 95.

actual protein complexes. For example, the affinity of specific interactions will be relevant to the stability of a biochemically defined protein complex and thus the conditions of isolation of the complex may underestimate the possible interactions. Secondly, a network may represent one protein binding several other proteins, whereas it may be that this anchor protein is only capable of binding one protein at a time in vivo.

Binary protein interactions can be linked together to predict large networks of proteins. For example in an analysis of 2,709 protein interactions (using the 2-hybrid method) in yeast *Saccharomyces cerevisiae* a single large network of 2,358 interactions between 1,538 proteins was made and the next largest network contained only 19 proteins (Schwikowski et al., 2000; reviewed in Tucker et al., 2001). Within the larger network emerged a regional architecture where local clusters of proteins corresponded to common functions. These networks represent potential interactions and although they contain errors (Von Mering et al., 2002), the confidence in a predicted association is often enhanced by multiple interactions between sets of proteins with similar functions (Mayer and Hieter, 2000).

The full value of protein interaction networks will only be realised when we go beyond simple (but useful) hypotheses on individual protein function toward an understanding of the properties of the networks themselves. Mathematical studies can be used to identify stability and robustness and thus focus attention on molecules (or nodes) that may prove to be useful targets for drugs or in diseases (Maslov and Sneppen, 2002). Outside the study of the networks themselves, it will be necessary to understand how these intracellular protein interaction networks are involved with function of the intact cell and its environment. The notion of hierarchies of organisation is a theme within the recently reinvigorated field of systems biology (Grant and Blackstock, 2001, Kitano, 2002).

Networks or maps of protein interactions have not been produced for neurons (an extract of an early version of an NRC network is shown in Figure 2). In the same way that these molecular maps are emerging as useful tools in simple cellular systems (such as yeast) a protein interaction map based on synapse biology may be useful for understanding synaptic biology. At first glance it may appear impractical to pursue this biology at the synapse rather than in yeast, however there are several compelling reasons to move toward this objective: First, we estimate that the synapse has a protein composition of around 1000-2000 proteins and there is abundant protein interaction data. Second, manipulation of these proteins using genetic and pharmacological tools is routine and the electrophysiological properties can be rigorously quantified (for example, see Migaud et al., 1998). Third, synaptic biology is important for behaviour of the intact animal and thus hypotheses based on our understanding of molecular networks could be tested *in vivo*.

We decided to develop a protein-protein interaction database (PPID) beginning with the proteomic datasets from the NRC. Here we describe a bioinformatics based strategy for building this network. This approach has general application outside synaptic biology and should be useful for the construction of any proteomic led protein interaction network.

3. A STRATEGY FOR CONSTRUCTION OF A PPID FROM THE NMDA RECEPTOR

A step-by-step guide beginning with proteomic approaches and progressing toward publication of an accessible database is summarised in Table 1.

Table 1. Strategy of PPID construction

1.	Isolation of large molecular networks form biological samples
2.	Identify molecules
3.	Collect molecular information
4.	Mine literature for interaction partners
5.	Collate information and cross-reference in PPID
6.	Re-iterate
7.	Publish

The steps in this strategy are described:

1. We developed novel techniques that allowed us to purify proteins from neuronal sources, such as brain and spinal cord, which left large molecular *in vivo* structures intact. These isolated complexes displayed an average molecular mass of 2 to 4 MDa, which corresponds roughly into complexes of approximately 20 to 40 proteins each if we assume an average molecular mass of 100 KDa per molecule (Husi and Grant, 2001).

2. The identification of molecules within these isolated complexes was achieved by standard Western blotting techniques as well as large-scale mass spectrometry (Husi et al., 2000). This resulted in an initial dataset of 77 proteins being part of the NMDA-Receptor complex (NRC). At this stage the list of NRC proteins consisted only of either accession numbers or names of the proteins.

3. Information on these identified proteins, such as sequence, known domains, etc., was obtained from the proteomic database SwissProt through the SRS (sequence retrieval system) interface. Novel molecules with no known function or described domains were analyzed by querying the ProDom database with the amino acid sequence of the protein in question in order to see whether part of the sequence fits known fingerprints of described domains. Furthermore, alternative names of the NRC proteins, as described in the SwissProt-entry, were also gathered to allow for subsequent searches of the published literature (see Table 2).

4. Information on the functional annotation of the proteins of interest was found by searching the literature. This was accomplished by querying the PubMed database using the names of the NRC-proteins either alone, using several names together, or in combination with trap-words such as 'binding', 'interaction' or 'association'. This step was very time consuming as it requires to at least scan through all of the abstracts found from the query, and automated approaches using programs are notorious for retrieving incorrect information.

Table 2. Links

PPID	http://www.PPID.mvm.ed.ac.uk
SwissProt	http://www.expasy.org
Unigene	http://www.ncbi.nlm.nih.gov/UniGene/
Medline	http://www.ncbi.nlm.nih.gov/
OMIM	http://www.ncbi.nlm.nih.gov/entrez/query.fcgi?db=Omim
EMBL/TrEMBL	http://www.ebi.ac.uk
ProDom	http://www.expasy.org

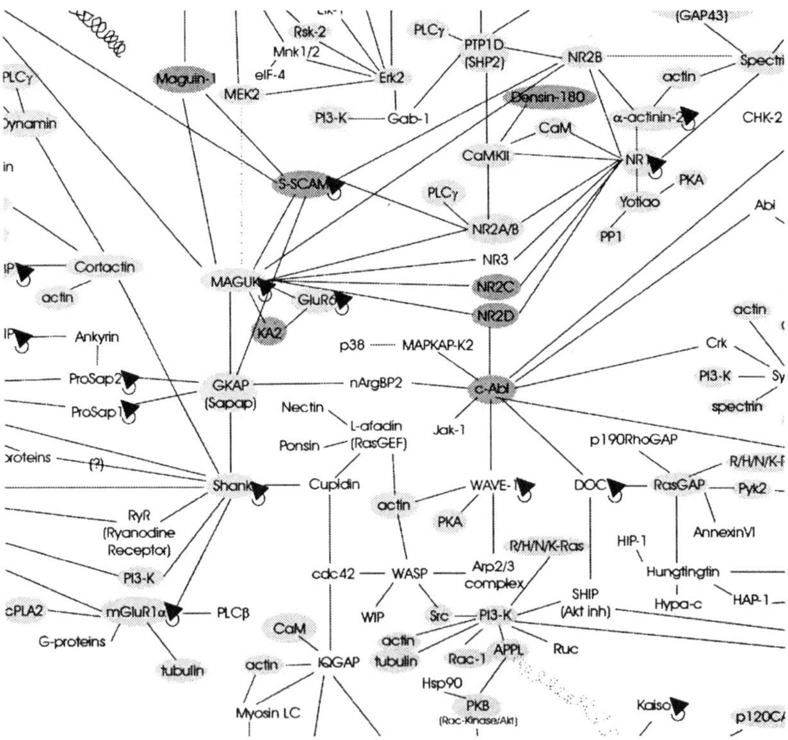

Figure 2. Figure 2. The extended NRC network.

Bioinformatic analysis and extended searches by Western blotting allowed to draw a hypothetical interaction map of molecules present in the NRC. Molecules highlighted in pink were reported to interact with connected components, and yellow highlighted proteins could be positively identified as being part of the extended NRC using Western blotting approaches. Their physical interaction profiles were also based on known findings. The map was collated manually and might not be a best-fit model. This map is an early version of the NRC network and later versions extend to 1900 proteins.

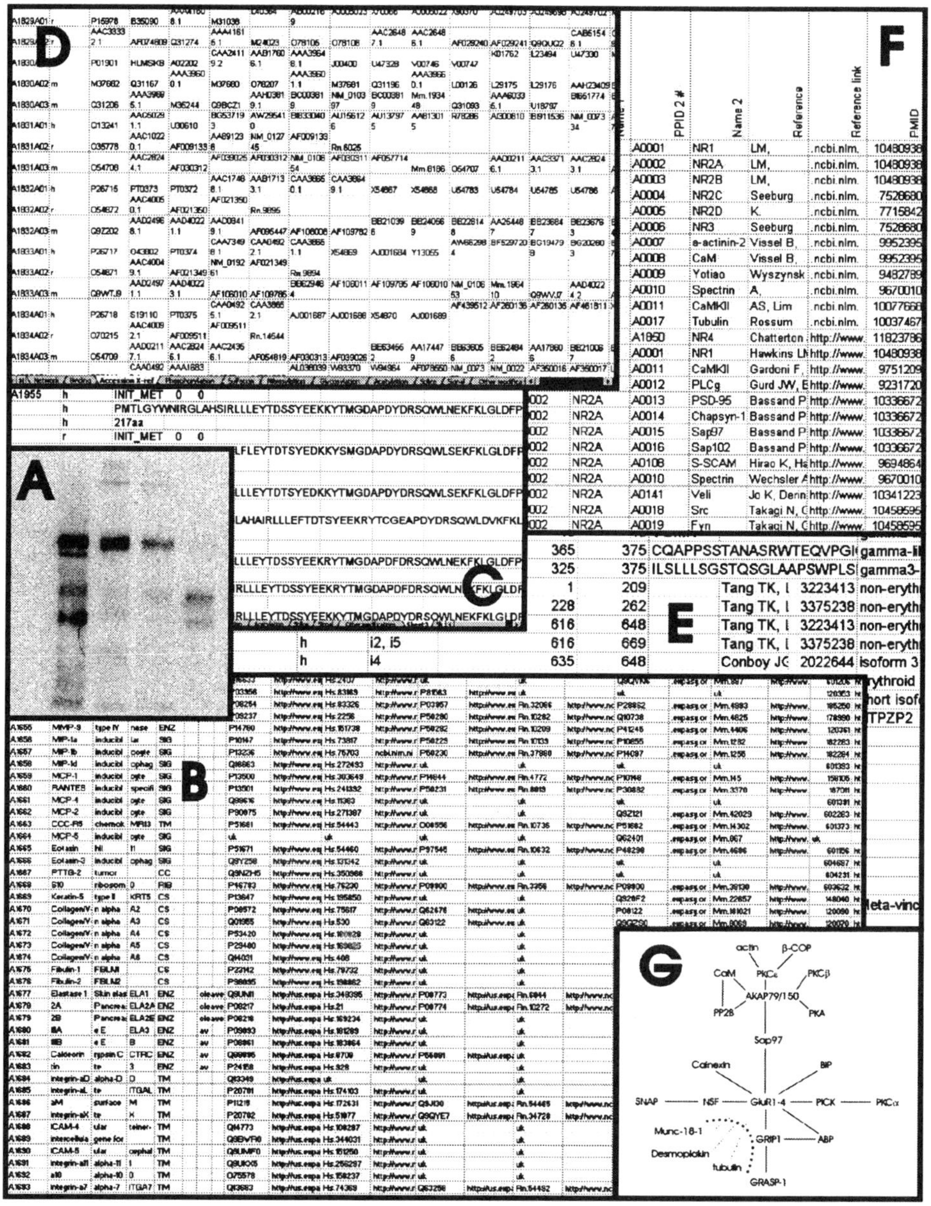

Figure 3. Snapshots of the PPID content

On-screen snapshots were taken from the various datasheets held in the database, ranging from screening results (A), species comparisons (B), sequence information (C), accession numbers (D), splice sites (E) to interaction data (F) as well as initial graphics files from manual cluster analysis (G).

5. All of the information gathered was assembled into a database holding information from our proteomic screens, sequence and domain information of each of the proteins listed, their names and a brief descriptor of the function, and protein-protein interactions published in the literature (see Figure 3). A unique internal accession number was assigned to each of the molecules, which is used throughout the database for cross-referencing and retrieval.

6. Protein-protein interactions described in the literature between proteins in the database and other proteins were annotated into the database, and steps 3 to 5 were carried out again with these added molecules. This re-iteration step was repeated several times and led to the current version of PPID, describing about 6000 molecules with more than 8000 described interactions. The database was then expanded to hold sequence and sequence-related information for the three species human, rat and mouse. Figure 3 shows snapshot fields of different database components.

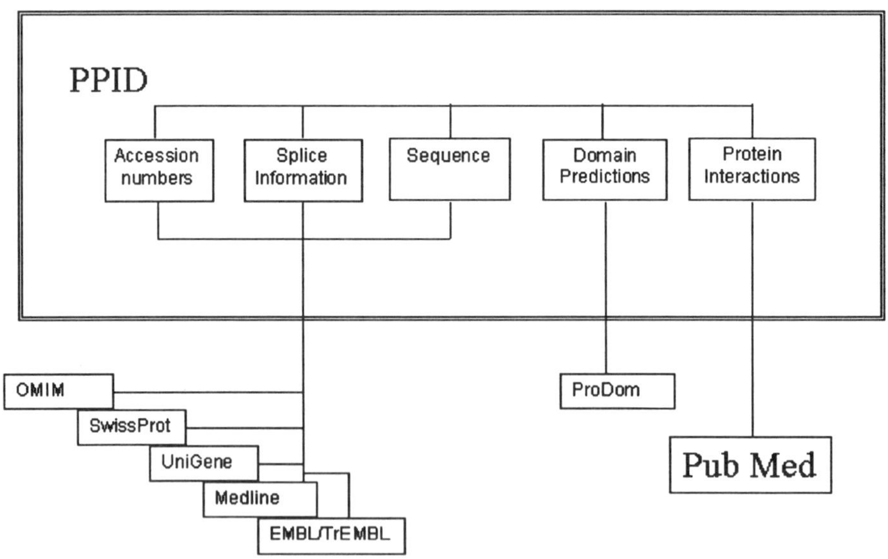

Figure 4. Structure of PPID

The internal structure of PPID is illustrated in the large box. An internal accession number with links to (human, rat, mouse) sequence and splice information is linked to protein structure information and the protein interaction data. Individual components of PPID are linked to external databases as illustrated: Online Mendelian Inheritance in Man (OMIM), accession numbers (SwissProt, EMBL, TrEMBL, Unigene), amino acid sequence (SwissProt, TrEMBL), splice information (literature (PubMed) and ClustalW sequence alignments) and interaction data (PubMed). Domain analysis (ProDom) and domain prediction (PPID internal) will be a future addition.

7. The database was then moved onto a UNIX platform to allow for greater flexibility and for future applications including online deployment and local compatibility. Additionally a user-friendly web interface will be developed.

4. DESCRIPTIONS AND APPLICATIONS OF PPID

PPID is designed to be both a versatile tool for analyzing new datasets as well as resource for studying organization of molecular networks. Intrinsic to the paradigm that molecules within living cells are not only compartmentalized but also associate into modules or complexes, and these modules are then assembled to become part of even larger complexes, is the need to handle information about many molecules (proteins, mRNAs and genes) simultaneously. Since these clusters of proteins require the handling of large datasets, we have tailored PPID toward integration with genomic and proteomics approaches. Below we will describe some of the features of PPID and hypothetical examples of applications (Figure 5).

The major role of PPID is to provide protein interaction data and thus a simple entry point is the name of an individual protein or gene. A typical simple query might be to list the known interacting proteins and provide links to their primary publication as well as information on the structure of their protein and information on the interaction domains. For each protein entry, amino acid sequence information is included together with known and novel splice-variants of each molecule as found by ClustalW sequence alignments. Graphical representation of protein interaction networks or tree diagrams will be produced (see example on http://www.PPID.mvm.ed.ac.uk).

Comparative genomics combined with genetics is an increasingly important vehicle for studying gene function. PPID provides unifying molecular entries across three species, namely human, rat and mouse. It is footed on sequence databases such as SwissProt, EMBL, TrEMBL (translated EMBL sequences) and Unigene and the literature database PubMed. A typical entry in PPID holds up to three 'general' entries for the three species, all protein and gene accession numbers associated with them (assembled from Blast2 searches of the databases) and the OMIM entry as maintained by Johns Hopkins University.

Querying the database will be possible by online browsing and batch-submission for large datasets holding accession number information. This type of data could include that generated using software such as Mascot for mass-spectrometry or microarray gene expression data. Cluster-analysis of the submitted datasets in the form of a graphical output will be developed.

Mutation and phenotype analysis may be facilitated with PPID. For example, a mutation in PSD-95 in mice (Migaud et al, 1998) might be expected to produce perturbations in the network of NRC proteins. The primary and secondary interacting proteins can be rapidly mined from PPID and assays designed to test them in vivo. A typical experiment might be to measure the levels of these proteins in the mutant mice. The quantitative data from these experiments could then be used to query the database to ask if changes in levels of specific proteins might reflect aspects of stability, robustness or redundancy. This information may also identify proteins that could be used in future mutagenesis strategies, where for example those proteins that change in response to a mutation would be candidates for experimental mutation.

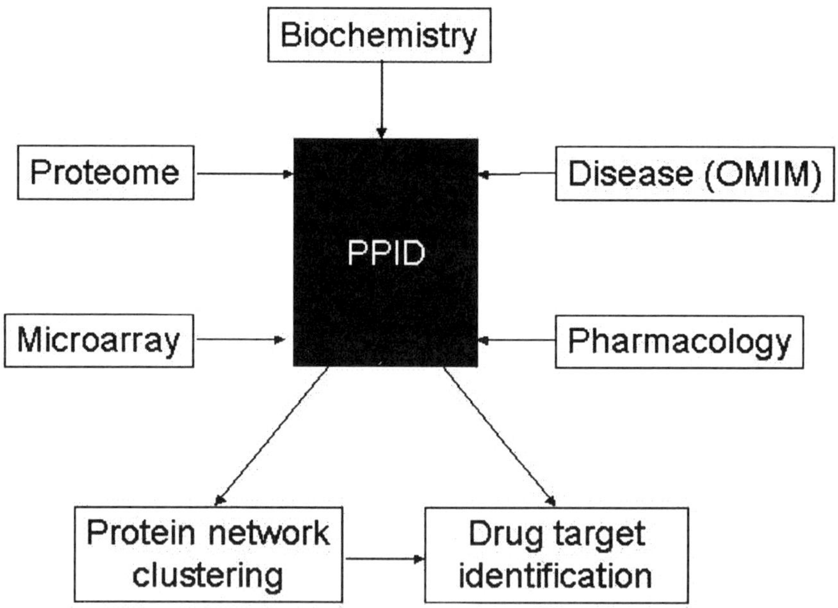

Figure 5. Application of PPID.

The schematic diagram shows how various research areas will benefit from a database like PPID by submitting single or batch queries and the potential output.

Studies of the proteins network are likely to have pharmacological applications including drug target identification. Some proteins are better suited for drug screening (e.g. enzymes rather than structural proteins) and although a genetic strategy could predict a good target, a potential binding protein predicted from PPID may have better prospects for medicinal chemistry.

5. FUTURE TRENDS

Other applications for PPID will emerge in the future due to the nature of the database and the way it was built in order to allow for maximum flexibility to add other fields such as transcriptional and translational profiles, known drug targets or even drugs themselves, or tissue distribution information, and developmental expression profiles.

The PPID is still under development and is an ongoing project with no real end-point. It will continue to grow by adding more entries and eventually also embrace information from other species such as chimp and other mammals. Future entries will also be the domain analysis of the sequences held in the database using both sequence-fingerprints held in the ProDom database and potential new domains by PPID-internal cross-comparisons. Another additional sequence-analysis will be the mapping of protein modifications according to ProSite and literature references.

We aim to build a database that encompasses all synaptic proteins and defines a subset of potential interactions. Clearly this dataset will link to protein interaction sets that are outwith the synapse and thereby be a useful tool for the general community. Because the function of synapses can be studied with exquisite quantitative precision and are readily genetically manipulated, this system and PPID may prove valuable for understanding the physiological relevant predictive value of protein interaction networks.

REFERENCES

Grant SG, Blackstock WP (2001) Proteomics in neuroscience: from protein to network. *J Neurosci* 21,8315-8318.

Husi H, Grant SG (2001) Isolation of 2000-kDa complexes of N-methyl-D-aspartate receptor and postsynaptic density 95 from mouse brain. *J Neurochem* 77,281-291.

Husi H, Ward MA, Choudhary JS, Blackstock WP, Grant SG (2000) Proteomic analysis of NMDA receptor-adhesion protein signaling complexes . *Nat Neurosci* 3,661-669.

Kitano H (2002) Systems biology: a brief overview. *Science* 295,1662-1664.

Maslov S, Sneppen K (2002) Specificity and stability in topology of protein networks. *Science* 296,910-913.

Mayer ML, Hieter P (2000) Protein networks-built by association. *Nat Biotechnol* 18,1242-1243.

Migaud M, Charlesworth P, Dempster M, Webster LC, Watabe AM, Makhinson M, He Y, Ramsay MF, Morris RG, Morrison JH, O'Dell TJ, Grant SG (1998) Enhanced long-term potentiation and impaired learning in mice with mutant postsynaptic density-95 protein. *Nature* 396,433-439.

Schwikowski B, Uetz P, Fields S (2000) A network of protein-protein interactions in yeast. *Nat Biotechnol* 18,1257-1261.

Tucker CL, Gera JF, Uetz P (2001) Towards an understanding of complex protein networks. *Trends Cell Biol* 11,102-106.

Von Mering C, Krause R, Snel B, Cornell M, Oliver SG, Fields S, Bork P (2002) Comparative assessment of large-scale data sets of protein protein interactions. *Nature* 417,399-403.

Chapter 5

Modeling The Dynamics of Second Messenger Pathways

Kim.T. Blackwell[1], Jeanette Hellgren Kotaleski[2]
[1]*Krasnow Institute of Advanced Studies and School of Computational Studies, George Mason University, Fairfax, VA 22030, USA*
[2]*Royal Institute of Technology and Karolinska Institute, S100-44 Stockholm, Sweden*
Correspondence to: avrama@gmu.edu

Abstract: Understanding the effect of neuromodulators requires modeling the dynamics of second messenger pathways and their effects on channel properties. Chemesis is software for modeling enzyme reactions, stoichiometric reactions, diffusion, and ligand gating of channels and is used in conjunction with GENESIS, the software program for neuronal modeling. This chapter explains the differential equations and corresponding Chemesis objects that implement these reactions. Glutamate stimulated IP_3 production and calcium waves are examples of the second messenger pathways that can be implemented in Chemesis.

Key words: diffusion, calcium, second messengers, enzymes, dynamics

1. INTRODUCTION

The importance of neuromodulators in neurological disorders such as Parkinson's Disease is generally accepted, (e.g. Albin et al. 1989; Blandini et al. 2000). Neuromodulators act through second messenger pathways, (e.g. Nishi et al. 1997; Choi et al. 1999) which influence neuronal behavior by their effect on membrane and channel properties, (e.g. West and Grace 2002; Levine et al. 1996; Kitai and Surmeier 1993; Surmeier and Kitai 1993). Most studies only evaluate steady state changes in channel properties, and, similarly, most compartmental models only include steady state effects of neuromodulation, (e.g. Seamans et al. 1998); but it is improbable that steady state concentrations occur *in vivo*. The release of neurotransmitters is a transient event, with rapid fluctuation in concentration mediated by release, re-uptake and degradation. Thus, to understand neural function in response to neuromodulators, it is necessary to study the dynamics of second messenger pathways and their effects on channel properties.

A technique to bridge this knowledge gap is computer modeling of the dynamics of second messenger pathways and their effects on membrane properties (e.g. Kotter et al. 1998; Kotter and Schirok 1999). Chemesis is software for modeling biochemical reaction pathways and is used in conjunction with GENESIS (Bower and Beeman 1998), a software program for neuronal modeling. Chemesis contains objects for describing enzyme reaction, stoichiometric reactions, diffusion, and ligand gating of channels. Creating models requires setting up the reaction pathways and assigning values for the parameters using standard GENESIS commands. Chemesis overlaps with Kinetikit, another set of GENESIS objects for simulating biochemical reactions, but Chemesis also contains objects for diffusion and calcium release.

The details of biochemical reaction pathways and kinetic parameters are maintained in several enzyme databases accessible via the internet such as ExPASy (http://www.expasy.ch/enzyme) and BRENDA (http://www.brenda.uni-koeln.de/.) For additional information it is necessary to refer to the original literature, much of which is published in the *Journal of Biological Chemistry*, http://www.jbc.org/. An explanation of how to extract reaction rate constants from the published experiments is provided in Chapter 10 of *The Book of Genesis* (Bower and Beeman 1998). Some of the neuron specific reaction pathways, with rate constants extracted from the literature, is provided in http://doqcs.ncbs.res.in/.

These databases, along with tools for modeling of biochemical reactions within compartmental models of neurons, allows the study of the dynamics of second messenger effects on neuronal function. The purpose of this chapter is to explain how to use Chemesis to model second messenger pathways, and to integrate these models with traditional compartmental models.

2. FROM CHEMISTRY TO MATHEMATICS

2.1 Equations Governing Biochemical Reactions

Biochemical reactions are interactions between substrate molecules to form product molecules, e.g.:

$$Subs1 \; + \; Subs2 \;\; \underset{k_b}{\overset{k_f}{\rightleftharpoons}} \;\; Prod \qquad (1)$$

The rate of the reaction is given by the forward rate constant, k_f, that describes the frequency with which product is formed, and the backward rate constant, k_b, which describes the frequency with which the substrates are regenerated. The quantity of each species in the reaction is determined from differential equations, which describe the change in quantity (or concentration) with time.

$$\frac{dProd}{dt} = k_f \cdot Subs1 \cdot Subs2 - k_b \cdot Prod \qquad \textbf{(2)}$$

A similar equation describes the change in substrate with time. These equations are commonly called "mass action kinetics", because they describe the behavior of large groups of molecules, and ignore the probabilistic interactions between individual molecules. Conservation of mass implies that, on the time scale of the reactions, there is no change in the total quantity of a particular molecule. In the example above, this implies that substrate1 and product are two forms of the same molecule; thus, they sum to a constant.

Most second messenger pathways are described as a cascade of such biochemical reactions. For example, the product in reaction 1 may be the substrate for the next reaction. Enzyme reactions are a special type of two step reaction, in which the enzyme is regenerated (and product formed) in the second step with rate V_{max}. Furthermore, the backward reaction rate for the second step is zero:

$$Enz + Subs \underset{k_b}{\overset{k_f}{\rightleftharpoons}} ES \overset{V_{\max}}{\rightarrow} Enz + Prod \qquad \textbf{(3)}$$

Equations describing this cascade of two reactions are written as for the first reaction:

$$\frac{dES}{dt} = k_f \cdot Enz \cdot Subs - (V_{\max} + k_b) \cdot ES$$

$$\frac{dProd}{dt} = V_{\max} \cdot ES \qquad \textbf{(4)}$$

If the reaction has certain properties, i.e. the enzyme substrate complex (*ES*) rapidly reaches equilibrium, then it is said to exhibit Michaelis-Menton kinetics (Schulz 1994). Given that the amount of free enzyme is the difference between the total amount of enzyme (E_{tot}) and the ES complex, then the equations simplify to:

$$\frac{dProd}{dt} = \frac{V_{max} \cdot E_{tot} \cdot Subs}{Subs + K_M}, \qquad where \quad K_M = \frac{V_{max} + k_b}{k_f} \tag{5}$$

2.2 Equations for Other Processes Affecting Concentration

Not all changes to molecules can be described using these biochemical reactions. In particular, movement of ions between compartments cannot be described using bimolecular reactions. One type of ion movement is flow from the extracellular space into the cell interior through channels such as the voltage gated calcium channel. This ion movement is described in terms of current flow, $I_{Ca}(V_M)$, with units of amperes. To determine the change in calcium per unit time (flux) produced by a given current flow, the following equation is used:

$$\Phi_I = \frac{-I_{Ca}(V_M)}{2\ F} \tag{6}$$

where F is Faraday's Constant, which specifies the moles of an ion per coulomb (charge), and Φ_I is the flux due to current flow.

A second type of movement is flow from the endoplasmic reticulum (ER) to the cell interior through calcium release channels. Similar to the flow of ions across ionic channels, the flow of ions across the membrane is governed by the permeability of the channel, and the driving force or difference between the inside and outside concentration. In contrast to the flow of ions across ionic channels, it is assumed that there is no potential difference between these compartments.

$$\Phi_R = P \cdot X_{open}^n \cdot ([Ca_{ER}^{++}] - [Ca^{++}]) \tag{7}$$

P, permeability of channel, is in units of volume/time; X_{open} is the open state of a calcium release channel, such as the ryanodine sensitive or IP_3 sensitive release channels, both of which depend on calcium concentration; n is an integer indicating how many independent subunits control the open state, and $[Ca^{++}_{ER}]$ $[Ca^{++}]$ are the calcium concentration in the ER and cytoplasm, respectively.

There are several models describing open state as a function of IP_3 and calcium. According to the model of the ryanodine receptor (RyR) channel (Tang and Othmer 1994), the RyR may reside in one of four different states, depending on the occupancy of the two calcium binding sites. The open state (R_{10}*) occurs when the excitatory site is bound. There are three closed states, occurring when either both calcium binding sites are unoccupied (R_{00}), the inhibitory calcium binding site is occupied (R_{01}), or both calcium binding sites are occupied (R_{11}):

$$
\begin{array}{ccccc}
 & & M_2 \cdot [Ca^{++}] & & \\
 & R_{00} & \rightleftharpoons & R_{01} & \\
 & & L_2 & & \\
M_1 \cdot [Ca^{++}] \downarrow\uparrow L_1 & & & & M_1 \cdot [Ca^{++}] \downarrow\uparrow L_1 \\
 & & M_2 \cdot [Ca^{++}] & & \\
 & R_{10}^{*} & \rightleftharpoons & R_{11} & \\
 & & L_2 & &
\end{array}
\qquad \textbf{(8)}
$$

where M_1, M_2, L_1 and L_2 are reaction rate constants.

Yet a third type of ion movement is diffusion from a region of high concentration to a region of low concentration. The latter does not really occur between compartments, but when modeling spatially extended structures, it is convenient to conceptualize such a structure as a series of discrete well-mixed compartments. I.e. within the compartment, no concentration gradients exist, and all the concentration gradient exists between adjacent compartments. This approximation of a spatially extended structure as a series of isoconcentration compartments is identical to that made for dendritic and axonal cables. The flux of ions between these pseudo compartments depends on the concentration gradient between compartments, the calcium diffusion coefficient (D_{Ca}) and also the geometry of the compartment. Thus, diffusional flux, Φ_D, increases with an increase in surface area ($area_j$) connecting compartments, and decreases as the length (distance separating the centers, $dist_j$) of compartments increases:

$$\Phi_D = -D_{Ca} \sum_j \frac{area_j \, ([Ca^{++}] - [Ca_j^{++}])}{dist_j} \qquad \textbf{(9)}$$

where the summation is over all adjacent compartments, j. The form of this equation allows for two- and three-dimensional diffusion.

The change in calcium for a compartment is the sum of the fluxes due to release, Φ_R, diffusion, Φ_D, ionic current, Φ_I, and bimolecular reactions with buffers or other molecules, Φ_B:

$$\frac{dCa^{++}}{dt} = \Phi_R + \Phi_D + \Phi_I + \Phi_B \qquad \textbf{(10)}$$

3. HOW TO USE CHEMESIS OBJECTS

Chemesis is a library of objects written in the GENESIS programming language (Bower and Beeman 1998) for use with GENESIS simulations. It can be downloaded from http://www.krasnow.gmu.edu/avrama/chemesis.html. A makefile is provided with instructions on compiling and installing the software on various UNIX platforms. Chemesis has objects for modeling the types of ion movement and biochemical reactions described above. Most second messenger pathways are described as a set of biochemical reactions, with diffusion of one or more species between compartments. The biochemical reactions are modeled using four Chemesis objects: rxnpool, conservepool, reaction, and enzyme; diffusion is modeled with an object called diffusion. For molecules such as calcium, which can flow into the cytosol through release channels in the ER, there are several additional objects: mmpump, cicr, and cicrflux. The following section explains the computations performed by each of these objects and their use with GENESIS commands.

3.1 Chemesis Objects for Bimolecular Reactions

Chemesis has two objects to keep track of the quantity and concentration of molecules. One of them is called rxnpool; the other is called conservepool. rxnpool accumulates the changes to molecule quantity due to various reactions, diffusion, etc. It then computes quantity from the equation $dC/dt = \sum A - \sum B \cdot C$, where A represents a change in quantity due to a reaction or other ion movement, and B represents a rate of change, e.g. degradation, which is proportional to the present quantity. The summation is over all processes that affect the quantity of the molecule, C. Conservepool is used to calculate the quantity and concentration of a molecule as the remainder after all other forms are accounted for: $C = Ctot - \sum C_i$.

Chemesis has two objects for biochemical reactions. One is called reaction; the other is called enzyme. Reaction takes as input the concentration or quantity of substrates and products (an arbitrary number of substrates and products can be specified), and calculates the rate of the forward reaction: the forward rate constant times the substrates, and the rate of the backward reaction: the backward rate constant times the products. Enzyme implements equation 5: it takes as input the concentration of substrate and enzyme, and calculates the change in the product molecule due to the enzyme reaction.

Standard GENESIS commands (Bower and Beeman 1998) are used to create a simulation of biochemical reactions. These commands are illustrated using the enzyme reaction that occurs after glutamate binds to the metabotropic glutamate recepter (mGluR) and catalyzes the production of the GTP bound form of the $G_{q\alpha}$ subunit, and the bimolecular reaction of active $G_{q\alpha}$ binding to PLC to activate it (Bockaert et al. 1993; Pin et al. 1994).

$$G_{\alpha\beta\gamma} + mGluR \rightleftharpoons GmGluR \xrightarrow{V_{max}} mGluR + G_{q\alpha}$$

$$PLC + G_{q\alpha} \underset{k_b}{\overset{k_f}{\rightleftharpoons}} PLC^* \tag{11}$$

1. Create pools of molecules using the GENESIS command *create*.
The syntax for this command is: create object_name element_name

```
create rxnpool mGluR
create rxnpool Gqa
create conservepool Gprot /* for Gαβγ */
create conservepool PLCtot  /* for PLC */
create rxnpool PLCstar
```

2. Create bimolecular and enzyme reactions using *create*:

```
create reaction react1
create enzyme enz1
```

3. Assign parameter values such as rate constants, compartment volumes (to calculate concentration from quantity and vice versa), and initial quantities using the genesis *setfield* command. The syntax for this command is:
setfield element_name field_name value

```
setfield PLCstar  volume 1e-15 Qinit 0
```

setfield react1 kf {kfvalue} kb {kbvalue}

In this command you have the option of specifying the value, or specifying a variable, in curly brackets, which has been given the value in an assignment statement: kfvalue = 3.5

4. Send molecule concentrations to the reactions using the GENESIS command *addmsg*. The syntax for this command is:
addmsg sending_object receiving_object message_type field_name

```
addmsg PLC react1 SUBSTRATE conc
addmsg Gqa react1 SUBSTRATE conc
addmsg PLCstar react1 PRODUCT conc

addmsg mGluR enz1 ENZYME conc
addmsg Gprot enz1 SUBSTRATE conc
```

5. Send the results of these calculations back to the rxnpools, for them to calculate their concentrations:

```
addmsg react1 Gqa RXN2 kbprods kfsubs
addmsg react1 PLCstar RXN2 kfsubs kbprods
addmsg enz1 Gqa MMPUMP kbprods
```

6. Last, send the messages to conservpool so it can calculate the amount of free substrate

```
addmsg Gqa Gprot CONC conc
addmsg PLCstar PLCtot CONC conc
addmsg PLCstar Gprot CONC conc
```

These commands implement the following equations:

$$\frac{dG_{q\alpha}}{dt} = \frac{V_{max} \cdot mGluR \cdot G_{\alpha\beta\gamma}}{G_{\alpha\beta\gamma} + K_M} - k_f \cdot PLC \cdot G_{q\alpha}$$

$$\frac{dPLC*}{dt} = k_f \cdot PLC \cdot G_{q\alpha} - k_b \cdot PLC* \qquad \textbf{(12)}$$

$$G_{\alpha\beta\gamma} = Gprot - G_{q\alpha} - PLC*$$

$$PLC = PLCtot - PLC*$$

To run the simulation standard GENESIS commands are used: *reset*, to initialize everything, then *step n*, to run the simulation for *n* time steps. The size of the time step is adjusted using the *setclock* and *useclock* commands.

3.2 Chemesis Objects for Diffusion and Flux Through Membranes

Chemesis has several objects for computing the flux between compartments. One object, called mmpump, is for calculating the amount of calcium (or optionally any other ion) pumped out of the cytosol by voltage independent pumps. This object implements the following equation:

$$\Phi_P = \frac{V_{pump} \cdot [Ca^{++}]^n}{[Ca^{++}]^n + K_D} \quad \textbf{(13)}$$

where n is the Hill coefficient, K_D is the concentration at which the pump rate is half maximal, and V_{pump} is the maximal rate of the pump. A second object is diffusion and implements Eqn 9. A third object, cicrflux, computes the flux across the membrane through a calcium release channel (Eqn 7). This object also can implement a constant flux, independent of channel state, such as a constant leakage. Lastly, the object cicr combines functions of both rxnpool and reaction to calculate the open state of calcium release channels. Cicr contain all the information about a single state of the release channel. It stores the reaction rates of the forward reactions, and calculates the fraction of molecules in that state. Both the model of De Young and Keizer (1992) for IP_3 induced calcium release (IICR), or Eqn 8 (Tang and Othmer 1994) for calcium induced calcium release (CICR) may be implemented with the cicr object.

Once again, standard GENESIS commands are used to create simulations of calcium dynamics. These commands are illustrated using a three compartment model of a soma with a single dendrite, and an ER compartment located within the soma compartment. The soma has a voltage dependent calcium channel. The ER has a rynodine receptor channel and a pump. Calcium is free to diffuse from the soma to the dendrite.

1. It is necessary to specify the dimensions of the compartments, and compute the volume and various surface areas given a certain geometry. For spherical compartment, radius and surface area are specified. For cylinderical compartments, radius, length, outer surface area and side surface area are specified. In the following example, it is assumed that compartments are

cylindrical. Furthermore, it is assumed that the ER volume is 0.185 times the soma volume. The syntax is: type variable_name {= variable_value}
type can be either float, int (for integer) or str (for a string). The variable can be set to a value as part of the command, or with an assignment statement later.

```
float soma_rad = 6e-6
float soma_len = 12e-6
float soma_vol = PI · soma_rad · soma_rad · soma_len
float soma_SA = 2 PI · soma_rad · soma_len
float soma_side = PI · soma_rad · soma_rad
```

SI units are used in this and subsequent examples. A similar set of commands is required for the dendrite and ER compartments.

2. Create pools of calcium molecules and assign values to fields that specify geometry

```
create rxnpool somaCa
create rxnpool CaER
create rxnpool dendrite Ca
```

In this following command, the \ is used to indicate a continuation of the command on the next line.

```
setfield somaCa volume {soma_vol} length {soma_len} \
    SAout {soma_SA} SAside {soma_side}
```

3. Create the diffusion object, set the value of the diffusion coefficient (in units of m^2/sec, and setup messages.

```
create diffusion dend_soma
setfifeld dend_soma D 0.006
```

The diffusion object expects two incoming messages, one from each of two adjacent compartments. Each message must contain three values: the concentration of the compartment, the area of the shared surface, and the length of the compartment. If the two adjacent compartments have different areas, the diffusion object uses the mean value of the two areas. The distance between two adjacent compartments is assumed to be the distance between the centers of the compartments, which is the mean of their lengths.

```
addmsg somaCa dend_soma CONC1 Conc SAside length
```

```
addmsg dendCa dend_soma CONC2 Conc SAside length
```

The diffusion object sends messages back to each compartment:

```
addmsg dend_soma somaCa DIFFUSION difconc2 difconc1
addmsg dend_soma dendCa DIFFUSION difconc1 difconc2
```

These messages have the values of flux within them, and are used by rxnpool to compute the new concentration. The two different pools get opposite messages because diffusion causes an increase in the concentration of one pool and a decrease in the concentration of the other.

4. Create an mmpump from the cytosol to the ER to represent the smooth endoplasmic reticulum Calcium-ATPase (SERCA). The mmpump formulation has the undesired side effect of eventually reducing calcium concentration to zero. To compensate, it is assumed that there is a slow leak of calcium from the ER, which is equal and opposite to the pump rate when calcium concentration is at its rest level. Thus, both an mmpump and a cicrflux (with a constant flux rate) are created:

```
create mmpump SERCA
setfield SERCA power 2 half_conc 0.1e-3 max_rate {Y}

create cicrflux leak
setfield leak maxflux {X} power 1
```

In this case Y is the maximal rate of pumping by the SERCA pump, and X is the permeability of the leak channel, and is calculated from the equation:

$$X = \frac{Y \cdot [C^{++}_{rest}]^2}{[C^{++}_{rest}]^2 + 0.0001^2} \tag{14}$$

where $[C^{++}_{rest}]$ is the resting value of calcium concentration, 0.0001 mM is the half concentration of the pump, and the factor of 2 is the power of the pump.

The mmpump object expects one incoming messages from the soma cytosolic compartment. It sends messages back to the soma, because it decreases the concentration in the soma, and to the ER, because it increases concentration in the ER. The leak object expects three incoming messages, one for soma cytosolic calcium, one for ER calcium, and one for open fraction. Sending the value of 1.0 indicates that permeability is constant. The leak sends messages back to the soma and ER using the rxnpool message DIFFUSION, because the values it sends have the same effect on concentration as a diffusion object.

```
addmsg somaCa SERCA CONC Conc
addmsg SERCA somaCa MMPUMP moles_out
addmsg SERCA CaER MMPUMP moles_in

addmsg somaCa leak CONC1 Conc
addmsg CaER leak CONC2 Conc
addmsg somaCa leak IP3R 1
addmsg leak somaCa DIFFUSION fluxconc2 fluxconc1
addmsg leak CaER DIFFUSION fluxconc1 fluxconc2
```

5. Set up calcium induced calcium release through the RyR channel by creating four calcium release objects, one for each state of the RyR. The commands for creating the R_{00} state are illustrated. Commands for the other three states are the same. The fields *alpha_state*, *beta_state* and *gamma_state* take values of 0 or 1, corresponding to the 0 or 1 subscripts in Eqn 8. *Alpha*, *beta* and *gamma* are the forward rate constants, M_1 and M_2 in Eqn 8. Both *alpha_state* and *alpha* are unused (i.e. 0) for the RyR, but they are used for the IP_3 receptor. The field *conserve* indicates whether this state is computed as 1 - sum of other states. *xinit* is the initial value.

```
create cicr R00
setfield R00 alpha_state 0 beta_state 0 gamma_state 0 \
alpha 0 beta {M1} gamma {M2} conserve 0  xinit {init00}
```

Each of the four states sends a message to the states to which it transitions containing the value of the backward rate constant of the transition and the fraction of channels in that state.

```
addmsg R10 R00 BSTATE beta fraction
addmsg R01 R00 GSTATE gamma fraction
addmsg R00 R10 BSTATE beta fraction
addmsg R11 R10 GSTATE gamma fraction
addmsg R11 R01 BSTATE beta fraction
addmsg R00 R01 GSTATE gamma fraction
```

The state which is conserved receives the message CONSERVE from all other states, e.g.:

```
addmsg R00 R11 CONSERVE fraction
```

In addition to these messages, each of the non-conserved states needs a calcium concentration message from somaCa, e.g.:

```
addmsg somaCa R00 CALCIUM Conc
```

Lastly, a cicrflux object is created to calculate the calcium flux from the ER to the soma. It requires messages not only from the somaCa and CaER objects, but also from the R10 object, which is the open state of the channel.

```
create cicrflux ryanflux
setfield ryanflux power {expon} maxflux {maxcond}

addmsg somaCa ryanflux CONC1 Conc
addmsg CaER ryanflux CONC2 Conc
addmsg R10 ryanflux IP3R fraction

addmsg ryanflux somaCa DIFFUSION fluxconc2 fluxconc1
addmsg ryanflux CaER DIFFUSION fluxconc1 fluxconc2
```

Combining ion movement such as diffusion with biochemical reactions makes it possible to model and simulate any second messenger pathway. The effect of second messengers on ionic channels, such as ligand gated or modulated channels, can be modeled by sending messages from the second messenger rxnpool to a GENESIS channel object.

4. EXAMPLES

The utility of Chemesis is illustrated with simulations of IP_3 production after stimulation of a metobtropic receptor on the head of a Purkinje cell spine (Finch and Augustine 1998). The biochemical reactions include those in Eqn 11, plus production of IP_3 from the active PLC enzyme acting on PIP_2:

$$PIP_2 + PLC^* \underset{k_{BPI}}{\overset{k_{FPI}}{\rightleftharpoons}} PLC\text{-}PI \xrightarrow{V_{maxPI}} PLC^* + IP_3 \qquad \textbf{(15)}$$

Degradation of IP_3 is included, and the equation describing Eqn 15 includes a volume term in the denominator to convert quantity to concentration:

$$\frac{d[IP_3]}{dt} = \frac{V_{maxPI} \cdot PLC^* \cdot [PIP_2]}{vol \cdot ([PIP_2] + K_{M-PI})} - D_{IP3} \cdot [IP_3] \qquad \textbf{(16)}$$

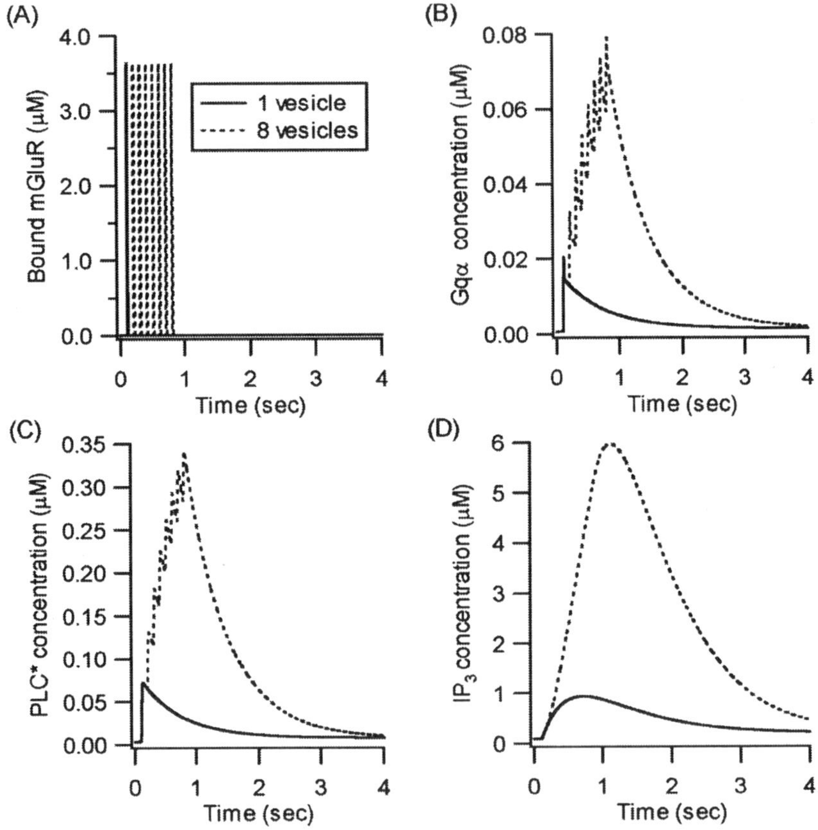

Figure 1 Concentration of molecules in biochemical reactions leading from mGluR activation to IP_3 production

Figure 1 illustrates the change in concentration of (A) bound mGluR, (B) $G_{q\alpha}$, (C) PLC* and (D) IP_3 in response to a single action potential and a short sequence of action potentials, each releasing a single vesicle of glutamate. Notice that IP_3 production is considerably slower than $G_{q\alpha}$ production and the contribution of individual vesicles is not visible.

Using Chemesis to Model Calcium Dynamics

Chemesis is not limited to single compartment models. Spatially extended structures may be modeled as a series of rxnpool compartments which communicate with each other via diffusion objects and with an ER via calcium release objects. Chemesis has been used to create multicompartment models to investigate calcium dynamics in classical conditioning (Blackwell 2002). The sea slug, *Hermissenda crassicornis,* is an animal model of classical conditioning and learns to associate light (the conditioned stimulus) with turbulence (the unconditioned stimulus). The memory of the associat on is stored in the type B photoreceptors and critically depends on an elevation of intracellular calcium concentration. Turbulence causes hair cells to release GABA onto the terminal branches of the type B photoreceptor and may produce a wave of calcium that propagates from the terminal branches to the soma. Chemesis objects were used to construct a model to determine if GABA stimulation can contribute a calcium signal that propagates along the 100 μm neurite separating the terminal branches from the soma. The model included the second messenger pathways from GABA stimulation to IP_3 production, and included release of calcium from stores through IP_3R and RyR channels. The neurite was divided into 100 isoconcentration compartments of 1 μm length.

Fig 2 illustrates the calcium concentration along the neurite in response to a 3 s (Left) and 30 ms (Right) GABA stimulation. The calcium wave propagates the entire length when RyR channels are present (A). With only IP_3R channels, the calcium wave does not propagate the entire distance (B), but only propagates the distance that IP_3 diffuses (C). Even a 30 ms GABA stimulus produces sufficient IP_3 to initiate a calcium wave.

To analyze the effect of calcium or IP_3 on membrane channels, the *addmsg* command is used to provide the second messenger concentration to a ligand gated channel, such as a calcium dependent potassium channel. Similarly, an *addmsg* command, from the calcium current to the calcium concentration rxnpool allows the traditional compartmental model to influence the second messenger pathway. Thus, the integration of second messenger pathways into GENESIS multi-compartmental neuronal models (Kotter and Schirok 1999), is extended and facilitated by using Chemesis.

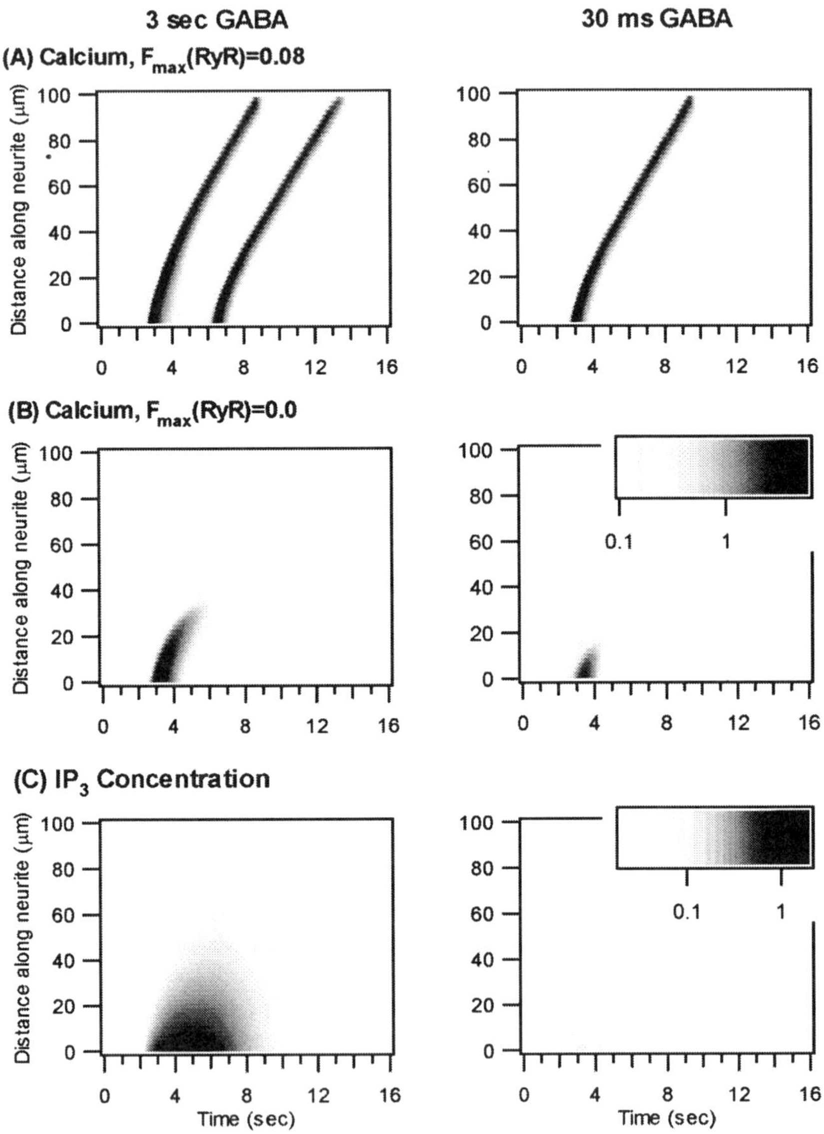

Figure 2 Calcium concentration (A and B) and IP_3 concentration (C) along type B photoreceptor neurite in response to GABA stimulation of terminal branches. Inset shows the concentration in μM corresponding to the gray shade of the figures.

REFERENCES

Albin, R. L., Young, A. B., and Penney, J. B. (1989) The functional anatomy of basal ganglia disorders. *Trends in Neuroscience* 12, 366-375

Blackwell, K. T. (2002) Calcium waves and closure of potassium channels in response to GABA stimulation in *Hermissenda* type B photoreceptors. *J.Neurophysiology* 87, 776-792

Blandini, F., Nappi, G., Tassorelli, C., and Martignoni, E. (2000) Functional changes of the basal ganglia circuitry in Parkinson's disease. *Progress in Neurobiology* 62, 63-88

Bockaert, J., Pin, J., and Fagni, L. (1993) Metabotropic glutamate receptors: an original family of G protein- coupled receptors. *Fundam.Clin.Pharmacol.* 7, 473-485

Bower, J. M. and Beeman, D. (1998) *The Book of Genesis: Exploring Realistic Neural Models with the GEneral NEural SImulation System.* Springer-Verlag. New York.

Choi, E. Y., Jeong, D., Won, K., Park, and Baik, J. H. (1999) G protein-mediated mitogen-activated protein kinase activation by two dopamine D2 receptors. *Biochemical and Biophysical Research Communications* 256, 33-40

De Young, G. W. and Keizer, J. (1992) A single-pool inositol 1,4,5-trisphosphate-receptor-based model for agonist-stimulated oscillations in Ca2+ concentration. *Proc.Natl.Acad.Sci.USA* 89, 9895-9899

Finch, E. A. and Augustine, G. J. (1998) Local calcium signalling by inositol-1,4,5-trisphosphate in Purkinje cell dendrites. *Nature* 396, 753-756

Kitai, S. T. and Surmeier, D. J. (1993) Cholinergic and dopaminergic modulation of potassium conductances in neostriatal neurons. *Adv.Neurol.* 60:40-52, 40-52

Kotter, R. and Schirok, D. (1999) Towards integration of biochemical and biophysical models of neuronal information processing: a case study in the nigro-striatal system. *Reviews in the Neurosciences* 10, 247-266

Kotter, R., Schirok, D., and Zilles, K. (1998) "Dopamine-mediated dephosphorylation of N/P type calcuim channels in striatal neurons: a quantitative model." In Holcombe, M. and Paton, R. eds. *Information Processing in Cells and Tissues.* Plenum Press. New York 95-106.

Levine, M. S., Altemus, K. L., Cepeda, C., Cromwell, H. C., Crawford, C., Ariano, M. A., Drago, J., Sibley, D. R., and Westphal, H. (1996) Modulatory actions of dopamine on NMDA receptor-mediated responses are reduced in D1A-deficient mutant mice. *J.Neuroscience* 16, 5870-5882

Nishi, A., Snyder, G. L., and Greengard, P. (1997) Bidirectional regulation of DARPP-32 phosphorylation by dopamine. *J.Neuroscience* 17, 8147-8155

Pin, J. P., Joly, C., Heinemann, S. F., and Bockaert, J. (1994) Domains involved in the specificity of G protein activation in phospholipase C-coupled metabotropic glutamate receptors. *EMBO J.* 13, 342-348

Schulz, A. R. (1994) *Enzyme Kinetics from Diastase to Multi-enzyme Systems.* Cambridge University Press. New York.

Seamans, J. K., Floresco, S. B., and Phillips, A. G. (1998) D1 Receptor modulation of hippocampal-prefrontal cortical circuits integrating spatial memory with executive functions in the rat. *Journal of Neuroscience* 18, 1613-1621

Surmeier, D. J. and Kitai, S. T. (1993) D1 and D2 dopamine receptor modulation of sodium and potassium currents in rat neostriatal neurons. *Progress in Brain Research* 99, 309-324

Tang, Y. and Othmer, H. G. (1994) A model of calcium dynamics in cardiac myocytes based on the kinetics of ryanodine-sensitive calcium channels. *Biophysical Journal* 67, 2223-2235

West, A. R. and Grace, A. A. (2002) Opposite influences of endogenous dopamine D1 and D2 receptor activation on activity states and electrophysiological properties of striatal neurons: studies combining in vivo intracellular recordings and reverse microdialysis. *J.Neuroscience* 22, 294-304

Chapter 6

Web-Based Neuronal Archives

Neuronal Morphometric and Electrotonic Analysis

Dennis A. Turner[1], Robert C. Cannon[2], and Giorgio A. Ascoli[3]
[1]Neurosurgery and Neurobiology, Duke University Medical Center and Durham VAMC, Durham, NC, USA; [2]Theoretical Biology, Born-Bunge Foundation, University of Antwerp, Wilrijk, Belgium; [3]Computational Neuroanatomy, Krasnow Institute for Advanced Study and Psychology, George Mason University, Fairfax, VA, USA.
Correspondence to: dennis.turner@duke.edu

Abstract: A practical guide is provided to web-based neuronal databases, and various forms of quantitative analysis of neurons and dendrites. Computational modeling and simulation of morphological and electrophysiological properties allow a complete and direct study of the structure-activity relationship in single cells. Large numbers of anatomically realistic synthetic neurons can also be assembled in real-scale networks, to complement reconstructed neurons. These constructs can provide a bridge between single-cell morphology and system-level stereology, as well as between subcellular biophysical properties and emergent network activity. Further, integrating together multiple databases of neurons may facilitate an overall goal of reconstruction and understanding of circuitry within the brain, approaching the brain as a collection of morphologically unique, interconnected neurons

Key words: Dendrites, Modelling, Synapses, Networks, Neuron Reconstruction, Neurolucida, Electrotonic Properties, Morphometric Analysis

1. INTRODUCTION

Dendritic appearance and the importance of dendritic function to neurons in the central nervous system have proven enigmatic, ever since the demonstration of Golgi stains for systematic visualization of neurons (Ascoli et al., 2001a; Turner et al., 1991). Computer-assisted reconstruction techniques are now widely available and in common use, facilitating a quantitative approach to both single neuron function and to brain models, using measured neurons as the primary structural platform. The analysis of dendrites has followed several different paths, depending on the analytic need and the level of resolution required. For example, dendrites have been compared between cell classes, during development and during pathological alterations in neuronal structure (Cannon et al., 1999). Due to the time and effort required for both individual neuron visualization and computer-aided reconstruction, such measured neurons are valuable, and have been placed in neuronal databases for open dissemination (Cannon et al, 1998; www.axiope.org; www.compneuro.org; cascade.utsa.edu/bjclab). Neuronal databases and enhancing potential uses of available neurons will promote optimal use of existing neurons, and to facilitate building neuronal, circuitry and brain models. The visualizations of real neurons can then also be compared critically to synthetic neurons, particularly examining the degree of differences between neurons of varying classes (Ascoli, 1999).

However, existing neuronal databases remain limited in accessibility, scope and usefulness, in spite of available tools. Use of measured neurons requires some background as to how the neurons were obtained, reconstructed and presented. This chapter outlines current approaches to labelling neurons *in vivo* and *in vitro*, as well as reconstruction techniques and limitations. The manipulation and analysis of these neurons will then be considered, together with issues concerning web-based storage and dissemination, giving examples from current websites. The creation and validation of synthetic neurons is also discussed, particularly in comparison to measured neurons obtained from biological sources.

2. NEURON LABELING

Measured neurons are now most frequently obtained using physiological injection of a dye (particularly biocytin) during sharp or patch intracellular recordings. A large amount of dye can be rapidly injected, and dendritic diffusion is enhanced in live neurons. These procedures require a physiological preparation, such as *in vivo* cerebral cortex, or slices (400 to 500 μm) maintained *in vitro*. The tissue is then fixed and processed for biocytin, usually resulting in an isolated neuron visualized within the tissue. This isolation facilitates the reconstruction, since there is no ambiguity about the identity of processes. Subsequent measurements are altered, however, by whatever processes of fixation and clearing are used. Most neurons available in the several web databases are derived from this type of visualization, usually from fixed tissue (Cannon et al., 1998; Pyapali and Turner, 1996). Appropriate correction factors may be applied to decrease possible histological or measurement distortions, if specific information is available (Horcholle-Bossavit et al., 2000).

3. CELL RECONSTRUCTION AND EDITING

Neuronal reconstruction methods have arisen from classic neuroanatomy, to portray the outline of neuronal structure. Traditional techniques include camera lucida tracings, and more recently computer-generated linear representations of structure. Preparing camera lucida tracings involves a compound microscope and a drawing tube, which allows the projection of an image of the microscopic specimen onto a sheet of paper. The systems currently in use allow the visualization of both the specimen and the paper through the microscope, so that a direct comparison can be made between the structure being traced and the drawing. Using a high magnification lens (typically a 100X plan-apochromat) fine details of dendritic structure can be achieved, to the level of resolution of light microscopy (~0.4 μm). The resulting planar representation of a neuron may subsequently be pieced together from multiple sheets.

The most common reconstruction system currently available, Neurolucida (Microbrightfield, Manchester, VT) uses a tracking apparatus for both planar and vertical measurements. This system consists of a motorized stage and focus mechanism, with a readout that can be transferred to a computerized database. The system keeps track of where the investigator is pointing/focusing in all three dimensions, and saves a log with length and orientation coordinates for all specified nodes. Dendritic diameters are measured using a variable spot cursor, which can be matched to the diameter at each node. The resulting neuronal structure contains a series of connected nodes, with calibrated, three-dimensional locations, and cylinders representing dendrites between the nodes. As with any approximation system, a proportionally large number of cylinders may be used to represent, for example, a tapering dendrite or a neuronal cell body, depending on the degree of fidelity desired. However, dendritic spines are rarely included or counted, which may result in significant loss of the actual surface contour, but can be added in later analysis (Turner, 1984).

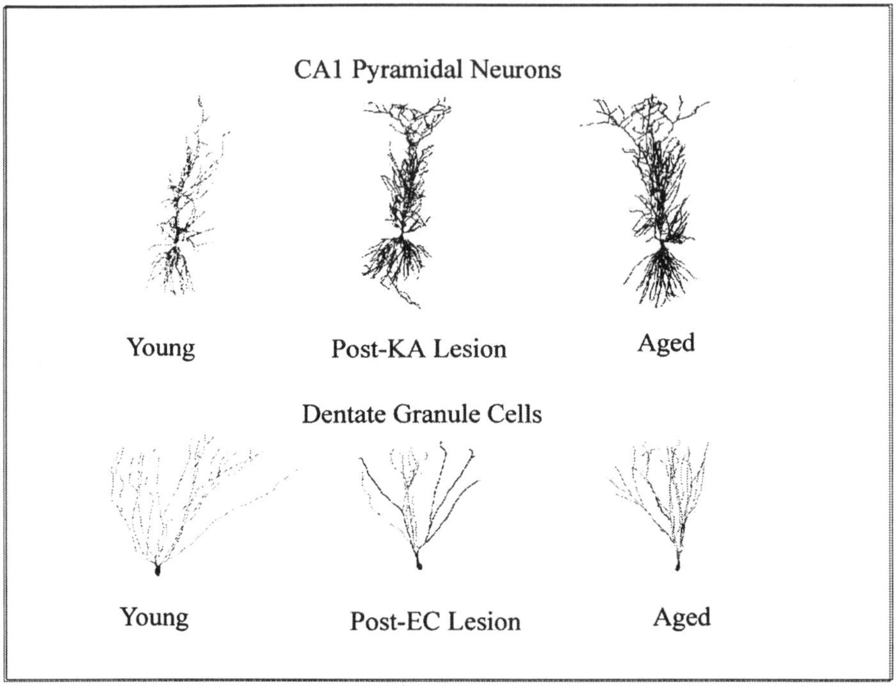

Figure 1. A montage of CA1 and dentate neurons after reconstruction.

Confocal microscopes share the advantage of short depth of field combined with the elimination of out of focus contamination, giving an almost ideal optical section of the tissue. Three-dimensional morphology of neurons stained with a fluorescent dye can be reconstructed without the need for computationally intensive out-of-focus image correction algorithms. Confocal techniques are especially helpful in that they can penetrate considerably thicker tissue slices than a conventional light microscope. Thus, full reconstructions may be derived from a single tissue slice, if this slice contains the entire neuronal structure. However, fading of the fluorescent dye with a series of sections sufficient for cubical voxels is severe, and connecting processes between voxels has proven difficult.

Shrinkage and distortion occurs during processing of nervous tissue, particularly during dehydration and fixation. This may result in a systematic underestimation of all the diameters and lengths measured from the preparation, creating a quandary for appropriate corrections. Limits on image resolution are determined by the magnification of the objective lens and its numerical aperture. Dendrites of CNS neurons have diameters close to the resolving power of the light microscope, and thus appear to have 'fuzzy' edges, which limit the accuracy of any diameter measurements. Since dendritic spines usually have spine widths less than the minimum resolving power, this also creates a blur or 'fuzzy' edge on the dendritic processes being measured, particularly where the spine density is high. Compared to dendritic lengths, dendritic diameters are more difficult to measure accurately, and yet are crucial for use in modelling dendritic function. If the dendritic contour is irregular or noncylindrical, then an inappropriate cylindrical dimension may be assumed. The subsequent measurement of dendritic length from the detailed tracings may be reasonably accurate (to within 10-15%). All of these concerns are reflected in caveats regarding available measured neurons in databases (Horcholle-Bossavit et al., 2000).

Once a cell is reconstructed, its digital file may contain several forms of errors, particularly if the arborization is highly complex. While intrinsic editors are available within Neurolucida, errors are usually not apparent until the reconstructed neuron is fully rendered and inspected. For example, very large datasets are not easily editable by Neurolucida, particularly for errors, which may render electrical modeling tools difficult, such as loops or apparent discontinuities. Therefore, the software tool CVAPP (Cannon et al, 1998; available in an updated version at www.axiope.org) provides an editor for electrical continuity and misplaced branches, specifically to overcome the limitations of the proprietary software. CVAPP includes excellent visualization tools, including panning, zooming and rotating, as well as shrinkage correction. CVAPP also offers several basic functions besides editing and viewing, including translating Neurolucida files into universal formats, or other formats appropriate for modeling programs, such as NEURON (Hines and Carnevale, 2001) or GENESIS (Bower and Beeman, 1998). This type of editor is critical for correcting errors incurred during reconstructions and visualizing the entire, complex neuron.

Another useful software tool to convert anatomical files of digitized neuronal reconstructions into compartmental modeling programs (including NEURON and its precursor, CABLE) is the program NTSCABLE, by J.C. Wathey (Salk Institute). NTSCABLE is available at cns.iaf.cnrs-gif.fr with exhaustive documentation. NTSCABLE reads and converts Neurolucida files and also older formats, including Eutectic and its variations (a complete description of these formats is beyond the scope of this chapter, and is included in Ascoli et al., 2001a). A similar program to NTSCABLE is Rogene M. Eichler West's Java class Neuromesher (available at www.science-renaissance.org/rogene. An additional program to convert anatomical files into simulation-ready formats is Lazarewicz' AWK routine swc2hoc (Lazarewicz et al., 2002a; www.neuroengineering.upenn.edu/~mlazarew.research.html). This program automatically simplifies neuronal morphology by maintaining the smallest amount of biophysically necessary geometrical and topological information, while merging data points with no variation in diameter or branch connectivity. Finally, it should be noted that any of the anatomical formats mentioned in the present chapter can be converted into a common standard by the software L-Measure, which is described in detail below.

4. ARCHIVING DIGITIZED NEURONS

4.1 Issues in Assembling Data Sets

Digitization of neuronal morphology is still a slow, time-consuming process. The results of this work are of interest both to those who undertake it, and to many others, such as computational modelers in the neuroscience community. It is therefore important to have an efficient system to facilitate data sharing and reuse. A priori, it might seem simplest to set up a centralized database of neuron structures, to which data providers can submit their work. This approach has been tried in several formats, notably in the Duke/Southampton cell archives (www.cns.soton.ac.uk; Cannon et al, 1998). However, no contributions were received, despite extensive use of the archives to retrieve data and software. There is now ample evidence that this approach simply does not work for intrinsically complex and hard-won data produced by individual researchers, as is the case for the majority of the data in neuroscience. The explanation is not hard to find: a centralized database is very convenient for data users (who have done no work); it is rewarding for the database constructors (who have done some work); and it requires considerable extra work for the data providers (who have already done most of the work). That is, the convenience and perceived intellectual property (and credit) in the result is heavily weighted against the data providers who, therefore, do not submit their data.

4.2 Currently Available Neuronal Archives

However, several laboratories do share anatomical neuron files with the rest of the scientific community, but most data are available only for the hippocampal formation. At present, three organized hippocampal cell archives are accessible through the web: Duke/Southampton (Turner Lab; Cannon et al, 1998), Claiborne Lab, and Gulyas' data. The Duke/Southampton archive currently contains 124 complete cells (all from the hippocampus) and it constitutes the largest public resource to date (www.cns.soton.ac.uk or www.compneuro.org/CDROM/nmorph). For each cell, the archive contains the anatomical file (in the standard SWC or Stockley-Wheal-Cannon format; Cannon et al., 1998; Ascoli et al., 2001a), a gif or jpg image of the neuron with calibration bar, a set of metadata (including cell type, authors and publication reference, species and age of the animal, type and date of preparation, shrinkage factor, and soma area), and a text file including additional comments and information by the authors (e.g. type of electrode used, input resistance and time constant, etc.). The formation of the Duke/Southampton archive was prompted by multiple users wanting enhanced access to measured neurons, primarily for modelling.

Cells in the Duke/Southampton archive are sorted by field/layer and type: 24 CA1 in vivo young pyramidal cells, 11 CA1 in vitro young pyramidal cells, 15 CA1 in vitro aged pyramidal cells, 7 CA1 in vitro pyramidal cells after kainate lesion, 19 dentate in vivo young granule cells, 7 dentate in vitro young granule cells, 6 dentate in vitro aged granule cells, 6 dentate in vitro granule cells after kainate lesion, 13 dentate in vitro young interneurons, and 18 CA3 in vivo young pyramidal cells. Cells are archived by their index number with the option to also display thumbnail images of each neuron. Finally, the whole collection (as well as the raw, unprocessed data) can be downloaded as a single zipped file. Data in this archive have already been used in several publications by laboratories other than the contributors/owners (see Cannon et al., 2002).

The archive of Brenda Claiborne's lab (cascade.utsa.edu/bjclab) contains 4 CA3 pyramidal cells, 7 CA1 pyramidal cells, and 43 dentate granule cells, reconstructed using an older, homemade system. For each neuron, the archive includes a thumbnail image with scale bar, the index number linked to the raw data, the age of the animal in days (ranging from 14 to 368), the sex, and the total dendritic length in microns. The archive is enriched by an extensive usage note, which specifies important metadata information (reconstruction system, accuracy, reference to publications). Finally, Attila Gulyas' archive (www.koki.hu/~gulyas/ca1cells) contains 18 CA1 pyramidal cells and 67 CA1 interneurons (further divided in three groups on the basis of their immunoreactivity). For each cell, the archive includes an index entry, a jpg image, and the anatomical files in virtual reality format, segment format, tree format (both of which are extensively described in the archive), and in the SWC standard. Metadata, including publication references, and additional information, such as synapse density from electron microscopy, are also linked to the archive.

In addition to these three web-based archives, two collections of hippocampal data are distributed by the owners, upon request: David Amaral's 24 CA3 and 23 CA1 pyramidal cells (Ishizuka et al., 1995), and German Barrionuevo's 6 CA3 pyramidal cells (Hentze et al., 1996). Both of these sets are in Eutectic format (see Ascoli et al., 2001a for description), and can be converted into other anatomical or modeling formats by available tools. Although this practical survey is limited to complete dendritic morphology reconstructed from optical microscopy, it is important to mention that the above described hippocampal data are complemented across scale by an ultrastructural archive based on electron microscopy (synapses.bu.edu/anatomy/ca1pyrmd). This archive contains 6 high-resolution apical dendritic segments from CA1 pyramidal cells. For each reconstruction, the archive includes information on the species, strain, sex, and age of the animal, a rotating gif image, the aligned image series, and the corresponding virtual reality model. There are currently hundreds of labs in the world using Neurolucida or another reconstruction system to measure their neurons, but few of these many additional cells are available publicly.

4.3 A New Approach: Axiope

Two of the biggest hurdles to data dissemination via centralized databases are the obligation to document data according to formats imposed by the database constructor (who almost certainly knows less about the data than the data provider), and the perceived loss of ownership, credit and control when data is transferred to the database. These issues can be overcome, however, with more recent and flexible technology using the web as a form of distributed database, and replacing centralized databases with servers which federate data from many distinct sources (Cannon et al., 2002). For example, Axiope (www.axiope.org), under development but available in a preliminary version, includes examples of federated cell morphology archives, including data originally located in the Duke/Southampton Archive (Cannon et al, 1998).

There are two main components to the system: a user-friendly data-publishing tool, and a server, which federates data from many sites. The data-publishing tool allows data-providers to document or annotate their data as they see fit, and then exports the data and meta-data to a website both as browsable web pages, and as XML. Annotation "as one sees fit" could involve defining and then filling in forms, or adopting forms used elsewhere for similar data. In most cases, the resulting structure will be complex, since ideally it should contain the same information as is required for the methods section of a publication. However, unlike a centralized database, the decision of what to include rests entirely with the data provider. The utility of the data to others is, of course, heavily dependent on the associated annotation, so providing relevant information is crucial.

Once data has been annotated and transferred to the provider's own web site, their involvement in the process is complete. It is the task of the data users and database experts to work out how best to federate the disparate sources of data so they can be exploited as a coherent entity. The current server software in the Axiope system accepts messages from the client tool whenever a new web site ("data site") is created. It then visits the site and extracts all the meta-data with standard http requests. The server software can then construct a searchable catalog of the contents of one site, and can attempt to merge metadata from different sites, which contain the same type of data. At its simplest, this involves spotting terms shared by the metadata on two sites (for example "keywords" is likely to appear in a great many records). In general, however, effective federation is not possible without either more standardization on the individual sites or the provision of further meta-data information to be used by the server to associate datasets.

The centralized approach to databases in effect demands that data providers standardize their meta-data. In the federated, distributed picture, an equivalent result can also be achieved by using information provided by the data users. Since they are the ones with the greatest incentive for the federation process to work well, they also have an incentive to provide the necessary information. The mechanism involves creation of new meta-data linking the meta-data on one site to that on another. This might, for example, be a synonym table to indicate that the field "area" in one set of records should be treated as equivalent to "region" in another. Current tools provide a small set of such functions for data users to improve the performance of a federating server. This is one of the most active areas of ongoing development, because the success of a federated distributed approach to databasing depends to a large extent on the breadth and consistency achievable by a federated cataloguing and search mechanism.

Although still in its infancy, using the Axiope system "off the shelf" already improves upon the results achieved in the Duke/Southampton archive, which involved considerable custom software development. Almost all the cost of using the system involves meta-data entry, which is the one inescapable component that can only be done by data providers. Further developments will add computational modules, providing enhanced visualization options or even complex searching based on derived properties, building on the original meta-data (see e.g. paragraph 5.1 below).

5. ANALYSIS OF DIGITIZED NEURONS

5.1 Measurements from Digitized Neurons

One of the advantages of storing reconstructed neurons in digital format is the possibility to extract any morphological information in a consistent, reliable, and automated way. The software accompanying the Neurolucida reconstruction system provides some morphological analysis. Recently, a more general, powerful, flexible, portable, and non-proprietary software tool to measure morphological parameters, called L-Measure, was publicly released (www.krasnow.gmu.edu/L-Neuron; Scorcioni and Ascoli, 2001). L-Measure consists of a C++ engine controlled by a user-friendly Java graphical interface (tested for Windows and Linux systems). Users select the cells to be analyzed, define the parameters of the analysis, and specify the parts of the dendrites that should be considered or excluded. The engine parses the digital files and returns the requested information as raw data columns (which can be imported in spreadsheet program or statistical packages) or as simple statistical summaries. L-Measure can read all known digital morphological formats (including Neurolucida files and the format adopted in the Duke/Southampton Archive). Any combination of more than 40 morphological parameters can be measured, including diameter, length, surface, volume, distance (Euclidian or along the path), angles, overall spread and extend of the trees, topological symmetry, number of termination and bifurcation events, etc. A typical L-Measure query, corresponding to an "advanced" Sholl analysis could be: "Plot the number of branches *of diameter greater than 1 μm* versus their Euclidian distance from the soma". A comprehensive analysis of hippocampal neurons using these types of analysis revealed large differences between hippocampal granule cells and pyramidal cells in particular (Cannon et al, 1999).

5.2 Simulations of Dendritic Growth and Branching

Growth and branching characteristics of neurons determine the three-dimensional appearance of dendrites, within externally imposed constraints. Quantitative analysis of such growth includes definition of likely branch angles, likelihood of branching at particular distances, branching complexity and whether dendrites behave as fractal entities in terms of organizational patterns (Cannon et al., 1999). A critical question in cellular and developmental neuroanatomy is how much of the structural complexity is captured by a given morphological analysis. An empirical answer may be provided by computational modeling. If synthetic neurons can be generated on the basis of the measured statistics, real and synthetic neurons of a given class can be quantitatively compared to search for any discrepancy, thus guiding the investigator towards a more complete description of the natural structures. L-Neuron (Ascoli and Krichmar, 2000) is one of the software tools that generates synthetic neurons in a format compatible with CVAPP, L-Measure, and common electrophysiological simulators.

L-Neuron (publicly available in Windows and Linux alpha versions) uses an iterative approach to defining synthetic neurons based on the local branch diameter (stochastically determining dendritic elongation, bifurcation, and termination) and three-dimensional directional biases (constraining dendritic orientation). Simulated neurons can be rendered and stored as digital files like the real, reconstructed neurons, and can thus be subjected to the same morphometric analysis (Ascoli et al., 2001b; Ascoli and Samsonovich, 2002). They can also be used as substrates for electrophysiological and biophysical simulations. Examples of virtual neurons generated with this approach can be found in the Virtual Neuromorphology Electronic Database at the L-Neuron web site (http://www.krasnow.gmu.edu/L-Neuron/). In addition to those implemented in L-Neuron, other algorithms have been developed by several laboratories to simulate dendritic morphology (Ascoli, 2002).

5.3 Modelling of Dendritic Function

An extensive discussion of the practical aspects of electrophysiological models and their computational simulations is obviously beyond the scope of the present chapter. Extended texts have been devoted to this topic, including excellent recent books (Bower and Beeman, 1998; Koch and Segev, 1998; De Schutter, 2000) and reviews (Hines and Carnevale, 2001). Practical aspects of biophysical simulations based on anatomical reconstructions such as those described in this chapter have also been recently discussed (Lazarewicz et al., 2002a). The following sections are kept at a general level to provide the non-expert reader with a functional context for web-based archives of neuronal morphology.

Detailed comparison of neuronal shape and function necessarily involves the use of detailed morphometric data. Since in most cases dendrites are roughly assumed to be cylindrical, cable models represent a first approximation. Dendritic dimensions and lengths are transformed into an electrical (electrotonic) image of a neuron. The electrotonic length values can be calculated for each dendritic path and for the whole neuron (Turner, 1984). Description of synaptic potential conduction may also be simulated onto discrete dendritic sites, for estimation of synaptic efficacy. The severe limitations of cable methods are the slow computation times, the limited number of input and output sites that may be included, and the lack of a format to include voltage-dependent properties (Turner et al., 1991, 1996).

The most common approach to estimating neuronal function is the use of compartmental models, which divide the neuronal structure into connected equipotential regions. These models can represent a wide variety of dendritic morphologies and spatially varying membrane properties (Migliore et al., 1995; Turner et al., 1991, 1996). If there are many small compartments the model can be regarded as a discrete numerical solution to the cable equations described above. Within each compartment the membrane and intracellular parameters can be lumped and the model then reduces to a system of (possibly nonlinear) differential equations. The inclusion of appropriate neuronal parameters is critical, but often difficult to specify quantitatively. There are many parameters for which good quantitative estimates are not available, thus increasing the need for extrapolation of data. These techniques offer the clear advantage of incorporating a number of nonlinear properties that cannot be included in the cable model format and also the inclusion of multiple, concurrent processes as well as multiple recording sites. The interaction between the distribution of nonlinear properties and the details of dendritic structure can result in extremely complex subcellular mechanisms of spike initiation and propagation (Lazarewicz et al., 2002b).

A very important (and often overlooked) aspect of morphologically realistic biophysical models of neuronal electrophysiology is the spatial resolution necessary to achieve numerical stability. The basic assumption of compartmental models is that membrane voltage is constant within each compartment. This assumption is used in the numerical approximation of the differential equations describing passive and active membrane properties. In reality, however, the voltage gradient is spatially continuous. The error introduced in the approximate numerical solution depends on the size of the model compartments: the smaller the compartment, the better the approximation. This relationship heavily depends on the type and complexity of the biophysical model, the specific kinetic equations, and the activity pattern under consideration. On the other hand, it is also important not to use *too small* a compartment size, which would slow down the simulation time as well as introduce the risk of numerical instability. While several methods are available to estimate the appropriate compartment size (Lazarewicz et al., 2002a), it is important to verify the stability of detailed biophysical models.

5.4 Structure-Activity Correlations

The parallel simulation of morphological and electrophysiological properties also allows a complete and direct investigation of the structure-activity relationship in single cells. One example is the simulation of synaptic potentials in CA1 hippocampal pyramidal neurons by cable modeling (Turner, 1984). Using this format of cable modeling a reasonable fit was obtained between model predictions and physiological estimates of dendritic EPSPs (Turner, 1988). In a recent study, Krichmar and coworkers loaded the same complex model of passive and active membrane properties onto all reconstructed CA3 pyramidal cell morphologies available at the Duke/Southampton Archive. Upon identical stimulation protocols, each individual cell produced a distinct simulated response. Overall, the electrophysiological behavior of these cells under simulated somatic current clamp conditions varied from quiescence to regular spiking, to complex spiking and irregular bursting (Krichmar et al., 2002). This broad range of firing characteristics (which is known for real CA3 pyramidal cells) was completely due in this modeling study to the morphological variability, since the biophysical and stimulation parameters were held constant for all cells (see also Migliore et al., 1995). The results demonstrated that dendritic size has a non-trivial effect on neuronal excitability (e.g. larger trees result in lower bursting rates but higher spiking rates within bursts), and that the specific dendritic branching pattern may influence burst duration. A similar study was carried out on a different subset of neurons from the Duke/Southampton archive, to determine the possible electrophysiological effects of cortical lesions (Nasuto et al., 2001).

Real neurons are both morphologically and biophysically heterogeneous. Thus, the direct experimental investigation of the influence of dendritic morphology on the electrophysiological behavior of neurons may require an exceedingly large number of recordings to isolate statistically the most significant correlation parameters. The use of computational modeling, together with archived digital reconstructions of real neurons, makes this type of studies possible. In addition, simulated neurons may be synthesized with extreme morphological characteristics and run through the biophysical simulations to verify the correlation trends between structural and electrophysiological properties. Finally, this approach may be extended to the study of the structure-activity relationship under more realistic conditions of synaptic stimulations, which are practically impossible to fully control in experimental preparations.

5.5 Extension to Network Modeling

Once validated in comparison to cohorts of measured neurons, large numbers of anatomically realistic synthetic neurons can also be assembled into real-scale networks. An essential step in this process is to define the appropriate surfaces and volumes in space that delimit the anatomical boundaries for the spatial distribution of neurons in the tissue. Real or synthetic neurons can then be virtually packed in space based on their preferential axes of orientation (e.g. Scorcioni et al., 2002). These constructs can provide a bridge between single-cell morphology and system-level stereology, as well as between subcellular biophysical properties and emergent network activity. There is the potential to build circuitry and regional brain models based on both a combination of measured neurons and synthetic neurons, with stochastic variability added to enhance the resemblance of simulated neurons to a large population of cells within the brain (Senft and Ascoli, 1999). Defining the innate characteristics of neuronal growth and associated physiological functioning should enhance the simulation of a large population of synthetic neurons, and their application towards realistic network and regional simulation.

6. DISCUSSION

Availability of experimental and simulated data on web-based archives will likely continue to increase over time. Concern about the quality of the primary data persists, particularly for reconstructed neurons obtained from physiological recordings. As more neurons become reconstructed, defining errors and morphological noise will become critical, to allow a more standard initial level on which to base subsequent modelling analyses. Specific definitions regarding how the cells were originally visualized, the type of tissue and location are included in currently available databases (http://www.compneuro.org/CDROM/nmorph/cellArchive.html; http://cascade.utsa.edu/bjclab/; http://www.koki.hu/~gulyas/ca1cells/). Some of the limitations of the available neurons and caveats on their use will be summarized briefly.

6.1 Limitations of Neuron Reconstructions

Most of the available neurons were visualized using biocytin injections, though the Claiborne lab cells were primarily obtained using HRP. Most cells have been processed in some way which engenders shrinkage, and often this shrinkage is differential in the plane of section (depth) versus across the plane. How and whether to correct for shrinkage is a difficult anatomical question, since the correction factors proposed (such as for the Duke/Southampton individual cells) were obtained from measurements on slices and sections, not on cells. Thus, from a modelling perspective, various approximations should be tried, rather than just assuming a fixed degree of shrinkage, which may not be well known.

Dendritic measurements are difficult, particularly in the hippocampus, with small diameter branches near the limit of light microscopy resolution. For example, Claiborne disclaims dendritic diameters for granule cells, whereas length measurements are much more accurate. How accurate proves difficult to calculate. Dendritic diameters are limited severely by size, resolution and the presence of a thick coating of barely perceptible dendritic spines. Methods to quantitatively add back a numerically accurate dendritic spine count have also been detailed (up to 25,000-40,000 per cell; Turner, 1984), which significantly increases dendritic surface area, and thus dramatically alters cells surface estimates, such as membrane resistivity.

Since most of the available neurons underwent physiological recording primarily for visualization and staining purposes, the quality and amount of physiological data available for each neuron is usually sparse. Though several archives do offer some basic physiology, most modellers have instead assumed more standard values for input resistance, time constant, etc., obtained primarily from high-quality physiological recordings. However, such standard values detract from the individual variability in cellular physiological properties. Until improved visualization techniques arise, which engender quality physiological recordings, neurons obtained primarily for staining may not be physiologically well characterized.

6.2 Usefulness of Cell Archives

The use of multiple, individual neurons for modelling incorporates inherent variability across neurons, and provides a realistic structural base for various types of structure-function analyses (Turner, 1984). The trend has clearly progressed towards the use of reconstructed neurons in modelling, as cell reconstructions have become more widespread, more neurons are available in archives, and computer modelling programs can now easily accommodate complex neurons. Since central archiving has not proven feasible or popular, then considerable hope may be placed in the use of federated data base structures, such as Axiope described above, wherein data is not stored centrally, but rather accessed from multiple servers. This type of system may overcome many of the objections of groups generating neuronal structures by the maintenance of authorship and control, and possibly could lead to thousands of neurons being available for public dissemination and subsequent analysis. The widespread availability of neurons may then lead to a cellular analysis of brain circuitry, rather than a systems-level approach commonly in use to approximate brain function.

ACKNOWLEDGMENTS

Supported by a Human Brain Project grant (NINDS/NIMH R01-NS39600 to GAA), and Veterans Affairs Merit Review Award (DAT).

REFERENCES

Ascoli, G.A. (1999) Progress and perspectives in computational neuroanatomy. Anat. Rec. 257, 195-207.

Ascoli, G.A. (2002) Neuroanatomical algorithms for dendritic modeling. Network, in press.

Ascoli, G.A. and Krichmar, J.L. (2000) L-Neuron: a modeling tool for the generation and parsimonious description of dendritic morphology. Neurocomputing 32-33, 1003-1011.

Ascoli, G.A., Krichmar, J.L., Nasuto, S.J. and Senft, S.L. (2001a) Generation, description and storage of dendritic morphology data. Phil. Trans. Royal Soc. London (Series B) 356, 1131-1145.

Ascoli, G.A., Krichmar, J.L., Scorcioni, R., Nasuto, S.J. and Senft, S.L. (2001b) Computer generation and quantitative morphometric analysis of virtual neurons. Anat. Embryol. 204, 283-301.

Ascoli, G.A. and Samsonovich, A. (2002) Bayesian morphometry of hippocampal cells suggests same-cell somatodendritic repulsion. Adv. Neural Proc. Syst, in press

Bower, J.M. and Beeman, D. (1998) The Book of GENESIS. Teleos Publishing, Los Angeles. Cannon, R.C., Howell, F.W., Goddard, N.H. and de Schutter, E. (in press) Non-curated distributed databases for experimental data and models in neuroscience. Network.

Cannon, R.C., Turner, D.A., Pyapali, G.K. and Wheal, H.V. (1998) On-line archive of reconstructed hippocampal neurons using CellViewer. J. Neurosci. Meth. 84, 49-54.

Cannon, R.C., Wheal, H.V. and Turner, D.A. (1999) Dendrites of classes of hippocampal neurons differ in structural complexity and branching. J. Comp. Neurol. 413, 619-633.

De Schutter, E. (2000) Computational Neuroscience: Realistic Modelling for Experimentalists. CRC Press, Boca Raton, FL.

Henze, D.A., Cameron, W.E., Barrionuevo, G. (1996) Dendritic morphology and its effects on the amplitude and rise-time of synaptic signals in hippocampal CA3 pyramidal cells. J Comp Neurol. 369, 331-44.

Hines, M.L. and Carnevale, N.T. (2001) NEURON: a tool for neuroscientists. The Neuroscientist 7, 23-135.

Horcholle-Bossavit, G., Gogan, P., Ivanov, Y., Korogod, S. and Tyc-Dumont, S. (2000) The problem of morphological noise in reconstructed dendritic arborizations. J. Neuroci. Meth. 95, 83-93.

Ishizuka, N., Cowan, W.M. and Amaral, D.G. (1995) A quantitative analysis of the dendritic organization of pyramidal cells in the rat hippocampus. J Comp Neurol. 362, 17-45.

Koch, C. and Segev, I. (1998) Methods in Neuronal Modeling: From Ions to Networks. 2nd edition. MIT Press, Cambridge, MA.

Krichmar, J.L., Nasuto, S.J., Scorcioni, R., Washington, S.D. and Ascoli, G.A. (2002) Effects of dendritic morphology on CA3 pyramidal cell electrophysiology: a simulation study. Brain Res., in press.

Lazarewicz, M.T., Boer-Iwema, S. and Ascoli, G.A. (2002a) Practical aspects in anatomically accurate simulations of neuronal electrophysiology. In Computational Neuroanatomy: Principles and Methods, ed. G. Ascoli. Humana Press, Totowa, NJ.

Lazarewicz, M.T., Migliore, M. and Ascoli, G.A. (2002b) A new bursting model of CA3 pyramidal cell physiology suggests multiple locations for spike initiation. Biosystems, in press.

Migliore, M., Cook, E.P., Jaffe, D.B., Turner, D.A. and Johnston, D. (1995) Computer simulations of morphologically reconstructed CA3 hippocampal neurons. J. Neurophysiol. 73, 1157-1168.

Mirsky, J.S., Nadkarni, P.M., Healy, M.D., Miller, P.L. and Shepherd, G.M. (1998) Database tools for integrating and searching membrane property data correlated with neuronal morphology. J. Neurosci. Meth. 82, 105-121.

Nasuto, S.J., Krichmar, J.L. and Ascoli, G.A. (2001) A computational study of the relationship between neuronal morphology and electrophysiology in an Alzheimer's Disease model. Neurocomputing 38-40, 1477-1487.

Pyapali, G.K. and Turner, D.A. (1996) Increased dendritic extent in CA1 hippocampal pyramidal cells from aged F344 rats. Neurobiol. Aging 17, 601-611.

Scorcioni, R. and Ascoli, G.A. (2001) Algorithmic extraction of morphological statistics from electronic archives of neuroanatomy. Lect. Notes Comp. Sci. 2084, 30-37.

Scorcioni, R., Boutiller, J.M. and Ascoli, G.A. (2002) A real scale model of the dentate gyrus based on single-cell reconstructions and 3D rendering of a brain atlas. Neurocomputing, in press.

Senft, S.L. and Ascoli, G.A. (1999) Reconstruction of brain networks by algorithmic amplification of morphometry data. Lect. Notes Comp. Sci. 1606, 25-33.
Turner, D.A. (1984) Conductance transients onto dendritic spines in a segmental cable model of CA1 and dentate hippocampal neurons. Biophys. J. 46, 85-96.
Turner, D.A. (1988) Waveform and amplitude characteristics of evoked responses to dendritic stimulation in guinea pig CA1 pyramidal cells. J. Physiol. (London) 395, 419-439.
Turner, D.A., Isaac, J., Chen, Y., Stockley, E.W. and Wheal, H.V. (1996) Analysis of dendritic synaptic sites in hippocampal CA1 pyramidal cells - variability and nonuniformity. In Excitatory Synaptic Transmission, ed. H.V. Wheal. Oxford Press, Oxford, pp. 171-189.
Turner, D.A., Wheal, H.V., Cole, H. and Stockley, E. (1991) Three-dimensional reconstructions and analysis of the cable properties of neurones. In Cellular Neurobiology, ed. J. Chad, H.V. Wheal. Oxford Press, Oxford, pp. 225-246.

Chapter 7

ModelDB: A Resource for Neuronal and Network Modeling

Andrew P. Davison[1], Thomas M. Morse[1], Michele Migliore[1,2], Luis Marenco[3], Gordon M. Shepherd[1] and Michael L. Hines[1]
[1]*Department of Neurobiology, Yale University School of Medicine, New Haven, CT, USA:*
[2]*Institute of Advanced Diagnostic Methodologies, National Research Council, Palermo, Italy:*
[3]*Center for Medical Informatics, Yale University School of Medicine, New Haven, CT, USA.*
Correspondence to: andrew.davison@yale.edu

Abstract: ModelDB is an online database (senselab.med.yale.edu/senselab/ModelDB) of published neuronal models, including models of ion channels, dendrites, axons, neurons, synapses and networks of neurons. Having ready access to the code for a model facilitates testing and verification of a model, re-use of model components to speed development of new models, and comparing a model to new experimental data. The database is useful for archiving models and for collaboration on modeling projects, and is a resource for teachers and students in both theoretical and experimental neuroscience. We describe here how to use the database: how to find specific models, how to obtain model code, how to run models, and how to contribute a model to the database.

Key words: neuronal models; neuroscience databases; verifying models; modeling tools; computational neuroscience; public contributions.

1. WHAT IS MODELDB?

ModelDB is an online database of published neuronal models (senselab.med.yale.edu/senselab/ModelDB). This includes models of ion channels, dendrites, axons, neurons, synapses and networks of neurons.

The core of each database entry is the code needed to run the model. The database can potentially contain models written in any programming language or simulation environment, although at present all the models run in either the NEURON (Hines and Carnevale, 1997; www.neuron.yale.edu) or GENESIS (Bower and Beeman, 1998; www.genesis-sim.org/GENESIS/)

simulation environments. If you have NEURON installed on your computer, you can run a NEURON model simulation directly from the web browser.

Each entry in the database also contains information about the model: the neuron types, ion channels, receptors and/or neurotransmitter release mechanisms that are included in the model; the reference for the paper in which the model was described; and the concepts underlying the model, i.e. the scientific motivation for the model or the scientific problems addressed in the associated paper. This information makes it possible either to retrieve readily a specific model or to search for models that satisfy a particular criterion, e.g. all models of CA1 pyramidal neurons (see *Searching for models*, below).

ModelDB is closely linked to NeuronDB, a database of neuronal properties (senselab.med.yale.edu/senselab/NeuronDB). NeuronDB provides a dynamically searchable database of three types of neuronal properties—voltage gated conductances, neurotransmitter receptors, and neurotransmitter substances—that are crucial for most current types of compartmental modeling. It contains tools that provide for integration of these properties in a given type of neuron and compartment, and for comparison of properties across different types of neurons and compartments. When analyzing a model of an olfactory bulb mitral cell in ModelDB, for example, it is easy to compare the model to the experimental knowledge about mitral cells available in NeuronDB.

The database also contains a few models that are not neuronal, but that use similar principles and may be of interest to neuroscientists, such as a model of a cardiac atrial cell (Courtemanche et al, 1998) and a model of networks of arterioles (Crane et al, 2001).

ModelDB, together with the other Senselab databases (senselab.med.yale.edu) use the EAV/CR model of data representation (Miller et al, 2001). This allows the structure of the databases to be easily extended as requirements evolve. For more information on EAV/CR see the chapter by Crasto et al in this same volume.

2. WHO IS MODELDB FOR?

ModelDB is intended to be useful to three groups of people: users of neuronal models, who may be computational or experimental neuroscientists; authors of computational neuronal models; and teachers/students.

2.1 Model users

Uses of ModelDB:

- Testing and verifying models. The essence of science is reproducibility, whether of experimental or computational results. Testing of a published model is required in deciding whether to accept or reject the conclusions of a modeling study and in evaluating the model's domain of validity.
- Re-using existing models. Using existing models and model components in developing a new model can greatly speed up the development process. Examples include taking an ion-channel model for a new model of a neuron, or taking an entire neuron model for use as a component in a network.
- Comparing a model to new experimental data. A model that was developed to investigate one scientific problem may be of use in approaching a different problem, e.g. exploring whether the model can explain a new experimental observation.

2.2 Model authors

Uses of ModelDB:

- Making a model available to the community. A model that is easily accessible, reproducible and well-tested is more likely to be frequently cited and to become the foundation of future research.
- Archiving a model. Changing computer platforms, operating systems and storage media can lead to model code becoming inaccessible. Similarly, as models are developed further, it can be difficult to recreate the exact model that was used in a published paper. ModelDB can be used as a repository for a 'standard' version of a model.
- Collaborating on a modeling project. ModelDB can be used to store private, unpublished models that are not accessible to the public but that can be accessed from any computer using an access code. This feature is useful when several authors are collaborating on a model.

2.3 Teachers and students

Uses of ModelDB:

- Neuroscience education. Understanding how neurons function is often much easier with a model than in an experimental preparation, as parameters may be easily changed and results easily graphed. ModelDB contains several models that were developed explicitly for this purpose, but even models that were intended for research rather than teaching may be useful. The close links between ModelDB and NeuronDB are especially useful for this purpose.

- For anyone trying to learn either the general principles of computational neuroscience or the details of a specific neuronal simulation environment it is valuable to have examples to study.

3. SEARCHING FOR MODELS

The starting point for a user who wishes to obtain and run a model is the ModelDB homepage (senselab.med.yale.edu/senselab/ModelDB/). From here, one can either browse the entire list of models or search for models which satisfy a criterion: models of a particular neuron type; models which include a particular component, such as a delayed rectifier potassium channel or NMDA receptor; models written by a particular author, models implemented in a particular simulator or programming language (e.g. NEURON, GENESIS, C++); or models which explore a concept of interest in computational neuroscience, such as parameter fitting or temporal pattern generation.

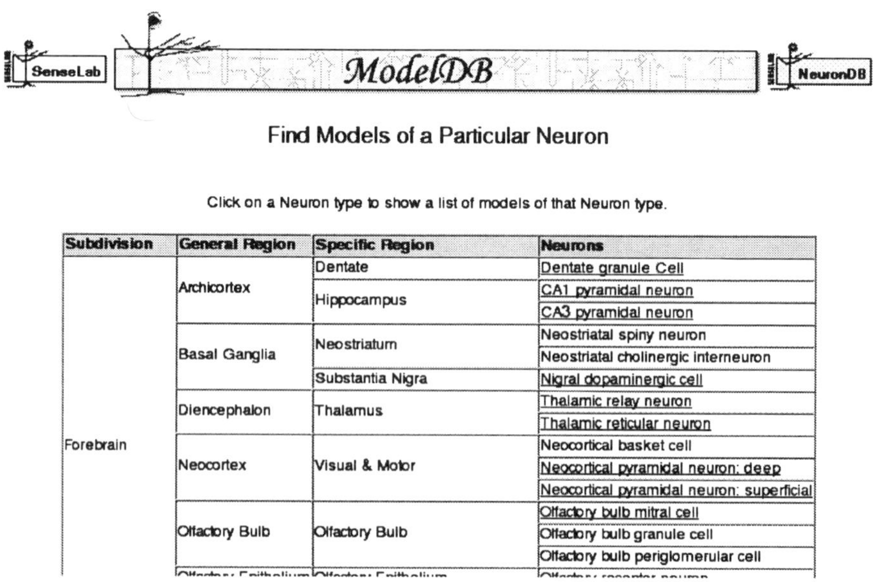
SenseLab ModelDB NeuronDB

Find Models of a Particular Neuron

Click on a Neuron type to show a list of models of that Neuron type.

Subdivision	General Region	Specific Region	Neurons
Forebrain	Archicortex	Dentate	Dentate granule Cell
		Hippocampus	CA1 pyramidal neuron
			CA3 pyramidal neuron
	Basal Ganglia	Neostriatum	Neostriatal spiny neuron
			Neostriatal cholinergic interneuron
		Substantia Nigra	Nigral dopaminergic cell
	Diencephalon	Thalamus	Thalamic relay neuron
			Thalamic reticular neuron
	Neocortex	Visual & Motor	Neocortical basket cell
			Neocortical pyramidal neuron: deep
			Neocortical pyramidal neuron: superficial
	Olfactory Bulb	Olfactory Bulb	Olfactory bulb mitral cell
			Olfactory bulb granule cell
			Olfactory bulb periglomerular cell
	Olfactory Epithelium	Olfactory Epithelium	Olfactory receptor neuron

Figure 1. Finding models of a particular neuron. This page shows all the neuron types contained in the ModelDB/NeuronDB. Clicking on a link brings up a list of models of that type (see Fig. 2).

SenseLab ModelDB NeuronDB

Models that contain the Neuron : *CA1 pyramidal neuron*

Models	Description
CA1 Pyramidal Neuron: Synaptic Scaling by London and Segev 2001	London and Segev (2001) discuss location dependent and location independent synaptic scaling in a model CA1 neuron with passive dendrites. The freely available text is followed by a critique by Maggee and Cook who comment that the London and Segev model is accurate and informative and however needs to be augmented by active channels in dendrites. Note: the zip files for this model are stored at the nature neuroscience website - Click above Supplementary Source Code in the readme.html in the model files
CA1 Pyramidal Neuron: slow Na+ inactivation, Migliore, 1996	Model files from the paper: M. Migliore, Modeling the attenuation and failure of action potentials in the dendrites of hippocampal neurons, Biophys. J. 71:2394-403 (1996). Please see the below readme file for installation and use instructions. Contact migliore@iaif.pa.cnr.it if you have any questions about the implementation of the model.
CA1 pyramidal neuron: Migliore et al 1999	Hippocampal CA1 pyramidal neuron model from the paper M.Migliore, D.A Hoffman, J.C. Magee and D. Johnston (1999) Role of an A-type K+ conductance in the back-propagation of action potentials in the dendrites of hippocampalpyramidal neurons, J. Comput. Neurosci. 7, 5-15. Instructions are provided in the below README file.Contact migliore@iaif.pa.cnr.it if you have any questions about the implementation of the model.
CA1 pyramidal neuron: Synaptic Scaling by Magee and Cook 2000	Jeffrey Magee and Erik Cook found evidence in experiments and modeling that support the hypothesis that an increase in synaptic conductance for synapses at larger distances from the soma is responsible for reducing the location dependence (relative to the soma) of synapses.
Dendritica by Vetter, Roth, and Hausser (2001)	Dendritica is a program package for relating dendritic geometry and signal propagation. The programs are based on those used for the simulations described in the following paper: Vetter, P., Roth, A. & Hausser, M. (2001). Action potential propagation in dendrites depends on dendritic morphology. Journal of Neurophysiology, 85: 926-937.

Figure 2. List of models of CA1 pyramidal neurons, with a short description of each model. Clicking on a model name shows that model (see Fig. 3).

As an example, clicking on 'Find models of a particular neuron type' leads to a page with a list of all the neuron types stored in NeuronDB/ModelDB, sorted by brain region (Fig. 1). Clicking on, for example, 'CA1 pyramidal neuron', leads to a page listing all the models of CA1 pyramidal neurons in ModelDB (Fig. 2). Clicking on a specific model, e.g. 'CA1 pyramidal neuron: Migliore et al 1999' brings up that specific model (Fig. 3).

4. USING MODELS

The files required to run a simulation, together with a 'README' file with a longer description of the model, are stored in a compressed archive in zip format ('zip file') and can be browsed from the model page. The zip file can be downloaded by clicking on the 'Download zip file' button. For models implemented in the NEURON simulator (Hines and Carnevale, 1997), the simulation can be run directly from the browser (by clicking on the 'Auto-launch' button), if NEURON is already installed on the host computer. Running a simulation usually reproduces one or more figures from the original paper(s) that first reported the model (e.g. Fig. 4).

The model description page contains tools for finding related models in the database. Clicking on a model property gives a list of all the other models in the database that have that property, e.g. in Fig. 3 clicking on 'I A' shows all other models containing this current. The citation tools allow relationships between models based on citations in published papers to be explored

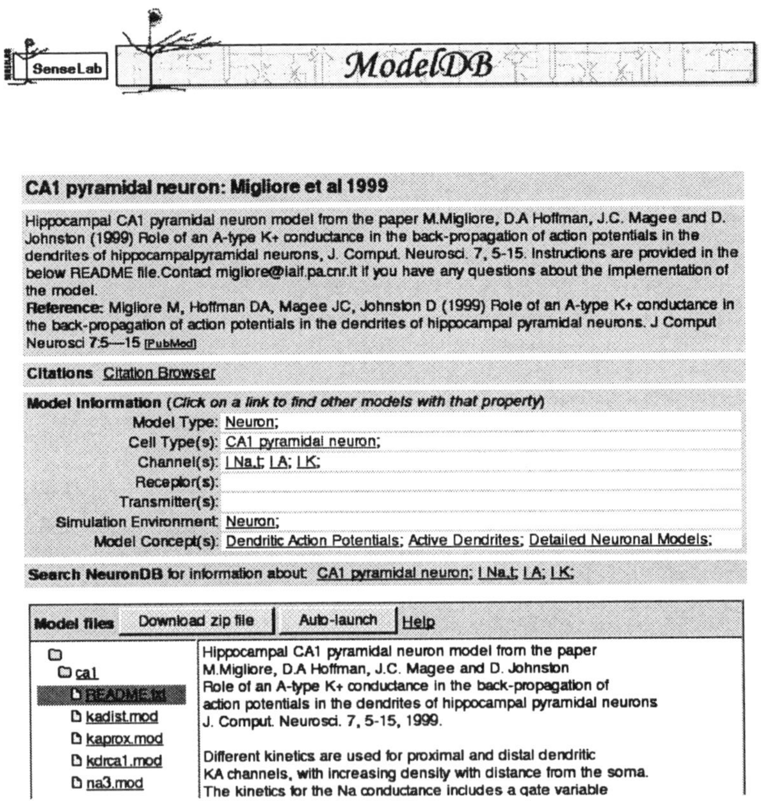

Figure 3. A specific model ('CA1 pyramidal neuron: Migliore et al 1999'). This page shows information about the model, links to find related models or to look up this neuron type in NeuronDB (see Fig. 5), and allows browsing all the code needed to run the model. The model code can be downloaded as a 'zip' archive or, if the NEURON simulator is installed on your computer, the model can be run directly from the browser by clicking on 'Auto-launch' (see Fig. 4).

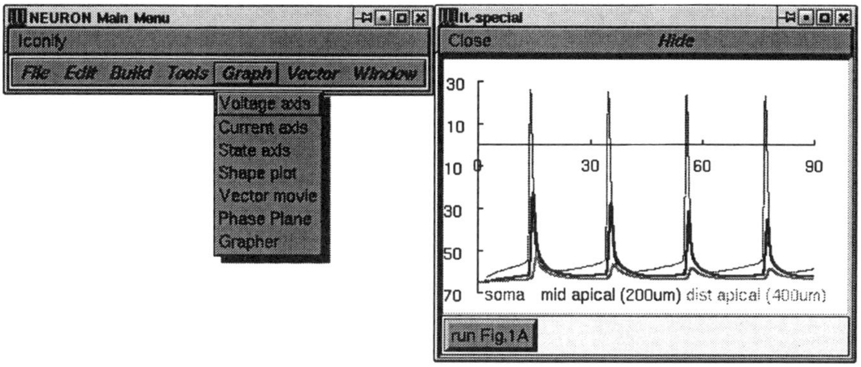

Figure 4. Running the 'CA1 pyramidal neuron: Migliore et al 1999' model in the NEURON simulation environment. The model is initially set up to show a specific figure from the paper, but the model can be explored in much more detail using the powerful tools available in NEURON.

There are close links between ModelDB and NeuronDB. For example the 'CA1 pyramidal neuron' link in the 'Search NeuronDB' section leads to the entry for this neuron type in NeuronDB (Fig. 5). This shows extensive information about the distribution of receptors, currents, and transmitters, derived from the experimental literature.

5. CONTRIBUTING A MODEL TO THE DATABASE

Contributing a model to ModelDB has been simplified as much as possible in order to encourage authors to contribute. One of the major factors inhibiting model submission, especially for beginners, is the worry that considerable time will be required to 'clean' the code and arrange for a public version of the model. It should be stressed, therefore, that the only mandatory requirement for contributing a model is that it must refer to a paper published in a peer-reviewed journal. Once the paper is published, interested authors may simply send the ModelDB administrator a complete set of the files required to reproduce a figure from the paper. In collaboration with the author, the administrator will check the files for consistency, confirm that the model reproduces at least one figure from the paper, and then add the model to the database.

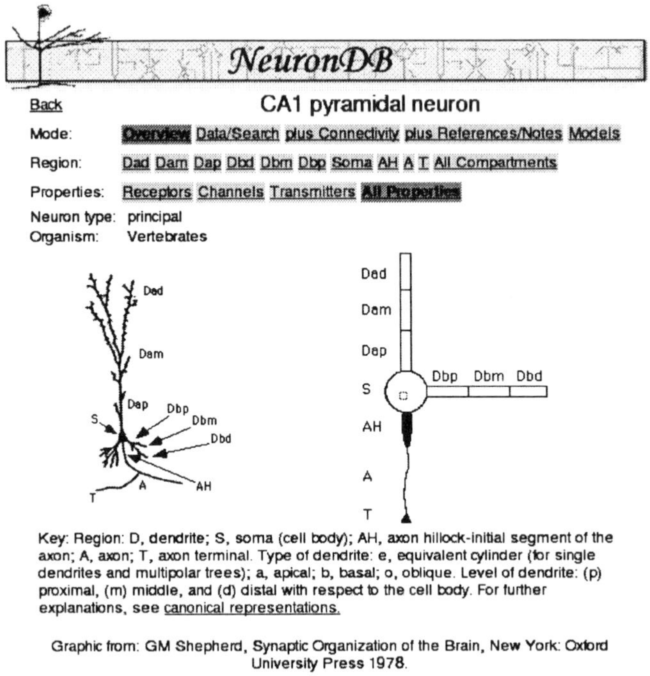

Figure 5. . An overview of the CA1 pyramidal neuron in NeuronDB. This page can be expanded to show information about the distribution of ion channels, neurotransmitters and receptors in the axon, soma and dendrites, together with references and annotations. There is also a link back to ModelDB.

Alternatively, users may prefer to add their model by themselves and maintain complete control of the entire process. In this case, upon request, the ModelDB administrator will activate a specific username/password to access a private space in the database. Authors then upload and test different ways to present a model. When the author is ready, s/he asks the ModelDB administrator to make her/his model public. Models already in the database can be used as templates to create new models by use of a 'Clone' button that appears when one is logged in as a private user. Even after a model has been made public, authors are welcome to refine the presentation of their model. We intend to add functionality allowing multiple versions of a single model (see *Future Developments*).

For a user who wishes to add a model themselves, the steps are as follows:

1. Create a zip file containing the model files and folders within a top-level folder. In addition to all the files that are required to run a sample simulation, the zip archive should include a file (called readme.txt or similar) with specific information and instructions on how to use the

model. To facilitate the use of model components as a starting point for new models and simulations, and in order to help and encourage the dissemination and use of the models, potential contributors should consider (although this is not required) simplifying, modularizing, and documenting their code as much as possible.

2. Log into ModelDB (once an account has been requested and created) and click on 'Add a model'. The 'Create a Model' web page is shown in Fig. 6. None of the fields is required (empty models can be created) and all fields are editable after the model has been created. The process can begin by editing the model name and then describing, in the notes, what is modeled and which paper the model is from. A quick look at several public models will suggest how to define the public model's name and how to format the 'Notes/References' field. Fill out the 'Properties' section, by selecting the model property values, such as each type of neuron, current, receptor, or neurotransmitter that is modeled within the model. All the relevant properties need to be selected in this step, in order for the new model to be found in searches for any of those specific properties. Some properties are hierarchical, e.g. if 'I Na,t', the common voltage-gated transient sodium channel, is selected in the list of Model Currents then subsequent searches for either 'I Na,t' or 'I Sodium' will find the new model. However if 'I Sodium' is selected then searches for 'I Na,t' will not find the new model since 'I Sodium' is more general than 'I Na,t' (above it in the hierarchy). In most browsers holding down the control key while clicking with the mouse allows one to select or deselect multiple items simultaneously.
3. Click the 'Create' button to enter the (private) model into the database. It will be accessible only by its owner at this time.
4. The final step is to upload the zip archive. Clicking on the 'Upload file' button will open a window that allows the name of the zip file to be entered directly or the file found by browsing in a graphical file manager. Every aspect of the model (name, notes, properties, and the archive file via re-uploading) can be edited as many times as necessary, until the contributor decides that the model is now ready to be made public.

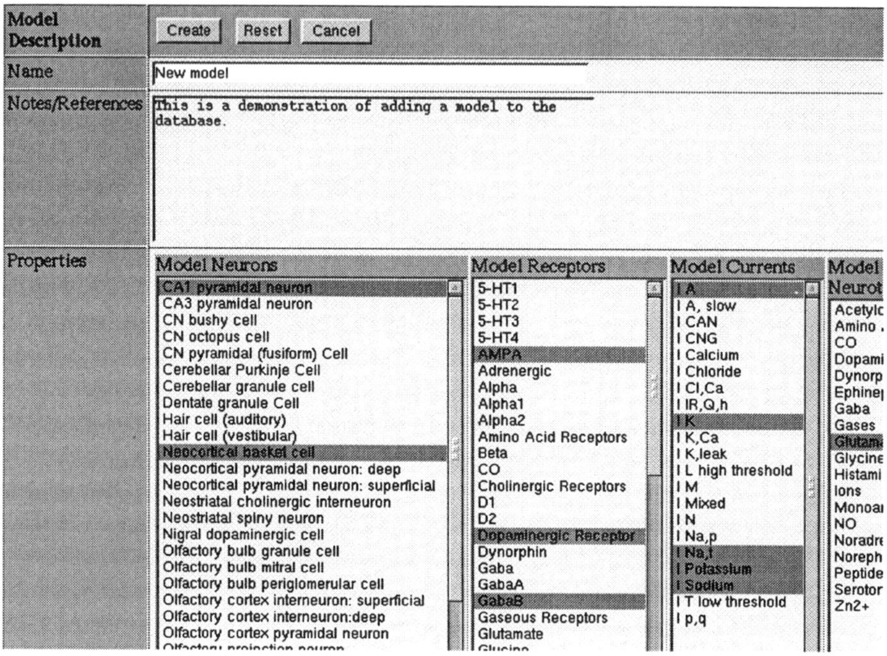

Figure 6. The 'Create Model' page. This shows part of the form used to add a model to the database.

Only ModelDB administrators are permitted to make models public. The model will be tested to ensure that it runs as intended, checked for consistency with the paper to which it refers, and then be made public. It should be stressed that only the general features (such as the graphical appearance) will be checked. The parameters used for the simulations will never be questioned, since it is expected that the author (and the peer-review process for the associated paper) take full responsibility for their values.

6. FUTURE DEVELOPMENTS

We will continue to expand the database by adding models and refining functionality. We encourage authors to contribute their models as described above. We also solicit comments on the ease-of-use and features of the site, to guide us in improving usability. The priorities for extended functionality are: developing the interoperability of our database with other databases and search engines; making it easier to reuse pieces of models, such as ion channel descriptions and morphological descriptions in data files, that now must be extracted by hand; and allowing multiple versions of a single model, so that models can be updated and/or refined without losing the original code.

ACKNOWLEDGEMENTS

We would like to thank the community of ModelDB users for their contributions, usability suggestions and patience. This work has been supported by the National Institute on Deafness and Other Communication Disorders, National Institute of Mental Health, National Institute of Neurological Disorders and Stroke, National Institute on Aging, and National Science Foundation (Human Brain Project), and a Multi University Research Initiative (Department of Defense).

REFERENCES

Bower, J.M. and Beeman, D. (1998) *The Book of GENESIS: Exploring Realistic Neural Models with the GEneral NEural SImulation System*. Springer-Verlag, New York.

Courtemanche, M., Ramirez, R.J., and Nattel, S. (1998) Ionic mechanisms underlying human atrial action potential properties: insights from a mathematical model. *Am. J. Physiol.* 275, H301-21.

Crane, G.J., Hines, M.L., and Neild, T.O. (2001) Simulating the spread of membrane potential changes in arteriolar networks. *Microcirculation* 8, 33-43.

Hines, M.L. and Carnevale, N.T. (1997) The NEURON simulation environment. *Neural Comput.* 9, 1179-209.

Migliore, M., Hoffman, D.A., Magee, J.C., and Johnston, D. (1999) Role of an A-type K+ conductance in the back-propagation of action potentials in the dendrites of hippocampal pyramidal neurons. *J. Comput. Neurosci.* 7, 5-15.

Miller, P.L., Nadkarni, P., Singer, M., Marenco, L., Hines, M., and Shepherd, G. (2001) Integration of multidisciplinary sensory data: a pilot model of the human brain project approach. *J. Am. Med. Inform. Assoc.* 8, 34-48.

Chapter 8

CoCoDat: Collation of Cortical Data on Neurons and Microcircuitry

Systematic Storage and Retrieval of Experimental Data for Biophysically Realistic Modeling

Jonas D.-Johnsen[1], Jürgen Maier[1], Klaas E. Stephan[3], and Rolf Kötter[1,2]
[1]*Computational | Systems | Neuroscience Group, C. & O. Vogt Brain Research Institute,*
[2]*Institute of Anatomy II, Heinrich Heine University, Düsseldorf, Germany.*
[3]*IME, Research Centre Jülich, Germany.*
Correspondence to: dyhrfjel@uni-duesseldorf.de, rk@hirn.uni-duesseldorf.de

Abstract: The CoCoDat (*Co*llation of *Co*rtical *Dat*a) database has been designed as a powerful tool facilitating realistic modelling of cortical neurons and local circuits. In such endeavors, collating and organizing quantitative single neuron and microcircuitry data from the literature is a time consuming task and currently available databases are specialized in either qualitative data (Mirsky et al. 1998) or reconstructed morphologies (Ascoli et al. 2001, Cannon et al. 1998). CoCoDat has been designed specifically to organize uninterpreted, quantitative data according to neurophysiological concepts and provides an advanced environment for storing and retrieving detailed data on morphology, electrophysiology and connectivity collated from the published literature. All experimental data are strictly linked to exact bibliographical references as well as detailed records of the procedures used in the experiments that produced the relevant data.

Key words: Databasing, microcircuitry, quantitative experimental data, modelling

1. BIOPHYSICAL MODELLING OF NEURONS AND MICROCIRCUITS

Modelling cortical microcircuits in a biophysically realistic manner requires a precise description of the cellular components and their interconnections. The single cell models are commonly based on the neuron morphology, which determines the passive membrane properties. Morphological subdomains are then equipped with distributions of ionic and synaptic conductances that, together with the morphology, determine the firing properties of the modelled cell. To collate, organize and select the experimental information necessary in the various steps of the modelling is generally very time consuming, and would be greatly facilitated by a dedicated tool for storing and grouping the experimental data according to the needs of the modeller.

2. STRUCTURE OF THE DATABASE

CoCoDat is implemented using the relational database approach with entries in different tables cross-linked through unique identifiers. These tables are organized into four major groups under the headings *Literature, Experimental Methodology, Mapping of Recording Site* and *Experimental Findings*. Each of the four groups contain several tables and are accessed through graphical user interface *forms*.

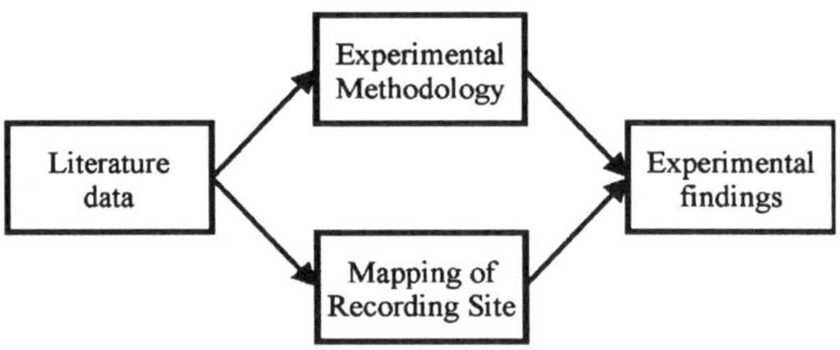

Figure 1. The relationship of the main data groups in CoCoDat.

All data are anchored in the detailed bibliographical data entered for the publication in question. After that the datastream diverges to records of methods and mapping respectively, before it reunites at the level of the actual quantitative data records. These are subdivided into the categories necessary for modelling and contain data on *Morphology, FiringProperties, IonicCurrents, IonicConductances, SynapticCurrents* and *Connectivity*. All information is entered exactly as given by the authors of the publication. The CoCoDat database is based on the principal structure used in the CoCoMac database of cortical connectivity in the macaque monkey (Stephan et al. 2001), and the two can easily be integrated.

3. MAPPING CONCEPTS

To associate the quantitative data in CoCoDat with a specific neuronal source, the exact site in the brain they were recorded from would ideally be registered by specifying the *BrainRegion*, *Layer*, *NeuronType* and *NeuronCompartment* of the *RecordingSite* given in the original publication. These four specifications can be understood as four levels of description of a *RecordingSite* and are named *BrainSites*. The electrophysiological literature rarely delineates the exact sites as it is done in tract-tracing studies, since precise definitions or techniques for unambigously determining their properties are not available. As an example, cell types and their properties are currently not defined in a principled manner but described according to heuristic criteria, e.g. IB (intrinsically bursting) vs. RS (regular spiking) pyramidal cells, irrespective of precise definitions. Therefore our *BrainSites* should be seen as prototypes, described by a generally accepted nomenclature on each of the four levels of description.

On each level of description, the *BrainSites* are arranged in a hierarchical manner, relating super- and substructures in conceptual parent/child relationships, as here illustrated for the compartment level (figure 2).

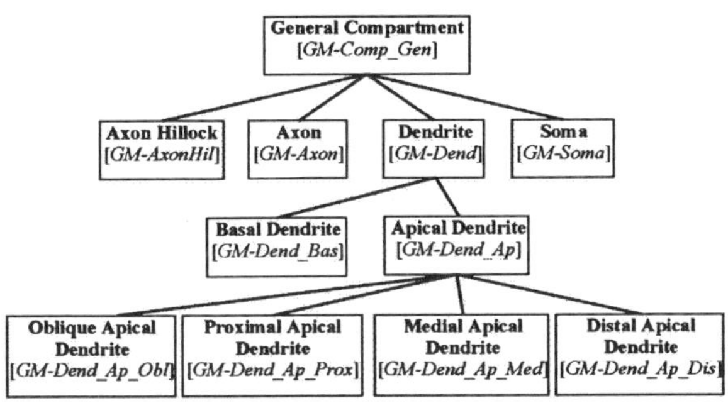

Figure 2. The hierarchy of *BrainSites* at the *NeuronCompartment* level. The abbreviation "GM" refers to the structures being organized in the *G*eneral *M*ap.

In the database, a *BrainSite* is given by a combination of the *BrainMap* it appears in (as these are prototypes we here use a prototypical General Map) and a shorthand acronym. If the *RecordingSite* is unspecified in one or more dimensions of description in the original publication, this *BrainSite* is specified as *general*. The *general*-designator resides at the top of the hierarchy at each level, as can be seen in figure 2. These hierarchies at the four levels of description are the basis of the powerful search routines and visualization methods that will be discussed below.

4. USING COCODAT

4.1 Obtaining and installing the database

The CoCoDat database is obtained by dowloading and un-zipping the file "CoCoDat_R1.0.zip" from the webpage http://www.cocomac.org/ under the heading "Databases". After unzipping the file to an appropriate directory, it can be opened using Microsoft ® Access 2000. By opening the database, the user is automatically taken to the the central CoCoDat *SwitchBoard* (figure 3), that gives access to all forms for entering and retrieving data.

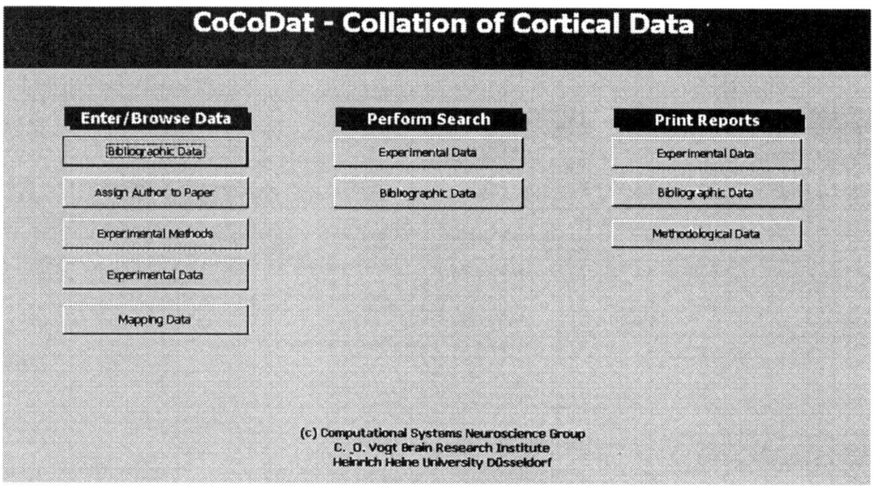

Figure 3. The CoCoDat main *SwitchBoard* gives access to all forms for entering, retrieving and printing the different classes of data.

Presently most of the collation has been targeted on data necessary for our modelling of local circuitry in the barrel cortex of the rat, with a primary focus on single cell properties of layer 5 pyramidal neurons (see below). The current contents (as of May 2002) stems from 37 publications with 76 records on experimental methodology and 65 neuronal *RecordingSites*. Linked to these are a total of 208 records of experimental data distributed as: *Morphology* (42 records), *FiringProperties* (52 records), *IonicCurrents* (68 records), *IonicConductances* (15 records), *SynapticCurrents* (11 records) and *Connectivity* (20 records).

4.2 Selecting parameters for CoCoDat-queries

By clicking "Experimental Data" under the heading "Perform Search" on the *SwitchBoard* the user opens the *SearchBoard* form for retrieving and displaying the experimental findings containined in CoCoDat. For each query to the database, the user can specify constraints on the neuronal structure(s), type of data and animal species to be included. The data can also be filtered by including a free-text search string that must be contained in the records. All parameters are combined to a final searchstring corresponding to the use of the Boolean AND operator.

Since all data are linked to *RecordingSites* entered exactly as in the original publications, a simple search specifying e.g. "dendrite" as *NeuronCompartment* would only return data from *RecordingSites* containing the exact matching *BrainSite* keyword, but not from more specific substructures like the apical or basal dendrites. Therefore the CoCoDat search routine expands the selected *BrainSites* to include all substructures on each of the 4 levels of description as defined in the previously described hierarchies. After the expansion of all terms, these are combined in an all-to-all fashion resulting in a list of *RecordingSites* used for the final query.

To illustrate the use of the database, we here consider the modelling of a layer 5 regular spiking pyramidal neuron from the sensorimotor cortex of the rat. Given that the modeller already has a suitable cell morphology (detailed or simplified) the next logical step is to equip the model with the appropriate voltage-gated currents. In this case the *BrainSites* constraining the query would be *GM-Ctx_SeM* (sensorimotor cortex), *GM-L_5* (layer 5), *GM-C_Pyr* (pyramidal neuron) and *GM-Comp_Gen* (general compartment). Leaving the *NeuronCompartment* as general will return data on all neuronal compartments after the expansion of the term, which is necessary to obtain data on all currents in the neuron. Combining these terms with *IonicCurrents* and *Rat* results in the query illustrated below.

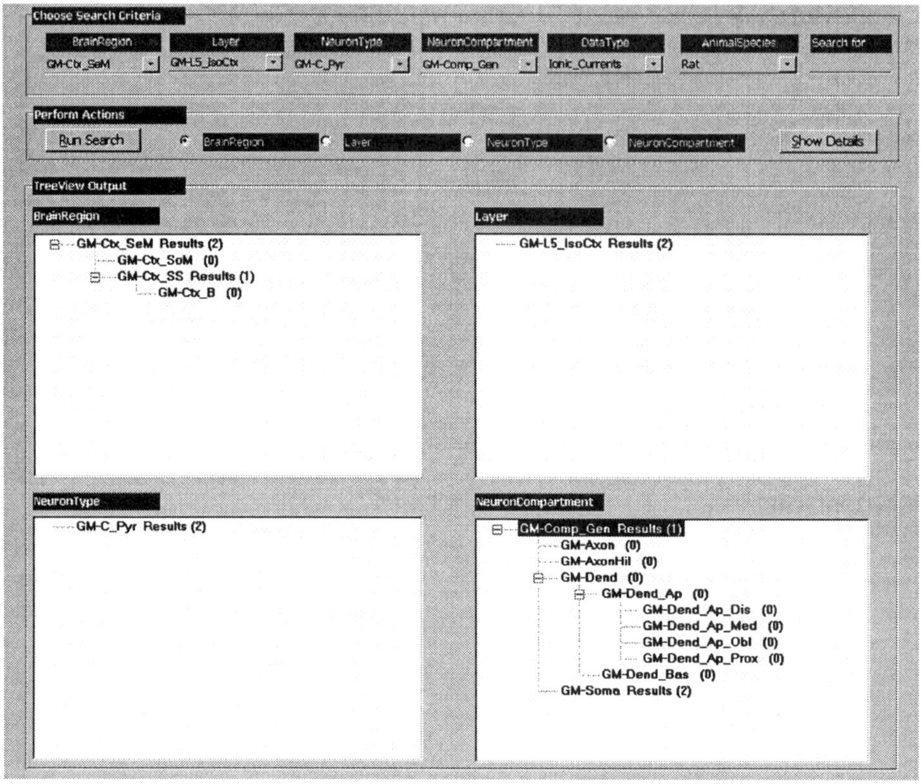

Figure 4. The CoCoDat *SearchBoard*: In the upper part of the board the criteria for extracting data on ionic currents in layer 5 pyramidal neurons from sensorimotor cortex in the rat are selected (here no free text criterion has been entered in the field "Search for").On the lower part the anatomical distribution of search results on all four levels of description is visualized in four panels.

4.3 Visualizing the search results

After selecting the appropriate criteria and clicking the "Run Search" button, the search routine expands all *BrainSites* as described above and the possible *RecordingSites* resulting from the all-to-all combination are checked against all existing sites in the database. For all matching *RecordingSites*, the existence of experimental data in the chosen category is verified and finally the specified animal species is checked against the records detailing the experimental methodology. If a search string has been specified, the resulting records are further filtered adding the entered criterion with a Boolean AND operator.

To visualize the anatomical origins of the resulting data, the four panels in the lower part of the *SearchBoard* (figure 4) show the distribution of returned records among the *BrainSites* constituting the *RecordingSites*. Each panel contains an illustration of the *Brainsite*-hierarchy rooted in the originally specified criteria. By clicking all "+"-signs appearing next to nodes in the hierarchy, the tree is expanded to show all substructures as already done in figure 4. The number of returned results at a given level in each hierarchy is given in parentheses. This method of visualization gives the user a way of instantly determining the distribution of returned results between the specified structures on the different levels of description.

In the case where data are not available for the exact structure one intends to model, the query returns an empty result. Here the hierarchical visualization becomes an even more powerful tool: By selecting *BrainSites* higher in the anatomical hierarchy on one or more levels of description (thus relaxing the constraints) in a new query, the user can allow more *RecordingSites* to be included. To determine, which of the results from the relaxed query are most appropriate for the model, the user can visually appraise the logical "distance" between the desired and available structures in the hierarchy, thus allowing for the selection of data most likely also to be appropriate for the modelled structure. The option to automatically perform a relaxation on one selected level of description until results are produced will be available in a future version of CoCoDat.

4.4 Examining and exporting the search results

In order to examine the retrieved data in detail, the user must first specify the level of organization according to which these should be grouped. This is done by selecting *BrainRegion*, *Layer*, *Neuron* or *NeuronCompartment* above the four panels in the lower part of the *SearchBoard*. By clicking "Display Details" a form summarizing the details is generated. In the current example, we choose to display the results grouped by *BrainRegion*. In the *SummaryForm* appearing, only the returned *BrainRegions* are visible at first, but by clicking the "+"-sign appearing to the left of each entry, a table containing the full information on the returned *RecordingSite* and brief description of the data appears. Here the user can also select, which results to export in printed form (figure 5).

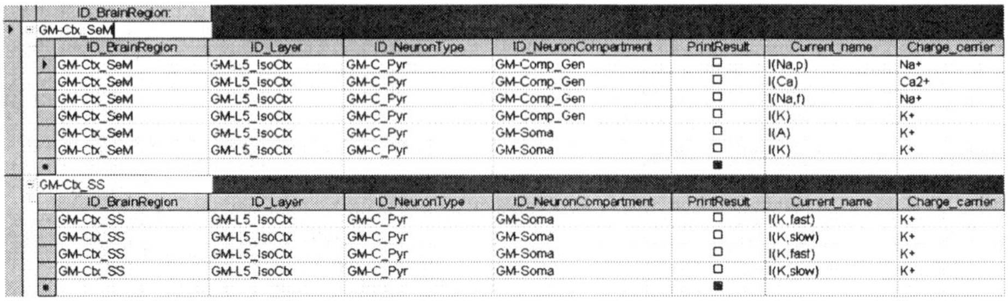

ID_BrainRegion:

GM-Ctx_SeM

ID_BrainRegion	ID_Layer	ID_NeuronType	ID_NeuronCompartment	PrintResult	Current_name	Charge_carrier
GM-Ctx_SeM	GM-L5_IsoCtx	GM-C_Pyr	GM-Comp_Gen	☐	I(Na,p)	Na+
GM-Ctx_SeM	GM-L5_IsoCtx	GM-C_Pyr	GM-Comp_Gen	☐	I(Ca)	Ca2+
GM-Ctx_SeM	GM-L5_IsoCtx	GM-C_Pyr	GM-Comp_Gen	☐	I(Na,f)	Na+
GM-Ctx_SeM	GM-L5_IsoCtx	GM-C_Pyr	GM-Comp_Gen	☐	I(K)	K+
GM-Ctx_SeM	GM-L5_IsoCtx	GM-C_Pyr	GM-Soma	☐	I(A)	K+
GM-Ctx_SeM	GM-L5_IsoCtx	GM-C_Pyr	GM-Soma	☐	I(K)	K+

GM-Ctx_SS

ID_BrainRegion	ID_Layer	ID_NeuronType	ID_NeuronCompartment	PrintResult	Current_name	Charge_carrier
GM-Ctx_SS	GM-L5_IsoCtx	GM-C_Pyr	GM-Soma	☐	I(K,fast)	K+
GM-Ctx_SS	GM-L5_IsoCtx	GM-C_Pyr	GM-Soma	☐	I(K,slow)	K+
GM-Ctx_SS	GM-L5_IsoCtx	GM-C_Pyr	GM-Soma	☐	I(K,fast)	K+
GM-Ctx_SS	GM-L5_IsoCtx	GM-C_Pyr	GM-Soma	☐	I(K,slow)	K+

Figure 5. The *SummaryForm*: Search results are grouped according to the selected level of description, here the *BrainRegion*, and can be seen in a summarized form after clicking the "+"-sign next to the initially appearing entries. By doubleclicking on an entry, the user can inspect the detailed experimental data (figure 6).

Using the summarized search results it is possible to quickly compare the data available from different *RecordingSites*. Along with the four *BrainSites* describing the *RecordingSite* the name of the "main feature" reported is given - in this example case the name of the voltage-gated current described as well as the charge carrier for the current in question. For the five other data categories, the display is identical apart from the feature reported.

By double-clicking the first column of the summarized results, the user can invoke a form containing the detailed quantitative experimental data reported along with precise bibliographical references to text and figures in the original publication (figure 6).

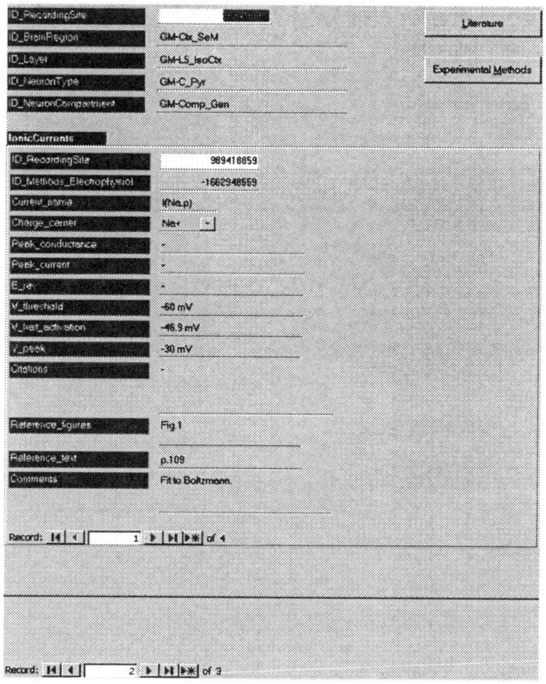

Figure 6. The detailed experimental information on a persistent sodium current ($I_{Na,p}$) recorded in a layer 5 pyramidal neuron from sensorimotor cortex in the rat. The associated bibliograpical and methodological data can be viewed by clicking the buttons appearing on the right.

Here the full information on the returned *RecordingSite* is repeated along with the quantitative data reported in the publication. To complement the quantitative data, most entries also contain textual quotes elaborating on the experimental data. Furthermore a field is reserved for comments entered by the person responsible for collating that particular dataset. The user can navigate between the records returned for a given *RecordingSite* by using the upper set of navigation arrows. Similarly, by using the lower set of navigation arrows, the user can switch between the returned *RecordingSites*.

By selecting "Literature" on a particular form containing the experimental findings, the user brings up the bibliographical information available on the publication. Here the specific data regarding the publication is registered, but also the abstract and a brief summary detailing in which of the six categories data from this publication has been entered.

The information on methodology used in the experiments generating the displayed data can be viewed by clicking "Experimental Methods". Having detailed information available regarding the methods used in the experiments enables the user to judge whether apparent contradictions between data obtained in different studies are actual contradictions or merely based on different experimental protocols. Furthermore, it is necessary to make sure that the data reported are obtained under circumstances comparable to those simulated in the modelling effort, or it may be necessary to correct for different ionic environments, temperatures or the effects of pharmacological manipulations. Table 1 contains a summary of all data-types in the *Literature*, *Experimental Methodology* and *Experimental Findings* groups.

Table 1. A summary of detailed data registered on *Literature, Experimental Methodology* and the six *Experimental Findings* categories.

Data group	Contents
Literature	ID, title, year of publication, publication type, abstract, keywords, comments and DB-collator ID, authors' initials and last names Page numbers, editors, book title, publisher, place of publishing, journal, volume and page numbers, links between literature IDs and authors
Experimental Methodology	Experimental preparation, solution, recording method, stimulation method, species, temperature, text- and figure references, comments, animal strain, age, sex and weight, pharmacological components and concentrations, slice orientation
Morphology	Morphological feature name, value, reconstruction and reconstruction method, citations, text- and figure references, comments
FiringProperties	Firing pattern type, action potential duration, input resistance, intracellular resistance, membrane time-constant, membrane resting potential, citations, text- and figure references, comments
IonicCurrents	Current name, charge-carrier, peak conductance, peak current, reversal potential, voltage threshold, half-activation voltage, peak voltage, citations, text- and figure references, comments
IonicConductances	Conductance name, charge-carrier, peak conductance, voltage threshold, half-activation voltage, peak voltage, citations, text- and figure references, comments
SynapticCurrents	Synapse type, peak conductance, peak current, peak potential, reversal potential, latency, citations, text- and figure references, comments
Connectivity	Target BrainRegion, Layer, NeuronType and Compartment, citations, text- and figure references, comments

To allow for later use or inspection of returned results, the user can select which records to include in automatically generated *reports* by activating the relevant tick boxes on the *SummaryForm* (figure 5). When returning to the main *SwitchBoard* (figure 3), reports of the detailed experimental data can be printed, as well as the bibliographic and methodological data associated with these, by clicking one of three buttons under the heading "Print Reports". These reports contain all data given in Table 1.

5. DISCUSSION

CoCoDat provides the user with a powerful and flexible tool for extracting desired datasets for biophysically realistical modelling. By allowing the user to specify all desired constraints and dynamically retrieving search results, CoCoDat provides customized extraction of information that is not limited by predefined groupings of data, but only the content of the database. Furthermore the selection of optimal data for a given model is facilitated by the visualization tools allowing for direct comparison of results from different *RecordingSites* along with detailed information on the experimental methods used for obtaining the data.

The use of CoCoDat is not limited to extracting subsets of the data already contained in the database at present. As can be seen on the CoCoDat *SwitchBoard* (figure 3), tools for entering data are implemented and available for the user in the left column. These forms resemble the forms used for displaying the detailed experimental data returned from the example search (figure 6). They provide a straightforward interface for entering more data for *BrainSites* already present in the database as well as adding sites that have not yet been described. A manual providing step-by-step instructions for efficiently collating and entering data in CoCoDat is provided when downloading the database from our website. Thus the user can customize a personal collection of data relevant to his or her work, as well as exchange collated data with other CoCoDat users in a predefined format.

As mentioned earlier, CoCoDat is based on the principal structure of the cortical connectivity database CoCoMac (Stephan et al. 2001), and the two can easily be integrated. This integration across levels of description (microcircuitry and connectivity of macroscopic areas) would provide the opportunity to access information on the anatomical connections in the macaque between e.g. two cortical areas as well as quantitative electrophysiological information on the neurons involved in the signalling, thus immediately giving a more complete picture of the nature and impact of the connection in question.

Another interesting possibility would be to integrate information on intracellular signalling into the CoCoDat database, or link information from already existing databases to the *BrainSites* in CoCoDat, possibly by using the XML capabilities provided in Microsoft Access. This would facilitate another kind of multi-level modelling, integrating information from the microcircuitry level with molecular processes in the single neurons constituting the local network.

Using XML output from CoCoDat would also make it possible to exchange information with the XML-based neuroscientific markup language NeuroML (Goddard et al. 2001), that suggests a common syntax and systematic ontology for describing entities used in neuronal modeling. Currently several simulation environments are being adapted to read/write descriptions in NeuroML, and the future possibility to incorporate collated experimental information directly in a modelling environment would further facilitate the use of detailed quantitative experimental data in modelling studies.

REFERENCES

Ascoli GA, Krichmar JL, Nasuto SJ, Senft SL (2001) Generation, description and storage of dendritic morphology data. *Phil. Trans. R. Soc. Lond. B* 356, 1131-45.

Cannon RC, Turner DA, Papyali G, Wheal HV (1998) An on-line archive of reconstructed hippocampal neurons. *J Neurosci Methods* 84(1-2), 49-54.

Goddard NH, Hucka M, Howell F, Cornelis H, Shankar K, Beeman D (2001) Towards NeuroML: model description methods for collaborative modelling in neuroscience. *Phil. Trans. R. Soc. Lond. B* 356, 1209-28.

Mirsky JS, Nadkarni PM, Healy MD, Miller PL, Shepherd GM (1998) Database tools for integrating and searching membrane property data correlated with neuronal morphology. *J Neurosci Methods* 82(1), 105-21.

Stephan KE, Kamper L, Bozkurt A, Burns GAPC, Young MP, Kötter R (2001) Advanced database methodology for the collation of connectivity data on the macaque brain (CoCoMac). *Phil. Trans. R. Soc. Lond. B* 356, 1159-1186.

Chapter 9

Computer Models and Analysis Tools for Neural Microcircuits

Thomas Natschläger[1], Henry Markram[2] and Wolfgang Maass[1]
[1] *Institute for Theoretical Computer Science, Technische Universität Graz, Austria*
[2] *Brain Mind Institute, Ecole Polytechnique Federale de Lausanne, Switzerland*
Correspondence to: tnatschl@igi.tu-graz.ac.at

Abstract: This chapter surveys web resources regarding computer models and analysis tools for neural microcircuits. In particular it describes the features of a new website (www.lsm.tugraz.at) that facilitates the creation of computer models for cortical neural microcircuits of various sizes and levels of detail, as well as tools for evaluating the computational power of these models in a Matlab-environment.

Key words: neural microcircuits, spiking neurons, computer simulations, real-time computing, learning algorithms, Matlab.

1. INTRODUCTION

A major problem in understanding brain-style computing is the identification of the *basic computational unit of the brain* (Zador, 2000). Neurons are obvious candidates. On the other hand there exist numerous emergent properties of recurrent circuits of neurons and synapses which cannot readily be explained on the basis of their components. This has motivated an alternative proposal: that stereotypical neural microcircuits which are repeated throughout the neocortex should be viewed as computational units of the brain (Shepherd, 1988), (Douglas and Martin, 1998). More detailed investigations of cortical neural microcircuits have provided further evidence that these circuits exhibit intriguing stereotypical features, but that these stereotypical circuit patterns are quite complex (Gupta et al., 2000), (Thomson et al., 2002), (Gupta et al., 2002). This additional complexity arises from the fact that stereotypical cortical microcircuits consist of a fairly large number of different types of neurons and synapses, which are combined in quite regular circuit patterns. For computational modeling this provides the additional challenge to explain how computations are organized in such complex microcircuits, and how the diversity of their components and the complexity of their connection patterns might contribute – or even be essential – for the virtually unlimited computational power of biological neural systems. Since at present it appears to be hopeless to analyze the computational power of realistic models for biological neural microcircuits with theoretical methods, computer simulations provide the most promising tool. This chapter discusses publicly available software and analysis tools that appear to be useful for such investigations.

Obviously there exists no agreement regarding the meaning of abstract concepts such as *computation* and *computational power*. In the following we mean with a computation simply an algorithm or circuit that assigns outputs to inputs. Evidence for the computational power of such circuit is provided by the complexity and diversity of associations of outputs to inputs that can be implemented on it. We will focus on the biological most common type of computations where the inputs to a computation are not static batch inputs, like in a computer, but rather *continuous streams of information*, such as for example afferent spike trains from neurons in peripheral neural systems. In addition we focus on the biologically most relevant case of *real-time computing*, where the system cannot afford to wait until a computation has terminated, but instead has to provide an output within a specific time period. More precisely, we focus on the case where the neural circuit has to respond to a stream of inputs with a stream of outputs, where the output at time t represents the best response to the stream of inputs up to time t that the circuit can provide at this time. In terms of concepts from computational theory this shifts the emphasis towards *online computations* and *anytime algorithms* (i.e., algorithms that output at any time the currently best guess at an appropriate response), and away from offline computations that are more common in computer applications (and which have frequently been used as paradigms for computations in biological neural systems).

2. WWW.LSM.TUGRAZ.AT

This website, which has been set up by the first author, contains a collection of Matlab[1] programs for constructing and simulating models for neural microcircuits, as well as Matlab programs for generating input streams and for analyzing the resulting circuit output for various benchmark tasks. We have chosen Matlab since it allows easy implementation of the analysis tools that we are using. Furthermore the powerful graphics offered by Matlab makes it easy to visualize the results. One drawback of choosing Matlab is that there is no readily available tool for the actual simulation of neural circuits. We solved this problem by writing our own simulator for leaky-integrate and fire neurons (in C) which can easily be accessed from within Matlab (by means of a MEX interface). However, with a conceptually very simple interface the user can use instead any other simulation software.

2.1 The Microcircuit Simulator

We provide a tool for simulating heterogeneous networks composed of leaky-integrate-and-fire neurons, or alternatively of nonspiking (sigmoidal) neurons. This simulator is written in C with an interface to Matlab (there is no standalone version). It is intended to simulate networks containing a few neurons, up to networks with a few thousand neurons and on the order of 100000 synapses.

Neuron models For the case of leaky-integrate-and-fire (LIF) neurons we implemented the standard model where the membrane potential V of a neuron is given by $\tau_m(dV/dt)=-(V-V_{rest})+R(I_{syn}(t)+I_{back})$, τ_m is the membrane time constant, V_{rest} is the resting membrane potential, R is the input resistance, $I_{syn}(t)$ is the current supplied by the synapses, and I_{back} is a non-specific background current. We have chosen the fixed values V_{rest}=0.0 and R=1.0. Alternatively the user can offset the voltages and scale the currents by the actual input resistance. If V exceeds the threshold voltage V_{thresh} it is reset to V_{reset} and held there during the absolute refractory period (of length τ_{refr}).

With regard to nonspiking neurons (which may be useful for analyzing population responses in larger circuits) we implemented a model of a sigmoidal neuron with leaky integration. The dynamics of such a neuron is given by $\tau(dx/dt)=-x+g(I_{syn}(t)+I_{back})$ where x is the output (in the range (0,1)) of the unit (which can be interpreted as an average firing rate over time or space), g is the logistic sigmoid function $g(z)=1/(1+\exp(-z))$, and τ some time

[1] Matlab is a registered trademark of *The Mathworks Inc.*, see www.mathworks.com for more information about Matlab.

constant (see e.g. chapter 7 in (Hertz et al., 1991)).

Synapse models Two types of synapses are implemented: static and dynamic synapses. While for static synapses the amplitude of each postsynaptic current (PSC) is the same, the amplitude of an PSC in the case of a dynamic synapse depends on the spike train that it has seen so far. For synapses transmitting spikes the time course of a PSC is modeled by $A \cdot \exp(-t/\tau_{syn})$, where τ_{syn} is the synaptic time constant and A is the synaptic strength (weight) which is constant for static synapses and given by the model described in (Markram et al., 1998) for dynamic synapses. For synapses transmitting analog values (such as the output of a sigmoidal neuron) static synapses are simply defined by their strength A (weight), whereas for dynamic synapses we implemented a continuous version of the dynamic synapse model for spiking neurons (Tsodyks et al., 1998).

Network inputs There are two forms of inputs which can be supplied to the simulated neural microcircuit: spike trains and analog currents, see the user manual for details.

Synaptic plasticity The simulator also supports spike time dependent plasticity, STDP, applying a similar model as in (Song et al., 2000). STDP can be modeled most easily by making the assumption that each pre- and postsynaptic spike pair contributes to synaptic modification independently and in a similar manner. Depending on the time difference $\Delta t = t_{pre} - t_{post}$ between pre- and postsynaptic spike the absolute synaptic strength is changed by an amount $L(\Delta t)$. A typical shape for the function $L(\Delta t)$ as found for synapses in neocortex layer 5 (Markram et al., 1997) is shown in Fig. 1. The current implementation allows for an arbitrary choice for the function $L(\Delta t)$, since various different shapes of it have been found for different synaptic connections (Abbott and Nelson, 2000). Synaptic strengthening and weakening are subject to constraints so that the synaptic strength does not go below zero or above a certain maximum value.

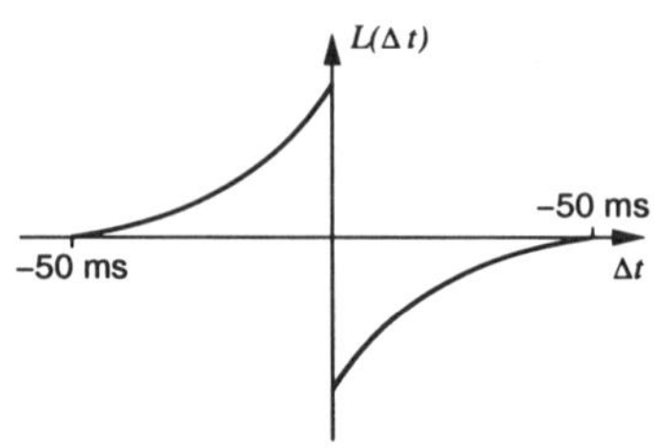

Figure 1: Spike time dependent plasticity (STDP) as found for synapses in neocortex layer 5 (Markram et al., 1997).

2.2 Tools for Creating Microcircuit Models

When one tries to build a realistic model of a stereotypical cortical microcircuit, one can rely on a number of clear rules that have been established by anatomical and neurophysiological research (Braitenberg et al, 1998), (Gupta et al, 2002), (Thomson et al, 2002). However many of these rules are of a statistical nature, and there are also various aspects where empirical data are still missing. In order to take this into account, the website provides tools not just for building a computer model of one particular neural microcircuit. It also allows the user to specify a probability distribution over neural microcircuits, and to draw random samples according to that distribution. For some questions the evaluation of their average computational power may be more realistic than the analysis of one particular circuit model (that necessarily also has several accidental features). As of this writing there are tools available for constructing multi-column circuits with a distribution of parameters that match those reported in (Gupta et al., 2000).

Connectivity within a single column In a single column the neurons are placed on a $N_x \times N_y \times N_z$ 3-dimensional grid. A certain fraction f_I of these neurons are randomly chosen to be inhibitory neurons. The synapses between these neurons are randomly created in the following way: the probability of a synaptic connection from neuron j to neuron i (as well as that of a synaptic connection from neuron i to neuron j) is defined by $C \cdot \exp(-D(i,j)/\lambda^2)$, where λ is a parameter which controls both the average number of connections and the average distance between neurons that are connected. $D(i,j)$ is the Euclidean distance between the neurons i and j (which are located on nodes of the $N_x \times N_y \times N_z$ grid). Depending on whether i and j are excitatory (E) or inhibitory (I), the value of C can be set to different values C_{EE}, C_{EI}, C_{IE} or C_{II}.

Synapse parameters A synapse can be chosen to be static or dynamic. In either case the associated scaling parameter A (the absolute synaptic strength) also depends on the type (E or I) of the pre- and postsynaptic neuron. To be precise: the four parameters A_{EE}, A_{EI}, A_{IE} or A_{II} describe the mean value of the absolute synaptic strength for a given type of connection (*EE*, *EI*, *IE*, *II*). The actual value of A_{ij} for a synaptic connection from neuron j to neuron i of type $T(ij)$ is chosen randomly from a gamma distribution with mean $A_{T(ij)}$ and a standard deviation of $A_{T(ij)} \cdot SH_A$. Dynamic synapses are described by the additional parameters U (use), D (depression time constant) and F (facilitation time constant) of the synapse model proposed in (Markram et al., 1998), which in general also depend on the type of the synapse. The actual values U_{ij}, D_{ij} and F_{ij} for a synaptic connection from neuron j to neuron i of type $T(ij)$ can be drawn from a Gaussian distribution (with appropriately resampled values for outliers) with user-specified mean values $U_{T(ij)}$, $D_{T(ij)}$, $F_{T(ij)}$ and standard deviations $SH_{udf} U_{T(ij)}$, $SH_{udf} D_{T(ij)}$, $SH_{udf} F_{T(ij)}$.

Intercolumn connectivity One can specify separately for each pair of columns the probability of connections between neurons in these two columns and the distribution of the synaptic values for these connections in an analogous manner.

2.3 Analyzing the Computational Power of Neural Microcircuit Models

The conceptual framework of a Liquid State Machine (LSM) (Maass et al., 2002) facilitates the analysis of the real-time computing capability of neural microcircuit models. It does not require a task-dependent construction of a neural circuit, and hence can be used to analyze computations on quite arbitrary "found" or constructed neural microcircuit models. It also does not require any a-priori decision regarding the "neural code" by which information is represented within the circuit.

A

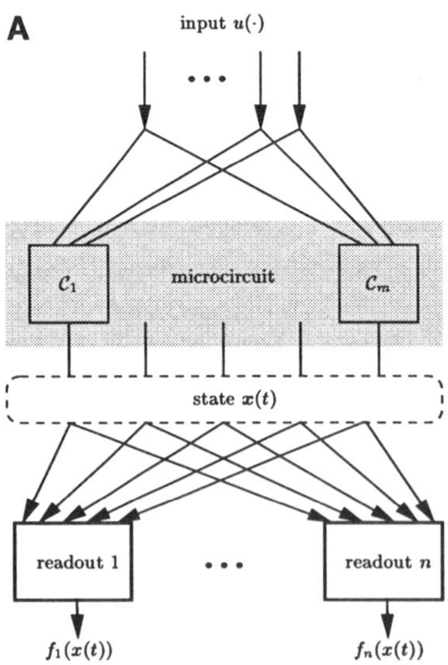

B

1.) Define the neural microcircuit to be analyzed

2.) Record states $x(t)$ of the microcircuit at various time points in response to numerous different (training) inputs $u(\cdot)$

3.) Apply a supervised learning algorithm to a set of training examples of the form $\langle x(t), y_u(t)\rangle$ to train a readout function f such that the actual outputs $f(x(t))$ are as close as possible to the target outputs $y_u(t)$.

Figure 2. **A** The Liquid State Machine (LSM). The recurrent microcircuit (liquid) transforms the input into states $\boldsymbol{x}(t)$, which are mapped by the memory-less readout functions $f_1, ..., f_n$ to the outputs $f_1(\boldsymbol{x}(t)), ..., f_n(\boldsymbol{x}(t))$. **B** A simple procedure for learning a given task defined by its target outputs $y_u(t)$ ($y_u(t)$ denotes the target output at time t if $u(\cdot)$ is given as input).

The basic idea is that a neural (recurrent) microcircuit may serve as an unbiased analog (fading) memory (informally referred to as "liquid") about current and preceding inputs to the circuit.

We refer to the vector of contributions of all the neurons in the microcircuit to the membrane potential at time t of a generic readout neuron as the *liquid state*[2] $\boldsymbol{x}(t)$ (see Fig. 2A). Note that this is all the information about the state of a microcircuit a readout neuron get access. In contrast to the finite state of a finite state machine the liquid state of an LSM need not be engineered for a particular task. It is assumed to vary continuously over time and to be sufficient sensitive and high-dimensional that it contains all information that may be needed for specific tasks.

The liquid state $\boldsymbol{x}(t)$ of a neural microcircuit can be transformed at any time t by a readout map f into some target output $f(\boldsymbol{x}(t))$ (which is in general given with a specific representation or neural code). We will argue that only the synapses of these readout neurons have to be adapted for a particular computational task. This requires that any two different input time series $u(s)$, $s \leq t$ and $v(s)$, $s \leq t$ which should produce different outputs at some subsequent time t put the recurrent circuit into two (significantly) different states $\boldsymbol{x}_u(t)$ and $\boldsymbol{x}_v(t)$ at time t. In other words: the current state $\boldsymbol{x}(t)$ of the microcircuit at time t has to hold all information about preceding inputs. If the liquid has this property it possible to train a memory-less readout to produce the desired output at time t. If one lets t vary, one can use the same principles to produce as output a desired time series or function of time t with the same readout unit (provided also the states $\boldsymbol{x}(t)$ and $\boldsymbol{x}(t')$ for different times t and t' are different for saliently different input histories).

Fig. 2B outlines the basic procedure for training a readout to perform a given task based on the ideas sketched above. One advantage of this approach is that it is not necessary to take any temporal aspects into account for the learning task, since all temporal processing is done implicitly in the recurrent circuit. Furthermore no a-priori decision is required regarding the neural code by which information about preceding inputs is encoded in the current liquid state of the circuit. Note also that one can easily implement several computations in parallel using the same recurrent circuit. One just has to train for each target output a separate readout neuron, which may all use the same recurrent circuit.

According to the theoretical analysis of this computational model, see (Maass et al., 2002), there are no a-priori limitations for the power of this model for real-time computing with fading memory. Of course one needs a larger circuit to implement computations that require more memory capacity or more noise-robust pattern discrimination. The theoretically predicted

[2] This terminology is motivated by the fact that in contrast to the *finite state* of a finite state machine this circuit output assumes continuous values and varies continuously over time.

universality of this computational model for neural microcircuits can not be tested by evaluating their performance for a single computational task. Instead, each microcircuit model should be tested on a large variety of computational benchmark tasks. Hence it is desirable that many users test the circuit models of the LSM-webpage on their favorite computational problem. Furthermore it seems advantageous to have a set of benchmark tests on which different models can be compared on a qualitative basis.

2.3.1 Benchmark Tasks

The tests for analyzing and comparing models of microcircuits that are currently available on the website all have the basic structure outlined in Fig. 2B. Hence to specify a test one just has to define a distribution of inputs and target outputs for these inputs. Then one collects a sufficiently large set of N_{train} training examples by recording the liquid states of the circuit for N_{train} inputs drawn from this distribution together with the associated target outputs. Subsequently a learning algorithm (see section 2.3.2) is applied to this set of training examples, and the performance of the trained readout module is assessed by measuring the error between the actual output and the target function on a distinct set of N_{test} test inputs. In the machine learning community it is common practice that the test inputs are generated by the same distribution as the training inputs.[3] If the error on this test set is small one may conclude that the readout is able to generalize to unseen inputs (and has not simply memorized the training examples, which is called overfitting in the machine learning community. In the following we describe some benchmark tasks which have already been used, and for which code is provided.

Classification of spike trains: The task is to output the class to which a spike train belongs. This task is motivated by the recent work of Hopfield and Brody (Hopfield and Brody, 2001), where they considered the task to classify spatio-temporal pattern of events in time (like spikes). In their case these events in time came from a simple procedure which converts single spoken words into spatio-temporal spike patterns (with at most one spike per channel). Our website provides code for the following generalization of the task: m arrays of d (e.g. m=10 and d=40 in (Hopfield and Brody, 2001)) Poisson spike trains of frequency f and length T_{max} (e.g. f=20Hz, T_{max}=0.5sec) are generated, and fixed as (spatio-temporal) templates 1 to m. For i=1 to m one generates jittered versions of template i by varying each spike in template i by a random drawn amount given by a Gaussian distribution with zero mean and a given STD; this STD is called *jitter*. The

[3] However the software lets the user specify also a distinct distribution.

task is to output the number of the template from which the spike train was actually generated.

Classification of segments of jittered spike trains: This is a more complex task which can be used to evaluate the fading memory capability of a specific microcircuit model. The goal is here to output the class to which a specific *time segment* of the input spike trains belongs. The input distribution is characterized as follows: All inputs are of length T_{max} (e.g. 1.0sec) and consist of n (e.g. 4) segments of length T_{max}/n each. For each segment m (e.g. m=2) templates are generated randomly (d Poisson spike trains with a frequency f as described in the preceding paragraph). The actual input spike trains are generated by choosing for each segment one of the m associated templates, and then generating a jittered version of it. One has in general n readout modules, whose task is to output for each of the n segments the number of the template from which the corresponding segment of the input was generated.

Fig. 3 shows the results for a microcircuit with λ=2 when tested on this task (d=1, n=4, m=2, f=20Hz, T_{max}=1sec). Two cases were investigated: a) with static synapses in the microcircuit and b) with dynamic synapses in the microcircuit. Note that for the case of static synapses the liquid state of the microcircuit contains substantially less information about preceding input segments than with dynamic synapses.

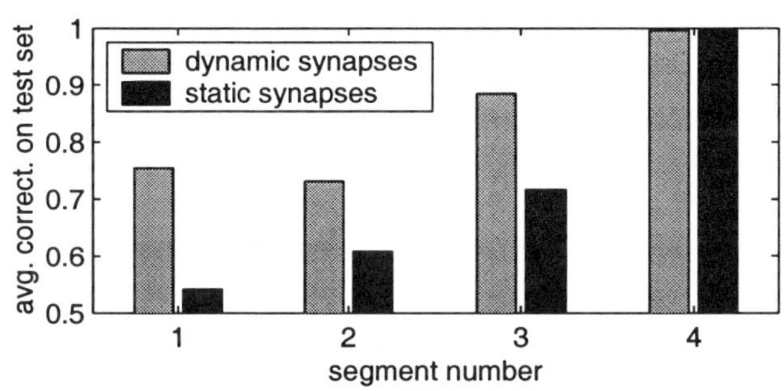

Figure 3: A result for the task "Classification of segments of jittered spike trains". 4 readout modules f_1 to f_4, each consisting of a single perceptron, were trained for this task. The readout module f_i was trained to output 1 at time t=1sec if the ith segment of the previously presented input spike train had been constructed from the corresponding template 1, and to output 0 at time t=1sec otherwise.

The avg. correctness (percentage of correct classification on an independent set of 500 inputs not used for training) is calculated as the average over 50 trials. In each trial a randomly connected circuit was constructed (1 column, λ=2) and the readout modules f_1 to f_4 were trained for perturbed versions (*jitter*=4ms) of randomly chosen Poisson spike trains

(f=20Hz) as templates. Such averaging is necessary to get statistically significant results. On the other hand a lot more CPU time is needed to get the simulations done. Since such trials are all independent they can easily be done in parallel. The benchmark tests provided at www.lsm.tugraz.at support this type of parallelism; see subsection 2.3.3.

Retrieval of delayed sum of rates: This is a more challenging memory task: at time t output the sum of rates of the input spike trains, averaged over a past time window. The input to the liquid consists of d Poisson spike trains. The rate of theses spike trains is modulated via a randomly chosen function. More precisely each input spike train has an instantaneous rate of $f_{max} \cdot r(t)$ where the random drawn function $r(t)$ is restricted to the interval (0,1). The function $r(t)$ is a sum of several frequency components f_i with independently chosen amplitudes and phase shifts; actually we choose the amplitude a_i of the sin part and the amplitude b_i of the cosine part for the modulation frequency f_i from a Gaussian distribution with zero mean and a variance of σ=1. In summary, $r(t)$ is of the form $r(t) = Au(t) + B$ with $u(t) = \Sigma_i\, a_i \sin(2\, \pi\, f_i\, t) + b_i \cos(2\, \pi\, f_i\, t)$ where A and B are some constants chosen such that $r(t)$ is in the range (0,1). The task is then to output at time t the sum of rates (normalized to the interval (0,1)) averaged over the interval (t-D-W,t-D) where W is the length of the interval and the "delay" D specifies how far in the past this interval lies. Fig. 4 shows results for two different microcircuits and for different values of the "delay" D. Note the significantly worse performance of the microcircuit with very sparse connectivity (λ=0.2). This shows that recurrent connections are necessary for the microcircuit to (temporally) integrate the stimulus.

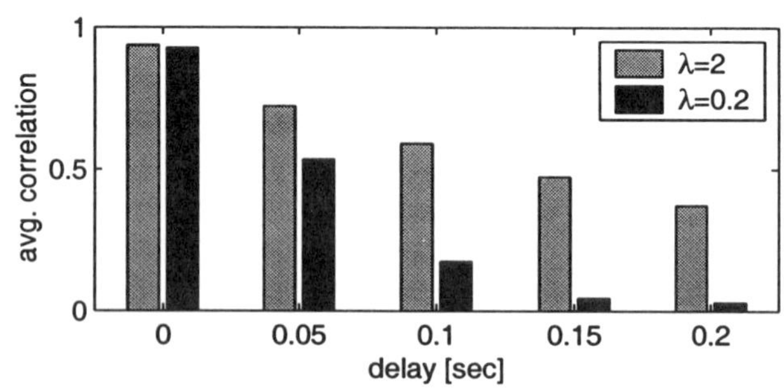

Figure 4: Results for the "Delayed Sum of Rates" task. 5 readout modules $f_0, f_{50}, f_{100}, f_{150}$, and f_{200} were trained (via linear regression) to map the liquid state of the circuit onto the scaled sum of rates for the five corresponding different values of D: 0ms, 50ms, 100ms, 150ms, and 200ms. The performance is measured as the average correlation between the actual output and the target output on an independent test set.

2.3.2 Available Learning Algorithms

All the previously described benchmark tasks employ some supervised learning algorithm in order to train the readout to map the liquid states of the microcircuit to the desired output. Note that in a biological context the target outputs may be provided by the environment (for example in the case of prediction tasks, or for function approximation in reinforcement learning), so that the learning algorithm is supervised only from the point of view of the local circuit, but unsupervised from the point of view of the system. In principle any supervised learning algorithm can be used for training readout modules. In fact it is rather straightforward to add a new algorithm to the set of readily available ones. In this section we will shortly describe the learning algorithms whose code is currently available on the website. The focus lies on learning algorithms that are commonly discussed in the context of neural computation. We do not go into the details of a particular algorithm, but just want to highlight some pros and cons. Excellent sources regarding details of these learning algorithms (except for the p-delta rule) are (Duda et al., 2001) and (Hertz et al., 1991).

The delta rule is one of the oldest and most studied neural learning algorithms. It is designed to train a single perceptron (or threshold gate) for a certain task. Due to the binary output of a perceptron it is limited to classification tasks. The appealing features of this algorithm are its simplicity and its online character which means that learning can be done on an example by example basis and it is not neccessary to have the whole batch of training examples available at once. The disadvantage of the delta-rule is that it does not converge to a stable solution if there is no perfect solution of the task at hand. (i.e. if the data is not linearly seperable).

Support vector machines (SVM) A SVM[4] is a more advanced algorithm for finding the parameters (weights) of a single perceptron applied to a virtual nonlinear projection of the data into a very high dimensional space (see (Vapnik, 1998) for details. The main idea is to choose among all the possible weight settings one which is least likely to overfit. An advantage of a SVM is that it is guaranteed to find the optimal weight setting (by means of a quadratic programming approach). The price to pay for this is the large computation time if the dimension and the size of the training data are substantial .

The p-delta learning rule The p-delta rule is a generalization of the delta

[4] In the case of a nonlinear SVM the input is first nonlinearly transformed in a high-dimensional feature space where then the optimal weight vector is determined.

rule that trains a population of perceptrons to adopt a given population response (in terms of the number of perceptrons that are above threshold). In contrast to other distributed neural learning algorithm it requires very little communication between the neural units (for details see (Auer et al., 2002)). The resulting readout function can also be implemented as a pool of unconnected spiking neurons. Obviously n perceptrons can output not only a binary value but a number between 0 and 1 (in discrete steps of size $1/n$). In fact, pools of perceptrons have the universal approximation property, i.e., they can approximate any given continuous function on any compact domain with any desired degree of precision. The p-delta learning algorithm is in some respect similar to support vector machines, since it explicitly tries to avoid overfitting. One possible drawback is that the user has to get a feeling for the numerous parameters which one has to set. However, by using the default parameters one usually gets quite good results.

Linear regression (least mean square) Linear regression is probably one of the simplest "learning" algorithms. It is the algorithm which one should try first since it is fast, there are no parameters to tune, one knows exactly what is computed and if one has enough data overfitting is unlikely. Furthermore it can be applied to binary (the output is simply passed through a threshold operation) as well to analog target functions

The back-propagation algorithm The back-propagation algorithm is a very popular algorithm for training multi-layer feed-forward artificial neural networks. Like the p-delta rule it has the advantage that it can in principle approximate very complicated target functions if one chooses the right network architecture (which is not a trivial problem). However, one has to be careful not to use too large networks in order to avoid overfitting. Regarding more details about back-propagation we refer the reader to (Hertz et al., 1991) and (Duda et al., 2001).

Which learning algorithm should I use? The answer strongly depends on the type of microcircuit that one wants to evaluate, the task one wants to implement, and last but not least on the type of results one is looking for. If one is able to get good performance with a very simple readout such as a simple perceptron (e.g. trained with the delta rule or linear regression) one can infer that the microcircuit not only carries out all temporal integration that is needed, but also projects its information about preceding inputs into a sufficiently high dimensional space to facilitate subsequent linear pattern recognition (in other words: in this case the microcircuit implements in addition a sufficiently nonlinear and high dimensional kernel, which plays a similar role as the kernel of a SVM).

2.3.3 Support for Parallel Processing

The software supports parallel execution of independent parts of the benchmark tasks. Such independent parts include simulation of a given microcircuit with different stimuli and training of several readout function for the same set of inputs (like in Fig. 3 and 4). Note that a single microcircuit will always be simulated at one processor. The following setup is required to be able to utilize the parallel processing feature: a cluster of Unix machines (we use Linux, Redhat 7.2 and SuSE 8.0) which share a common file system (via NFS) and where the software packages Matlab, the Parallel Virtual Machine (PVM) Library (http://www.csm.ornl.gov/pvm/), and the Parallel Matlab Toolbox (Svahn, 2001)[5] are installed.

2.4 Contributing to www.lsm.tugraz.at

The microcircuit models, benchmark tasks and learning algorithms described so far should just serve as a starting point for this website. We would like to encourage everybody to contribute his/her models, simulation tools, data, algorithms and last but not least links to related webpages to www.lsm.tugraz.at.

As of this writing there is no automated way for doing this. But any material which is potentially related/interesting can be sent by e-mail (lsm@igi.tu-graz.ac) to the maintainers of the website.

2.5 Comparison to Other Tools for Analyzing Neural Microcircuits

There are many other tools available for the simulation and analysis of the type of networks which are currently supported via the software available at www.lsm.tugraz.at. A quite comprehensive list can be found at www.hirnforschung.net/cneuro/cneuro_software.htm. The goal of most of these tools is to provide some kind of general purpose simulation and analysis environment. In principle it would have been possible to implement all code provided at www.lsm.tugraz.at for example within GENESIS (www.genesis-sim.org) or NEURON (www.neuron.yale.edu), to name two prominent representatives of this class of simulators. The advantages would have been that these systems are designed to simulate biologically realistic neurons and one does not have to worry about the particularities like numerical integration. However, we found that most of these systems have disadvantages in comparison with the standard tool Matlab when one

[5] Note that each Matlab process which is spawned uses one full Matlab licence.

implements the analysis tools (for example the learning algorithms) that are needed to evaluate the computational power of microcircuit models. Another (not major) issue was simulation speed: general purpose simulators are usually slower than specialized tools (due to the overhead needed for generality).

The software describe in this chapter is limited to moderate sized models and does not scale to really large models. One project aiming for the simulation of large scale models is the NeoCortexSimulator developed at the Goodman Brain Computation Lab (see http://brain.cs.unr.edu). The goal of this project is to emulate a multi-columnar brain of up to 1 million compartmental neurons by the year 2003. To achieve this performance the simulation software makes use of the supercomputing Beowulf network of the University of Nevada, Reno.

3. OUTLOOK

As soon as more experimental data regarding the connectivity and anatomy of neural microcircuits will become publicly available, we plan to add software to the website for creating neural microcircuits that reflect these new data. However for building such more complex microcircuit models it may turn out that one needs a more abstract language for describing the neural microcircuit. One project which tries to establish such a description language for neural systems (which is independent of the software which is actually used to simulate the circuit) is NeuroML, see http://www.neuroml.org.

In the near future we will provide on our website additional analysis tools that are based on concepts from information theory. These will complement the application of learning algorithms by a direct analysis of the mutual information between the liquid state (or other internal variables) of a microcircuit with the preceding circuit input. The estimation of this mutual information is not a trivial problem, since one has to deal with the problem of undersampling (see chapter 10: A practical guide to information analysis of spike train by S. Panzeri, R.S. Petersen and S.R. Schultz). However we think that this is a valuable way to analyze functional properties of neural microcircuits.

We also plan to make publicly available new software which allows the application of microcircuit models to simple sensory processing tasks, such as the ones discussed in (Legenstein et al., 2002). There neural microcircuit models were trained to predict movements of different objects, that move within an unlimited number of combinations of speed, angle, and offset over a simulated visual field. This approach provides interfaces for using simulated models of biological neural microcircuits in actually sensory processing tasks, for example for controlling a mobile robot.

The software currently available is based on the extremely simple LSM architecture shown in Fig. 2A. For a biologically more realistic scenario we will also allow for feedback from the readout back into the microcircuit. In fact, neurons within the liquid circuit may simultaneously serve as readout neurons, thereby blurring the distinction between a microcircuit and its readout. To explore this more complex scenario we are currently developing the necessary software tools. Another simplification made so far is that we are only looking at one microcircuit – more precisely one microcircuit and its associated readouts - at a time. In the spirit of the discussion about microcircuits as potential computational units of the brains it is obvious that one also wants to look at hierarchies or even recurrent networks of such units. Software for such higher level networks also needs to be developed.

Last but not least some kind of graphical user interface (GUI) is needed, which should make the available tools more easily accessible. We are looking forward to feedback and software contributions from the users.

REFERENCES

Abbott, L. F., and Nelson, S. B. (2000) Synaptic plasticity: taming the beast. *Nature Neurosci.* 3, 1178-1183.

Auer, P., Burgsteiner, H., and Maass, W. (2002) Reducing communication for distributed learning in neural networks. *Proc. ICANN'2002*. Online available as # 127 on http://www.igi.tugraz.at/maass/publications.html.

Braitenberg, V., and Schuez, A. (1998) *Cortex: Statistics and Geometry of Neuronal Connectivity*, 2nd ed., Springer Verlag, Berlin.

Douglas, R., and Martin, K. (1998) Neocortex. In: *The Synaptic Organization of the Brain*. G. M. Shepherd, Ed., Oxford University Press, 459-509.

Duda, R. O., Hart, P. E., and Stork, D. G. (2001) *Pattern Classification*, 2nd ed., John Wiley & Sons, New York.

Gupta, A., Wang, Y., and Markram, H. (2000) Organizing principles for a diversity of GABAergic interneurons and synapses in the neocortex. *Science* 287, 273-278.

Gupta, A., Silberber, G., Toledo-Rodriguez, M., Wu, C. Z., Wang, Y., and Markram, H. (2002, in press) Organizing principles of neocortical microcircuits. *Cellular and Molecular Life Sciences*.

Hertz, J., Krogh, A., and Palmer, R. G. (1991) *Introduction to the Theory of Neural Computation*. Addison-Wesley, Redwood City, Ca.

Hopfield, J. J., and Brody, C. D. (2001) What is a moment? Transient synchrony as a collective mechanism for spatio-temporal integration. *Proc. Natl. Acad. Sci.*, USA, 89(3), 1282.

Legenstein, R. A., Maass, W., and Markram, H. (2002) Input prediction and autonomous movement analysis in recurrent circuits of spiking neurons, submitted for publication. Online available as # 140 on http://www.igi.tugraz.at/maass/publications.html.

Maass, W., Natschläger, T., and Markram, H. (2002, in press) Real-time computing without stable states: A new framework for neural computation based on perturbations. *Neural Computation*. Online available as # 130 on http://www.igi.tugraz.at/maass/publications.html.

Markram, H., Lubke, J., Frotscher, M., and Sakmann, B. (1997) Regulation of synaptic efficacy by coincidence of postsynaptic APs and EPSPs. *Science* 275, 213-215.

Markram, H., Wang, Y., and Tsodyks, M. (1998) Differential signaling via the same axon of neocortical pyramidal neurons. *Proc. Natl. Acad. Sci.*, 95, 5323-5328.

Markram, H., Ofer, M., Natschläger, T., Maass, W. (2002, in press) Temporal integration in neocortical microcircuits. *Cerebral Cortex*.

Shepherd, G. M. (1988) A basic circuit for cortical organization. In: *Perspectives in Memory Research*, M. Gazzaniga, Ed., MIT-Press, 93-134.

Song, S., Miller, K., and Abbott, L. F. (2000) Competitive Hebbian learning through spike-timing-dependent synaptic plasticity. *Nature Neurosci.* 3, 919-926.

Svahn, E. (2001) *Parallel Matlab Toolbox: User Documentation.* Masters Thesis, Chalmers University of Technology, Sweden. To get a copy of the toolbox contact E. Svahn (email: ersva@igb.polymtl.ca, d96svahn@dtek.chalmers.se) or get it via ftp://ftp-at.e-technik.uni-rostock.de/pub/pm/.

Thomson, A., West, D. C., Wang, Y., and Bannister, A. P. (2002, in press) Synaptic connections and small circuits involving excitatory and inhibitory neurons in layers 2 to 5 of adult rat and cat neocortex: triple intracellular recordings and biocytin-labelling in vitro. *Cerebral Cortex*.

Tsodyks, M., Pawelzik, K., and Markram, H. (1998) Neural networks with dynamic synapses. *Neural Computation* 10, 821-835.

Vapnik, V. N. (1998) *Statistical Learning Theory*. John Wiley, New York.

Zador, A. (2000) The basic unit of computation. *Nature Neurosci.* 3, 1167.

Chapter 10

A Practical Guide to Information Analysis of Spike Trains

Gianni Pola[1], Simon R. Schultz[2], Rasmus S. Petersen[3] and Stefano Panzeri[1]
[1]Department of Psychology, University of Newcastle upon Tyne, The Henry Wellcome Building for Neuroecology, Medical School, Framlington Place, Newcastle upon Tyne, NE2 4HH, UK
[2]Howard Hughes Medical Institute and Center for Neural Science, New York University, 4 Washington Place, New York, New York 10003
[3]Cognitive Neuroscience Sector, International School for Advanced Studies, Via Beirut 2/4, 34014 Trieste, Italy
Correspondence to: stefano.panzeri@ncl.ac.uk

Abstract: Information Theory enables different candidate coding strategies to be quantified and compared, and hence is a natural framework for studying neural coding. The main difficulty is that estimates of information from experimental data are prone to systematic *sampling error*. In this chapter, we present a step-by-step guide to how this error can be addressed, and reliable information estimates obtained.

Key words: Information Theory, Spike Timing, Neural Coding.

1. INTRODUCTION

Almost all sensory messages and motor commands are encoded as temporal patterns of action potentials (spikes), often distributed over thousands of neurons. A fundamental problem in neuroscience is to understand the nature of this neural code. Ideally, as argued by Rieke et al (1996), we would like a dictionary that, given some ensemble of spike trains, tells us what sensory signal has occurred or what motor command is about to be executed. As a first step, we need to know what kind of code we are dealing with. For example, is the precise timing of spikes important, or is it just the number of spikes that matters? A powerful approach is to treat the brain as a communication channel, and to seek to quantify and compare the "information" available in different candidate codes. Information Theory provides a mathematical framework for quantifying information using rigorous measures such as mutual information (Shannon, 1948; Cover and Thomas, 1991). This approach has recently become a very popular way of studying neural coding (Borst and Theunissen, 1999).

Mutual information quantifies how well an ideal observer reading out the neuronal responses on a given trial is able to determine the stimulus identity. It is a functional of the distribution of joint probabilities between each possible stimulus and each possible response. In a typical neurophysiological study, the experimenters repeatedly present certain stimuli to the animal, whilst recording the spike emission times. To measure the mutual information that such responses convey about the stimulus set, it is necessary to use the recorded data to *estimate* these joint probabilities. The number of trials that can feasibly be collected is limited, and the probability estimates are therefore necessarily imprecise. The key difficulty in applying Information Theory is that these estimation errors lead to an *upward bias* in the mutual information, and since this bias gets worse for more complex codes, there is a very real danger of spurious results.

In this chapter we will focus on two approaches developed and used by the authors to address this problem. The *bias correction* approach is to explicitly estimate the bias, so that it can be subtracted. The *series expansion approach* (Panzeri and Schultz, 2001) is to approximate the mutual information by an expression that depends only on individual-cell mean firing rates and pair-wise correlations between spikes. Since these quantities are much easier to estimate accurately from limited data than the full response probabilities, the series expansion has much better sampling properties than "brute force" estimation of the mutual information. These two approaches are complementary, and indeed we often use them together.

With the aid of this chapter and of the discussed references, the reader should be able to compute basic information quantities (Section 2), subtract bias estimates to improve their accuracy with limited data samples (Sections 3 and 4), and implement the series expansion formalism (Section 5).

2. SHANNON INFORMATION: DEFINITIONS

Consider a time period of duration T, associated with a dynamic or static sensory stimulus, during which the activity of C cells is observed. The neuronal population response to the stimulus in this post-stimulus time window is described by a vector r - each element $r_1, \ldots, r_C$ of the vector indicating the response of an individual cell. The response of each cell can be described in a number of ways depending on the experimental questions to be addressed. For example, the experimenter might be interested in a *spike count code*. In this case r_c would simply be the number of spikes emitted by cell c in the post-stimulus time window $[0,T]$ on a given trial (see Fig 1). Or else, the experimenter might wish to investigate a *spike timing code*. In this case the response r_c would be a sequence of spike arrival times $\{t^c_j\}$, t^c_j being the time of the j-th spike emitted by the c-th neuron in a given trial. In this case the responses of the cell ensemble are sampled into L time-bins with a finite time precision dt (i.e. $T=Ldt$), small enough to contain at most one spike. Each cell response can be represented by a binary word[1] of length L, the entries of which represent the occupation number of each time-bin (see Fig 1). The response of the entire neuronal population can be represented by a binary word of length LC.

For a given choice of code, following Shannon (1948), we can write down the mutual information transmitted by the population response about the whole set of stimuli as

$$I = \sum_s P(s) \sum_r P(r \mid s) \log_2 \frac{P(r \mid s)}{P(r)} \qquad (1).$$

Summation is over all possible population responses r and stimuli s. $P(r|s)$ is the probability of response r given presentation of stimulus s, and $P(r)$ is its average across all stimulus presentations. The probability $P(r|s)$ is determined experimentally by repeating each stimulus in exactly the same way on many trials, while recording the neuronal responses. The probability

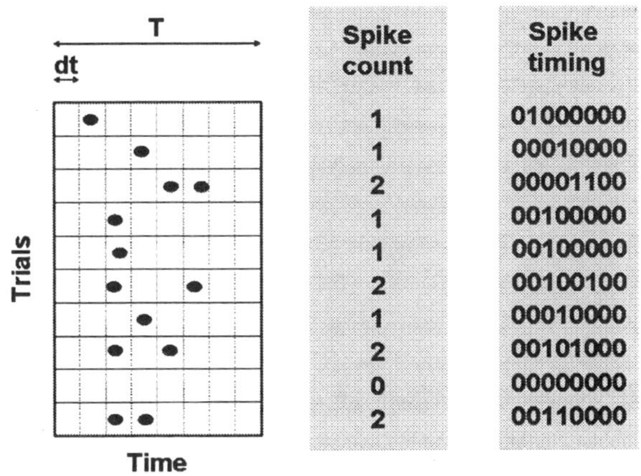

Figure 1. Spike count and spike timing codes. Each trial consists of a number of bins, which can either be occupied by an action potential (indicated by a dot), or not. The way in which this pattern is turned into a number is determined by the *code* – two examples are shown at right.

[1] A binary word is a sequence of two different symbols. In this context we are assuming that there is at most one spike in each time-bin, hence we are considering a $\{0,1\}$ sequence.

$P(s)$ is usually chosen by the experimenter.

A set of neurons can only encode stimuli effectively if each stimulus evokes a distinct response. Mutual information quantifies how informative these distinct response distributions are: it tells us how well an ideal observer of neuronal responses can discriminate between all the different stimuli, based on single trial responses. Mutual information is actually a "Kullback-Leibler distance" between the response distribution for each particular stimulus and the "overall" distribution of responses. This measures how well any given stimulus can be "picked out from the crowd" – see Cover and Thomas (1991).

3. WHAT IS THE BIAS PROBLEM?

Shannon's mutual information (Eq. (1)) is defined in terms of the probabilities of neuronal responses to the stimuli. In order to estimate the mutual information conveyed by a neuron, for example, the key step is to *estimate* these probabilities from the available neurophysiological data. If we had an infinite amount of data, we could measure the true probabilities precisely but, in practice, our estimates will necessarily differ from the true values. As explained below, this leads directly to the key issue to the practical application of Information Theory.

Consider, for example, a "neuron" that responds identically to each of two stimuli. On each trial, the neuron fires from *0* to *9* spikes with equal probability, and this is the same for both stimuli. In this case, $P(r|s) = 0.1$ for all responses and both stimuli, and Eq. (1) tells us that the mutual information is precisely zero. However, consider the effect of limited sampling. Fig 2a shows the probabilities estimated from a random sample of *10* trials per stimulus. The estimated probabilities differ markedly from *0.1* and, as a result, the estimated mutual information is *0.419 bits*. The distribution of information estimates obtained by repeating this calculation for *5000* different samples of *10* trials is shown in the right panel. Although the exact result varies from sample to sample, there is a large, consistent, bias. The lower panels illustrate the effect of increasing the number of trials. With $N=100$, the fluctuations in the estimated probabilities are smaller. There is still a bias, but it is much lower (*0.046 bits*). With $N=1000$, the estimated probabilities are very near to their true values, and the bias has almost disappeared (*0.004 bits*).

In this example, the magnitude of the bias is inversely proportional to the number of trials per stimulus. In fact, this is a general characteristic of information bias (Miller 1955; Treves and Panzeri 1995; Panzeri and Treves 1996; Victor 2000). A second general characteristic is that the bias magnitude depends on the number of different response categories that are considered. With a spike count code (e.g., the simulated neuron in Fig 2), the number of response bins required scales approximately linearly with spike count. With spike timing codes and population codes, the number of different response categories scales exponentially with the number of neurons and with the number of time bins, and hence the bias problem is much more severe.

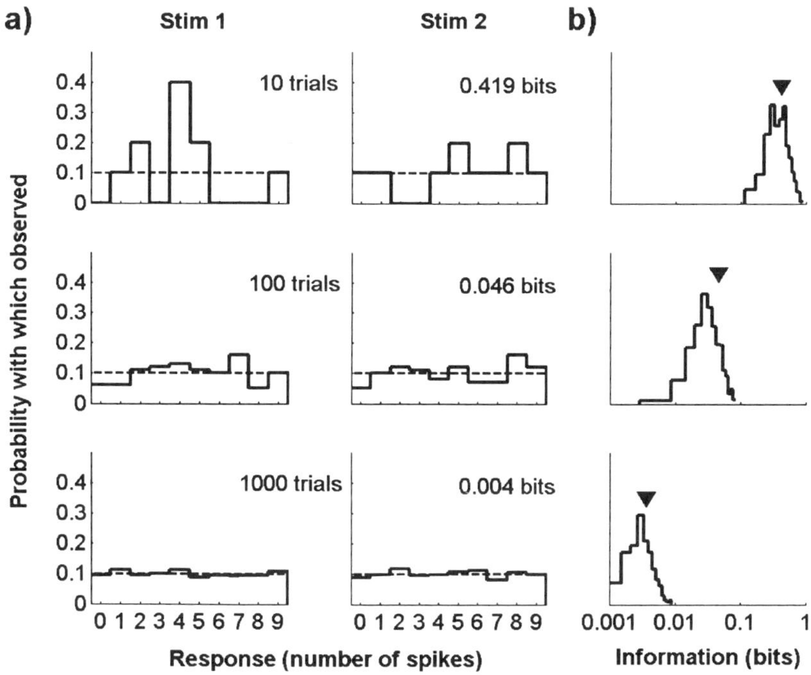

Figure 2. The effect of limited sampling. An uninformative neuron was simulated, which responded on each trial with a uniform distribution of spike counts ranging from *0* to *9*, regardless of which of two stimuli was presented. (a) Examples of response probability histograms for *10, 100* and *1000* trials. Left: responses to Stimulus 1; Right: responses to Stimulus 2; the dashed line is the true (underlying) response distribution. Also shown are the mutual information values corresponding to each pair of distributions, obtained from Eq. (1). (b) The distribution of information values obtained with a sample of *10, 100* and *1000* trials per stimulus (see main text). As the number of trials increases, the information bias shifts towards zero and the standard deviation decreases. The arrows in panel (b) indicate the values of the information of the corresponding example in panel (a).

3.1 Calculating the bias correction and estimating the number of relevant bins

We define the *bias* of the mutual information to be the difference between an average of the estimated information I_N across repeated samples of N trials (across all stimuli) and the true information I computed with the true underlying probability distributions

$$Bias(I) = < I_N[P_N(\cdot)] >_N - I[P(\cdot)] \quad (2),$$

where $< \ldots >_N$ stands for an average performed over all possible outcomes of the N trials (Treves and Panzeri, 1995) and I_N stands for the information computed starting with the estimated probabilities. Remarkably, provided that the number of trials is not too small (we make this more precise below), this expected bias is well-approximated by a simple expression. This expression depends only on the total number of trials and on the number of so-called *relevant* response categories. A given stimulus will evoke different responses with different probabilities: some will have high probability, some low probability, and others will have zero probability. A response is *relevant* if its probability, conditional to the stimulus, is non-zero. R_s is the number of responses that are relevant to stimulus s; R is the number of responses that are relevant considering all the stimuli together. The formula for the bias (Panzeri and Treves, 1996) is:

$$Bias(I) = -\frac{1}{2N \ln 2}(R - 1 - \sum_s (R_s - 1)) \quad (3).$$

To evaluate this expression, the key step is to estimate the number of relevant responses. This is non-trivial, since, with real data, we only have an *estimate* of the response probabilities, and hence only an estimate of the number of relevant responses. We will discuss three approaches to this problem.

The simplest approach is to approximate R_s by the *total* number of response categories R^{tot}, which we call the "full" count. In this case, the bias is:

$$Bias(I) = \frac{1}{2N \ln 2}(S - 1)(R^{tot} - 1) \quad (4),$$

where S is the number of the stimuli. Since $R^{tot} \geq R_s$, the full count bias is an upper bound on the actual bias. If all responses occur, the bound is right, but if the responses are sparse, the bias will be significantly over-estimated.

Another simple approach is to approximate R_s by the number of responses that were observed at least once – the "naïve" count. This will be a *lower* bound on the actual number of relevant bins, and hence the naïve count bias will be a lower bound on the actual bias.

These ideas are illustrated in Fig 3. We generated up to *8192* trials for the responses of a pair of neurons to each of three stimuli. If no bias correction is applied (solid line), the information decreases as the number of trials increases, as expected. The figure shows the effect of correcting the information using both the full count and naïve count estimates of R_s. When the number of trials is small, there is a big difference, which gradually shrinks as the N increases. The true mutual information must lie between the full and naïve curves.

Although these simple strategies are useful for bracketing the true information, we would like a method for making a better estimate of the number of relevant responses. In many cases, good results can be obtained by using a Bayesian procedure for estimating R_s (Panzeri and Treves 1996).

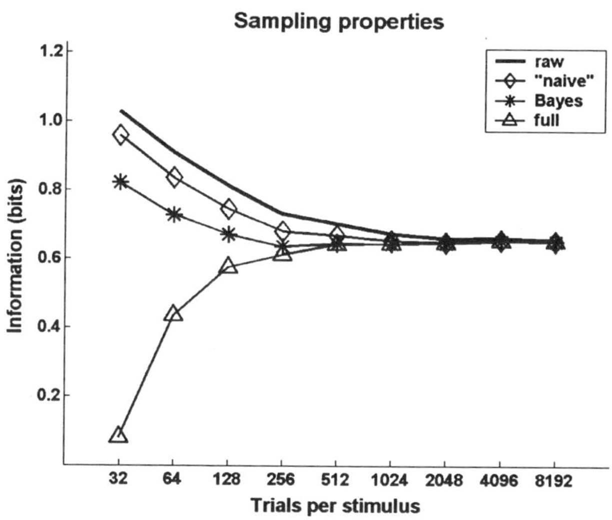

Figure 3. Correcting the bias with different subtraction procedures. In this example we considered the information carried by the joint spike timing of simulated neuronal pairs. We simulated three stimuli and we considered three time bins per cell. Hence the total number of response bins R^{tot} is equal to *64*. The information is plotted as a function of the number of simulated trials per stimulus. Results are averaged over *100* simulations for each fixed number of trials per stimulus. The "raw" information obtained without any bias correction is compared with the values obtained by using three different ways to count the number of relevant bins in the bias correction formula Eq. (3): the "naïve" count, the "full" count, and the "Bayes" count. It is clear that the "Bayes" count produces more accurate estimates than the other procedures, particularly at low number of trials.

In brief, the Bayes counting procedure of Panzeri and Treves (1996) is based on the empirical idea to use some simple prior information about the response probability distribution (such as "the prior probability of bins in which many responses were observed may be higher than that of bins in which no response was observed") to guess the number of relevant bins from the data. According to this procedure, the most likely number of relevant bins given the data and the prior is the one that reproduces more closely the number of response bins with non-zero probability observed experimentally from N trials. (Details on how to implement this procedure are reported in Appendix A). Results of applying this method are also shown in Fig 3. It can be seen that the Bayes-corrected information falls between the bounds given by the full and naïve estimates. The Bayes method has been extensively tested on various datasets and we apply it routinely in our studies of neural coding (Panzeri et al 2001; Petersen et al 2002). Although the Bayes count procedure is usually very effective, there are two potential problems with it. The first problem is that it can go wrong when the assumptions made for the priors (see Appendix A) are violated. The second problem is that the Bayes count procedure, in the form presented here, does not allow to estimate the residual error in the information evaluation. This last problem was to a good extent overcome by the more advanced and recent work of Nemenman et al (2002). A rule of thumb for the amount of data required to compute accurately the information using the "Bayes" procedure is as follows (Panzeri and Treves 1996; Pola et al 2002). When there are more trials per stimulus (at least *2* to *4* times) than population response classes R^{tot}, then the mutual information can be reasonably well corrected by the Bayes "bias" correction term and the estimated information is computed with reasonable accuracy.

We finally note that another method to empirically determine the value of the bias correction term of Eq. (3) from the data was proposed in Strong et al (1998). The procedure of Strong and colleagues consists in dividing the dataset into sub-samples and to determine empirically a best fit polynomial curve to the scaling of information with the data size. The linear term of the fit could be used to estimate the magnitude of the bias term in Eq. (3). We have tested this interesting empirical algorithm extensively, and found that it generally works less well than the Bayes procedure for estimating the bias, although it works generally much better than the "naïve" procedure.

4. SERIES EXPANSION APPROACH

Since the bias increases with the number of response categories R^{tot}, it is a particular problem when estimating the information carried by spike timing codes or population codes. Even using the correction procedures discussed above, it is usually possible to make useful information estimates only for very short word lengths. However, it is possible to achieve considerably better resolution if the number of spikes per stimulus is small. Since this is the case for many neurophysiological preparations (see, for example, Rieke et al 1996), we developed the *series expansion approach* to address such situations. The idea is that, when the number of spikes in the response time window is small, the mutual information can be well-approximated by a power series expansion in the time window that depends only on low order firing statistics. Hence, the series expansion method tends to be much less susceptible to sampling bias than "brute force" estimation from Eq. (1) and, for a given number of trials, permits information to be estimated at greater temporal resolution. Provided the assumptions of the method are met (see below), we have found a second-order approximation to provide accurate results in a number of cases. The derivation of the equations has been presented elsewhere (Panzeri et al 1999, Panzeri and Schultz 2001), as have its applications to neurophysiological data (Panzeri et al 2001, Petersen et al 2001, Petroni 2002). Our aims here are to explain how to correct the series expansion for bias and how to verify if its assumptions are satisfied.

4.1 The series expansion of the Mutual Information

The series expansion method mainly assumes that (i) the time window is short enough (or rates are low enough) so that there are very few spikes per time window (ii) Spikes are not locked to one another with infinite time precision – in other words the conditional probability of a spike occurring in a time bin given the presence of other patterns of spikes scales for small dt proportionally to dt plus higher order terms. If these assumptions are satisfied (see Panzeri and Schultz 2001 for a more precise formulation), the mutual information can be approximated by a Taylor series expansion. To second order:

$$I \approx It + Itta + Ittb + Ittc \quad (5).$$

An important feature of the method is that each of these terms quantifies different types of coding. It and $Itta$ together express information that the neurons convey by virtue of the timing of *individual* spikes. Any extra contribution due to correlations between spikes is given by $Ittb$ and $Ittc$. For precise details, see Panzeri and Schultz (2001); for a more intuitive description, see Petersen et al (2001). The four terms of the second-order series expansion are as follows:

$$It = \sum_{a,i} \left\langle \overline{n}_{ais} \log_2 \frac{\overline{n}_{ais}}{\left\langle \overline{n}_{ais'} \right\rangle_{s'}} \right\rangle_s \quad (6),$$

$$Itta = \frac{1}{2} \sum_{a,b,i,j} \left[CS_{aibj} \left(1 - \log_2 \frac{CS_{aibj}}{MS_{ai} MS_{bj}} \right) - ECS_{aibj} \right] \quad (7),$$

$$Ittb = -\frac{1}{2} \sum_{a,b,i,j} \left\langle CN_{aibjs} - ECN_{aibjs} \right\rangle \log_2 \frac{CS_{aibj}}{ECS_{aibj}} \quad (8),$$

$$Ittc = \frac{1}{2} \sum_{a,b,i,j} \left\langle CN_{aibjs} \log_2 \left[\frac{CN_{aibjs}}{ECN_{aibjs}} \div \frac{\left\langle CN_{aibjs'} \right\rangle_{s'}}{\left\langle ECN_{aibjs'} \right\rangle_{s'}} \right] \right\rangle_{s'} \quad (9),$$

where n_{ais} is the response in time bin i of cell a to stimulus s on a particular trial. The bar means an average over trials, thus $\overline{n}_{ais}$ is simply the corresponding Post-Stimulus-Time-Histogram (PSTH) normalized in units of mean number of spike per time bin. The angle brackets $<\ldots>_s$ denote an average over stimuli, weighted by the stimulus probabilities $P(s)$. $MS_{ai} = \left\langle \overline{n}_{ais} \right\rangle_s$ is the average of the PSTH over stimuli for time bin i of cell a. $CS_{aibj} = \left\langle \overline{n}_{ais} \overline{n}_{bjs} \right\rangle_s$ is the *signal correlation* between time bin i of cell a and bin j of cell b; $ECS_{aibj} = MS_{ai} MS_{bj}$ is the expected value of CS_{aibj} for PSTHs that are uncorrelated across stimuli. CN_{aibjs} *(noise correlation)* is the *joint* PSTH of bin i of cell a and bin j of cell b given stimulus s. It is equal to $n_{ais} n_{bjs}$, unless a=b and i=j, in which case it is zero. $ECN_{aibjs} = \overline{n}_{ais} \overline{n}_{bjs}$ is the expected value of CN_{aibjs} for statistically independent spikes.

4.2 Bias correction for the series expansion

However, it is useful to estimate the bias for each of the four terms of the expansion separately. The first important point is that to a very good approximation, the bias comes only from *It* and *Ittc*. The rationale for correcting bias in these terms remains that of Eq. (3). For the first term, *It*, the number of responses is simply the total number of bins in all the (single cell) PSTHs. Thus $R_{tot}=CL$. R_s is the number of non-zero PSTH bins. For the fourth term, *Ittc*, the number of responses is the number of pairwise combinations of time bins. This has two components – an auto-correlation component from within single cells, and a cross-correlation component from across pairs of cells. Auto-correlation contributes $CL(L-1)/2$ responses, cross-correlation $L^2C(C-1)/2$ responses. For a single cell, *Ittc* is characterised by $L(L-1)/2$ responses. For a pair of cells, it is characterised by $L^2 + L(L-1)$ responses.

The key point is that, whereas the number of response categories in the brute force method scales *exponentially* with the number of time bins, it scales only *quadratically* with the second order series expansion. Thus, for a given amount of data, the series expansion permits an analysis at higher temporal resolution. For example, by considering the responses of one cell with *10* time bins, some two hundred trials per stimulus should be sufficient to have a good estimation of the information with the Bayes count procedure (out of two thousands for a "brute force" evaluation directly from the mutual information formuale, see Section 2 and 3).

How does the series expansion method compare to the brute force method on real data? We analysed the mutual information conveyed in the responses of single cells recorded from rat barrel cortex in response to stimulation of each of *9* whiskers on the rat's snout. In this case, *50* trials per stimulus were available. We considered the time window [*0-80 ms*] post-stimulus (*0*= time of whisker deflection) and subdivided it into *10 ms* bins. Results of applying the two methods are shown in Fig 4. It is evident that the "brute force" estimation of the full temporal information diverges rapidly after the first four to five time bins. This is due to failure of corrections for finite sampling, as expected by the rule of thumb for sampling corrections discussed above. In fact while at *40 ms* the number of the response bins is

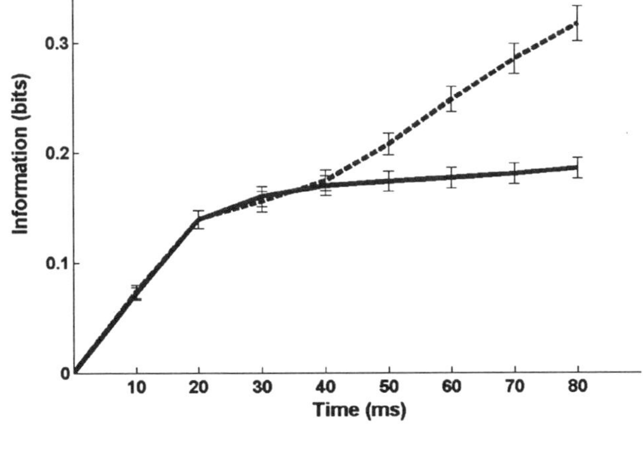

Figure 4. Information analysis of neuronal responses of *106* single units recorded in the barrel-column D2 of the rat somatosensory cortex (see Panzeri et al. (2001) for full details). Both series expansion and "brute force" method were applied. Results averaged across all cells and are plotted as a function of post-stimulus time. The information is about which of 9 whiskers (centred around whisker D2) was stimulated.

$2^4=16$, at *80 ms* the response bins are $2^8=256$. It is clear that *50* trials per stimulus could be enough to estimate the spike timing information until *40 ms* but they do not give credible results for larger time windows. The spike time series expansion is a close match to the "brute force" estimator up to *40* to *50 ms*. This result illustrates the benefit obtained by the superior scaling properties of the series expansion, compared to "brute force" information estimates. A detailed comparison of the sampling properties of brute force and series expansion approaches is reported in Schultz and Panzeri (2001).

4.3 Checking the method

The series expansion approach is an approximation: it is based on some assumptions. So it is necessary to evaluate if the results are valid. Here we discuss how to perform some important checks to validate the results. First, it is necessary to verify that the approximation is sensible: the mathematical assumptions of the method should be satisfied. Second, in order to get an idea of how precise the approximation, it is important to compare the "brute force" estimation of data-robust information quantities (such as spike count information or total response entropies) to the corresponding estimates obtained with the series expansion.

Suppose that we have applied the series expansion method to some data and obtained a certain result. How can we validate whether the results are well-sampled? We suggest that the following points should be checked:

- Is the stimulus average firing rate in all time windows of interest less than one?
- Is CS_{abij} / ECS_{abij} not divergent at any time resolution dt?
- Do brute force and series expansion estimates match for spike count? Usually, enough data is available with which to estimate information in the spike count using the brute force method. Since the series expansion for spike *count* makes the same assumptions as that for spike timing, a useful check is to compare the two spike count estimates. An example for rat barrel cortex data is shown in Fig 5a.
- Do brute force and series expansion estimates match for the response entropy? The response entropy in spike timing for each cell is well sampled especially with a large number of stimuli: in fact its bias properties depend on the total number of trials available and not on the trials per stimulus (see Cover and Thomas 1991). So this check can address directly the hypothesis of the series expansion in the spike timing. An example for rat barrel cortex data is shown in Fig 5b.

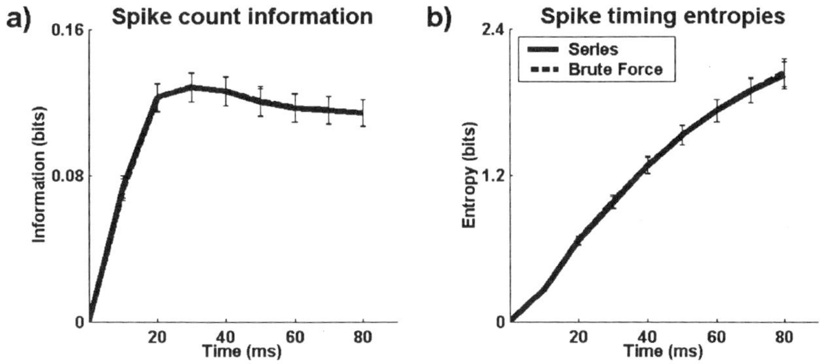

Figure 5. Checking the method on single cells from rat somatosensory cortex (a) A comparison of spike count information with series expansion and "brute force" method. (b) A comparison of spike timing entropy information with series expansion and "brute force" method.

5. SUMMARY

The main practical issue in making accurate information estimations from neurophysiological data is that these estimates are prone to systematic bias. In this chapter we have reviewed some practical aspects on how to solve or at least alleviate this problem, by discussing how to apply the bias subtraction procedure and by presenting the series expansion approach. These methods have permitted a detailed quantitative analysis of the role of both spike timing and spike correlations in cortical sensory coding (Panzeri et al 2001; Petersen et al 2001). Although these methods have been applied so far mainly to real experimental data, where data sampling problem is particularly severe, it is conceivable that they will be fruitfully applied in the future to the analysis of simulated spike trains generated by computer models of neuronal microcircuits (see Chapter 9: Computer models and analysis tools for neural microcircuits, by Natschläger, Markram and Maas), contributing thereby to a better understanding of the relationship between neuronal structures and neuronal coding.

ACKNOWLEDGEMENTS

We thank Mathew Diamond and Misha Lebedev for valuable scientific collaboration and for making available to us the neuronal data that we used as an illustration. This research was supported by EU, BBSRC, Telethon Foundation, J.S. McDonnell Foundation, Wellcome Trust grant 066372/Z/01/Z, MURST, CNR and Regione Friuli Venezia Giulia. SRS was supported by HHMI and SP by an MRC Research Fellowship. Some matlab codes that implement the procedures described here are available at http://www.staff.ncl.ac.uk/stefano.panzeri/software.html

APPENDIX A

In this Appendix we review the Bayesian procedure to count the number of relevant response bins that was developed by Panzeri and Treves (1996). With respect to what presented in Panzeri and Treves (1996), we try to describe more clearly the practical implementation of this empirical procedure.

In brief, the Bayesian counting procedure is based on the idea of using Bayes' theorem to "guess" the number of relevant bins from the data. The number of relevant bins, i.e. the number of response bins with non-zero probability, is one of the parameters of the prior distribution of probabilities. This parameter is varied in the allowed range, and then the most likely value for number of relevant bins is chosen to be the one that predicts an expected number of occupied bins (calculated from the Bayes estimate of the probabilities) that matches the experimentally observed value.

We focus for simplicity on how to compute the number of bins R_s relevant to a certain stimulus *s*, and we discuss at the end how this generalizes to the calculation of the number of bins *R* relevant to the unconditional response probability. In the following we denote by R_s^{naive} the "naïve" count of the relevant bins, i.e. the number of different response bins observed at least once from *N* trials (see main text), and by R_s^{tot} the total number of response bins. Obviously, the true number of relevant bins R_s is bound between R_s^{naive} and R_s^{tot}. The procedure for the Bayesian estimation of R_s is as follows.

- We select a particular value of R_s, starting from $R_s = R_s^{naive}$, and then increasing by one R_s at each recursive step until the procedure converges (see below);
- For the selected value of R_s, we construct a Bayesian estimate of the true response probabilities $P^B(r|s)$ given the experimental frequencies. This is done as follows. The prior distribution of response probabilities is constant across all the R_s - R_s^{naive} that were non-occupied in the experiment. The sum of the total prior probability allocated across all non-occupied bins is denoted by Γ. The prior probability is also assumed to be constant across experimentally occupied bin. Of course the total prior probability allocated across the occupied bins is $1\text{-}\Gamma$. With these flat choice for priors, one can obtain the following expression for the Bayesian estimates of the probabilities as a function of R_s and Γ (derivation given in Wolpert and Wolf 1995):

$$P^B(r \mid s) = \frac{n+1}{N_s + R_s^{naive}}(1-\Gamma) \quad \text{if } r \text{ is "occupied"}$$

$$P^B(r\,|\,s) = \frac{\Gamma}{R_s - R_s^{naive}} \quad \text{if } r \text{ is not "occupied"} \quad \text{(A1)},$$

where n is the number of events observed in the particular bin r.

- For each , we compute the expected value of the number of occupied bins:

$$\left\langle R_s^{naive} \right\rangle = \sum_r \left\{ 1 - \left(1 - P^B(r\,|\,s)\right)^{N_s} \right\} \quad \text{(A2)};$$

- We then compare this expected value with the one compared when R_s was one less. We stop the procedure when increasing R_s does not bring the expected number of bins any closer to the observed value of the number of occupied bins, or else when R_s reaches its maximum allowed value R^{tot}.
- After stopping the procedure, we choose as our estimate of the number of relevant bins the value of R_s that gives the expected number of bins which is closest to the experimental value of the number of occupied bins to stimulus s.

One arbitrary point of the procedure is the choice of the factor Γ that describes the flat priors in both the occupied and non-occupied response bins. We tried a range of different choices, and found that they work with roughly similar performance as long as the prior probability per bin is higher for occupied than for non-occupied bins. The particular choice of Γ that we used in Panzeri and Treves (1996), as well as in several other papers (e.g. Pola et al 2002; Petersen et al 2002; Schultz and Panzeri 2001), is as follows. We chose Γ so that the Bayes estimate of the response probability $P^B(r|s)$ of a non-occupied bin satisfies the following condition: the probability of that bin being empty in N_s trials is N_s / R_s times larger than the probability of being occupied in the same number of N_s trials. This requirement reflects the idea that if all observed responses are concentrated in a few bins only (i.e. high N_s/R_s), the probability in the non-occupied bins should be less than the probability assigned by a prior function constant across all R_s bins. The resulting equation for Γ is:

$$\Gamma = \left(R_s - R_s^{naive}\right)\left[1 - \left(\frac{N_s}{N_s + R_s^{naive}}\right)^{1/N_s}\right] \quad \text{(A3)}.$$

The procedure to evaluate the stimulus-unconditional number of bins R is as above, the only difference being that the quantities have to be computed from all N trials, and not from only the N_s trials specific to stimulus s.

REFERENCES

Borst, A. and Theunissen, F. E. (1999) Information theory and neural coding. *Nature Neuroscience* 2, 947-957.

Cover, T. M. and Thomas, J. A. (1991) *Elements of information theory*, Wiley, New York.

Miller, G. A. (1955) *Note on the bias on information estimates. Information Theory in Psychology; Problems and Methods II-B*, 95-100.

Nemenman, I., Shafee, F. and Bialek, W. (2002) *Entropy and Inference, Revisited* in Advances in Neural Information Processing Systems 14. Dietterich, T. G., Becker, S. and Z. Ghahramani (Eds.), MIT Press, Cambridge, MA.

Panzeri, S., Petersen, R. S., Schultz, S. R., Lebedev M. and Diamond, M. E. (2001) The role of spike timing in the coding of stimulus location in rat somatosensory cortex. *Neuron* 29, 769-777.

Panzeri, S. and Treves, A. (1996) Analytical estimates of limited sampling in different information measures. *Network* 7, 87-107.

Panzeri, S. and Schultz, S. R. (2001) A unified approach to the study of temporal, correlational and rate coding. *Neural Comput.* 13, 1311-1349.

Panzeri, S., Schultz, S. R., Treves, A. and Rolls, E. T. (1999) Correlations and the encoding of information in the nervous system. *Proc. R. Soc. Lond. B Biol. Sci.* 266, 1001-1012.

Petersen, R. S., Panzeri, S. and Diamond, M. E. (2001) Population coding of stimulus location in rat somatosensory cortex. *Neuron* 32, 503-514.

Petroni, F. (2002) *A computational analysis of the functional role of timing of action potentials in the cerebral cortex.* Unpublished doctoral dissertation, University of Newcastle upon Tyne.

Pola, G., Thiele, A., K. –P. Hoffmann and Panzeri, S. (2002) An exact method to quantify the information transmitted by different mechanisms of correlational coding. *Network: Special issue on statistical analysis of spike trains and information theory*, submitted.

Rieke, F., Warland, D., de Ruyter van Stevenick, R. R. and Bialek, W. (1996) *Spikes: Exploring the neural code,* MIT Press, Cambridge, MA.

Schultz, S. R. and Panzeri, S. (2001) Temporal correlations and neural spike train entropy. *Phys. Rev. Lett.* 86, 5823-5826.

Shannon, C. E. (1948) A mathematical theory of communication. *AT&T Bell Labs. Tech. J.,* 27, 379-423.

Strong, S.P., Koberle R., de Ruyter van Stevenink, R.R. and Bialek, W. (1998) Entropy and information in neuronal spike trains. *Phys. Rev. Lett.* 80, 197-200.

Treves, A. and Panzeri, S. (1995) The upward bias in measures of information derived from limited data samples. *Neural Comput.* 7, 399-407.

Victor, J. D. (2000) Asymptotic bias in information estimates and the exponential (bell) polynomials. *Neural Comput.* 12, 2797-2804.

Wolpert, D. H. and Wolf , D. R. (1995) Estimating functions of probability distributions from a finite set of samples. *Phys. Rev. E*, 52, 6841-54.

Chapter 11

An Introduction to CoCoMac-Online

The Online-Interface of the Primate Connectivity Database CoCoMac

Lars Kamper[1], Ahmet Bozkurt[1], Konrad Rybacki[1], Andreas Geissler[1], Ina Gerken[1], Klaas E. Stephan[1], and Rolf Kötter[1, 2]
[1]Computational | Systems | Neuroscience Group, C. & O. Vogt Brain Research Institute, and [2]Institute of Anatomy II, Heinrich Heine University, 40225 Düsseldorf, Germany
Correspondence to: kamper@uni-duesseldorf.de, rk@hirn.uni-duesseldorf.de

Abstract: The CoCoMac database is a collation of published data from tracing and mapping experiments in the macaque brain. It contains detailed connectivity data (connection strength, laminar patterns, experimental methods) as well as information about the relations between areas in different brain maps. We currently provide information from more than 300 articles.

Here we introduce our new online interface (*CoCoMac-Online*), which is a powerful search tool providing systematic access to the collated information.

CoCoMac-Online presents the data from three points of view: literature, inter-map relations, and connectivity. The search can be carried out manually, and data are returned via e-mail or http protocol (e.g. in a web browser) in two formats (HTML and XML). In addition, we provide an interface for automated database queries, which is used by other applications, for example, graphical front-ends, mapping tools and modelling environments.

In this article we present the concepts and functionality of the text interface and demonstrate its practical use with walk-through examples for the literature access. Further documentation, search options and online access to the largest collation of primate connectivity data are available from www.cocomac.org.

Key words: tract tracing, data mining, macaque monkey

1. INTRODUCTION

Anatomical connectivity data are relevant for many aspects of neuroscience research. They provide the basis for: (i) the interpretation of the measured activation pattern in functional imaging studies, such as PET and fMRI (Tononi & Edelman, 2000; Northoff et al., 2000); (ii) the connections between simulated components of anatomical realistic computer models (Kamper et al., 2002; Kötter et al., 2002); (iii) network analyses with multivariate statistics (Bozkurt et al., 2001a, 2002; Kötter et al., 2001) and (iv) theories about the pathogenesis of diseases like schizophrenia (Friston, 1999) or Alzheimer's disease (Bozkurt et al., 2001b).

Over the last years, neuroscience databases have evolved from rather simple and separated collations of published data, to powerful repositories that include sophisticated tools for data analysis (for a survey, see Kötter, 2001). For example, our main database CoCoMac-Tracer (www.cocomac.org and Stephan et al., 2001) contains systematically collated data from partitioning schemes and anatomical tracing studies of the macaque monkey brain. Currently (May 2002), this comprises more than 25,000 experimental findings from more than 1,700 tracer injections as well as 5,700 relations between more than 300 brain maps.

Most databases so far have been restricted to one data modality, however, and efforts to cross-link them have only been started (e.g. Burns, 2001). Linking these databases by standard data transfer protocols and making them available on the internet is highly desirable for the neuroscientific community since the databases contain information that are relevant to many neuroscientists in many different contexts.

CoCoMac-Online is a prime example of a comprehensive database with interfaces for manual and automated search and retrieval of data for a wide range of applications in neuroscience.

2. CONCEPT

2.1 Search categories

CoCoMac-Online provides access to connectivity and mapping data from three points of view (*search categories*): literature, inter-map relations and connections. Adapted to the needs of the respective category the interface offers different functions and options. The literature category is designed for a bibliographical search; here it is possible to search using keywords and specific criteria (listed in Table 1). In contrast, our search-interface for connections and relations allows to specify brain areas and to search for their connections, synonyms and relations to other specific partitioning schemes.

2.2 Data representation

Data were collated using operational rules for data representation, including abbreviations for the name of brain sites (*BrainSite-ID*), for the spatial extent of label in injected or labelled brain sites (extension codes, *EC*), for the extensional relation between brain sites from different maps (relation codes, *RC*) and for the precision of data representation (precision of description codes, *PDC*). Detailed descriptions of these concepts are provided in Stephan et al., 2001 and in our manual for data entry at www.cocomac.org. To guide the user to the requested level of detail extensive help options have been implemented. Acronyms and abbreviations used are explained by "mouse over" context sensitive help functions and, in more detail, in a new browser window, by clicking on the respective abbreviation. Furthermore, a general and unified description to the CoCoMac methodology is available as a help text as well.

3. TECHNICAL DETAILS

3.1 Computational requirements

As most other web applications, CoCoMac-Online does not require much performance from the user's computer system. A standard computer with an Internet browser and a display resolution of minimum 1024 x 768 dots will be sufficient. Although the display of CoCoMac-Online is optimised for the Microsoft Internet Explorer (v6), it is compatible with other standard browsers, such as Netscape (v6). As some queries can lead to large amounts of requested data we advise that CoCoMac-Online should be accessed through high bandwidth Internet connections.

3.2 Data output

The standard input and output format will be HyperText Markup Language (HTML) displayed by the user's Internet browser. To generate a hardcopy of the displayed results, CoCoMac-Online provides a printer-ready HTML format.

In addition, it is possible to import connectivity data from CoCoMac into other tools or applications. Data is exported from CoCoMac in Extensible Markup Language (XML) format. The data files are structured according to the CoCoMac XML schema, which is accessible at www.cocomac.org as well. Furthermore, CoCoMac-Online can be queried through an extended-URL syntax, independent of the graphical user interface displayed in the browser. The input by extended URL and the output by XML enable CoCoMac-Online to cooperate and communicate with other applications and databases on the basis of standard protocols. Currently, two other applications possess direct links to CoCoMac-Online:

(i) "Catacomb" by Robert Cannon, a workbench for developing biological network models that provides a Java-based graphical interface to view connectivity data in an intuitive fashion on the basis of a selected brain map (http://www.compneuro.org/catacomb/).
(ii) Van Essen et al. have integrated CoCoMac-Data in their tool "CARET" (Computerized Anatomical Reconstruction and Editing Toolkit). It allows projecting connectivity data obtained from CoCoMac-Online onto 2D and 3D brain maps (http://stp.wustl.edu/caret.html).

At the moment, both Catacomb and CARET focus on visual data in the partitioning scheme of Felleman and Van Essen (1991).

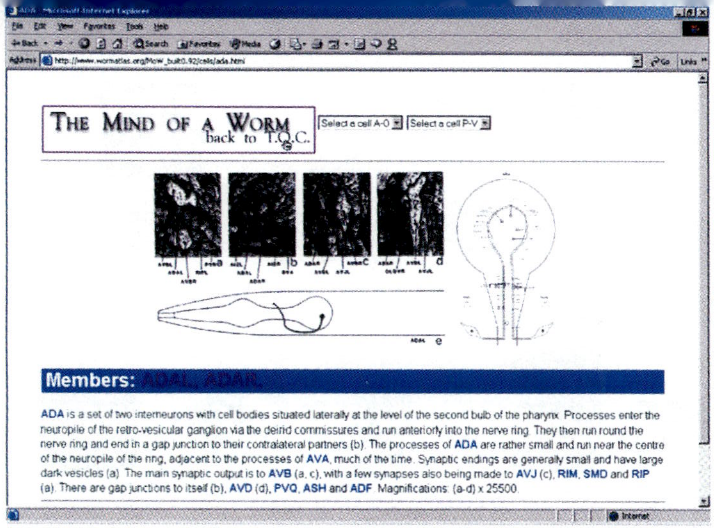

Chapter 1, Figure 1, p. 3

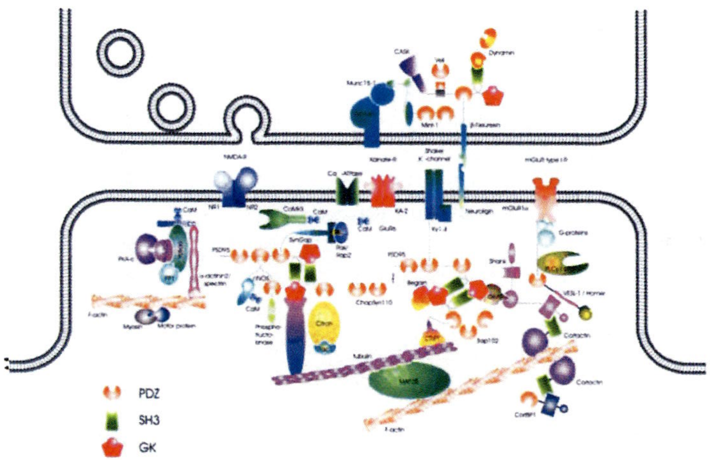

Chapter 4, Figure 1, p. 53

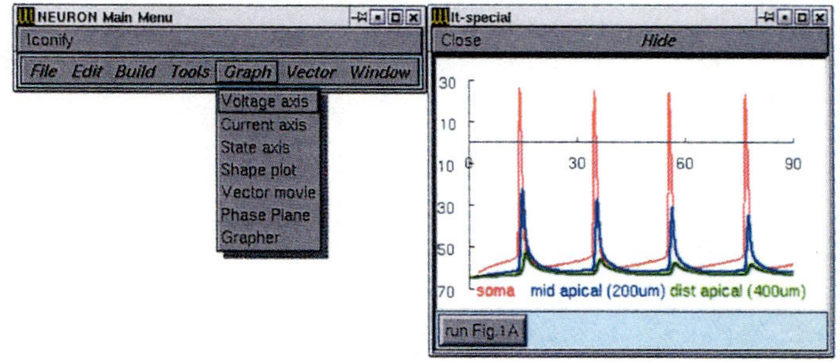

Chapter 7, Figure 4, p. 105

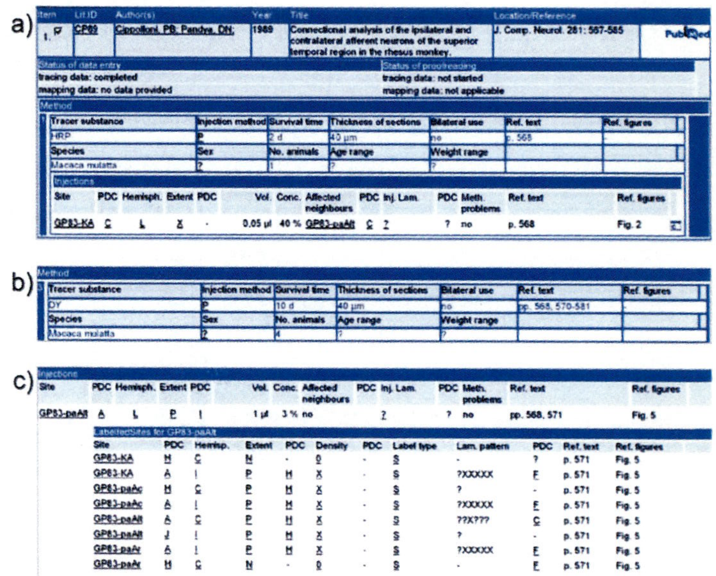

Chapter 11, Figure 3, p. 165

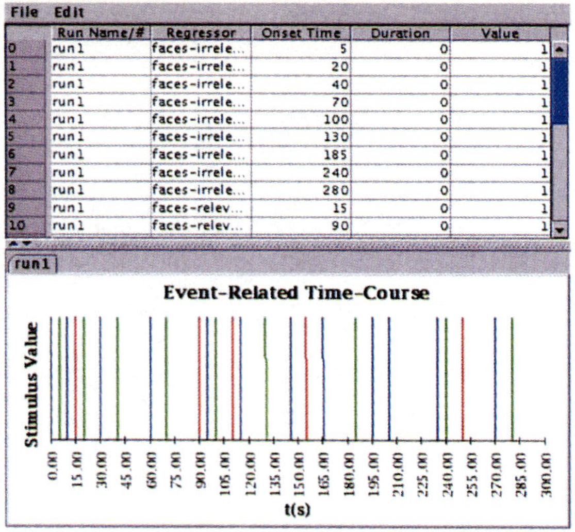

Chapter 15, Figure 3, p. 229

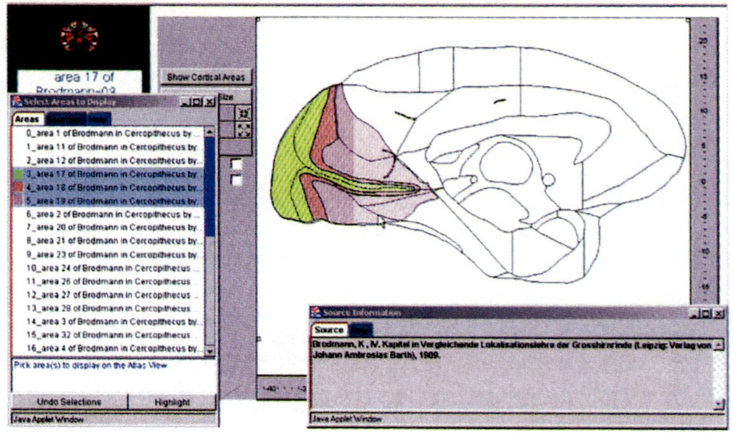

Chapter 18, Figure 5, p. 266

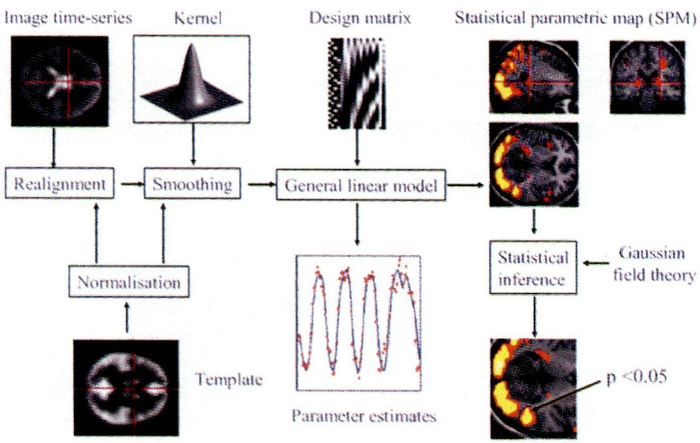

Chapter 16, Figure 1, p. 240

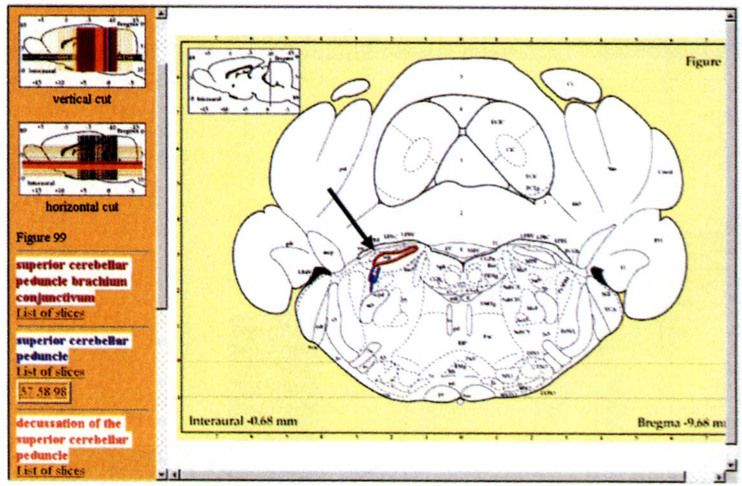

Chapter 19, Figure 6, p. 283

4. LITERATURE SEARCH

4.1 General information

4.1.1 Keyword search and search criteria

The literature category of our online interface offers two alternative ways for querying original data from published tracing studies, either by user chosen keywords or by predefined criteria. The latter can be grouped into (i) standard bibliographical criteria (e.g., names of authors, year of print, and PubMed-ID), (ii) method-related criteria (e.g. tracer substance), and (iii) CoCoMac-specific criteria (e.g., the representation of the brain areas as subcortical nuclei or isocortical brain areas). Each criterion is linked to a list box of corresponding search options. Multiple search options of different search criteria can be combined by the Boolean operators AND, OR, NOT. Table 1 gives a list of available search-criteria and the associated search-options.

Table 1. [List of search criteria in the Literature search. Together with a brief explanation of the associated "search options" and an exemplary selection.]

Search criteria	Associated search options
Keyword	Text box for free text search based on abstracts and CoCoMac keywords
Author	List box with all authors of all articles provided (e.g. Barbas, H.)
Year	List box with all publishing years from all provided tracing studies, additionally time ranges can be selected (e.g. 1976 – 2001)
Title	List box with all titles of all provided tracing studies (e.g. "Thalamic connections of the cortex of the superior temporal sulcus in the Rhesus monkey")
Journal (exact reference)	List box of all exact references (e.g. Science, volume 198, p315-317)
Journals	List box of all journals (e.g. Cerebral Cortex; Brain)
PMID	List box of all PubMed-ID's from all provided tracing studies
LiteratureType	Tick boxes for provided studies in journals, books, or book chapters
DataType	Tick boxes to search for studies that contain mapping data or connectivity data
BrainMap	List box of all partitioning schemes contained in CoCoMac (e.g. Olszewski, 1952 [O52]; Walker, 1940 [W40])
BrainSite	List box of acronyms for all brain areas provided in CoCoMac (e.g. supplementary motor area [SMA]; primary visual cortex [V1])
BrainSiteClass	List box to search for specific classes of brain structures (e.g. isocortical brain areas [Area_IsoCtx]; subcortical nuclei [Nucl_SubCtx])
TracerSubstance	List box of all tracer substances from all provided tracing studies (e.g. diamidino yellow [DY]; horseradish peroxidase [HRP])

4.1.2 Literature output list

The Literature output list displays general bibliographic information for articles that match the particular search, like title, authors, year, journal and a specific link to the entry in PubMed (PMID), if available. For books and book chapters we provide information about the editor and publisher (see Fig. 1, for an example of our literature output list).

The user can select articles of interest to display all available information on (i) brain sites, (ii) their inter-map relations, (iii) the tracing methods used, and (iv) injected and (v) labelled areas referred to in the article. The output list can be modified to the particular needs of the user by the option bar (see Fig. 1). For example, it is possible to order the entries by the name of the author, the year of publication, or the name of the journal and to display detailed information for selected articles at different levels of resolution (as shown in examples 2 and 3). In the standard output list, the names of all specified brain areas, the status of data entry and the status of proofreading are displayed. By default, information on experimental methods, e.g. tracer substances and macaque species, segregate data of the same article into independent method-related sections. Within each of these sections, the injection sites of the respective method are listed.

The relations between different partitioning schemes, described in the selected articles, can be displayed by marking the respective tick box (*Relations*) in the option bar. The labelled brain sites resulting from each injection are either shown in separate windows (see Fig. 2) or interspersed within the list of injections (see Fig. 3).

Every user has the option to submit own comments for each article contained in CoCoMac. These comments are subsequently accessible to all users.

4.2 Examples for the Literature-Search

4.2.1 Example 1

In our first example, we use the literature search to obtain an overview of the information available for the prefrontal cortex in CoCoMac, with a focus on standard partitioning schemes. Using the keyword search with the query term "prefrontal", CoCoMac-Online generates an output list comprising all relevant articles. Note that the keyword search generally indexes not only the title and abstract of the provided articles, but also the journal name, the year of publication and BrainSites in our database.

By default, the list is ordered by the year of publication in a descending order so that the most recent articles are shown at the top. Using the option bar, the output list can be reordered by publication year in an ascending fashion: now the oldest publications appear at the top (Fig. 1), which include some standard partitioning schemes.

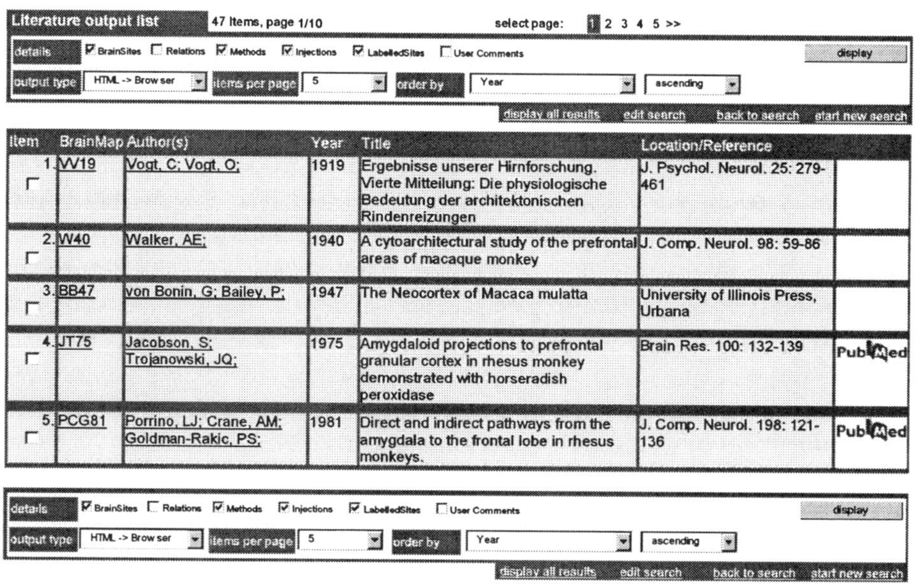

Figure 1. Literature output list, ordered by "Year" and "ascending".

The second entry now presents the bibliographic data for the article "A cytoarchitectural study of the prefrontal areas of macaque monkey " by Walker, AE (1940) that appeared in "Journal of comparative Neurology" (volume 98, pages 59-86). The CoCoMac specific *BrainMap-ID* for this article is "W40", an acronym that is composed of the first letter of the author's surname and two digits for the year of print (1940). Note that for publication years before 1975 PubMed includes no abstracts.

Entry Nr. 5 with the *BrainMap-ID* "PCG81" refers to a tracing study published in 1981 by the authors Porrino, Crane and Goldmann-Rakic. Again, the user gets a quick overview of the bibliographical information of this article (*year of publication*, *title* and *reference*). In this case the PubMed link is available by a mouse click on the PubMed icon, which opens a new browser window with the abstract from PubMed.

In the next step, one or more articles can be selected, and the resolution of the displayed details can be modified using the option bar (explained in example 3). Alternatively, the user selects one article by clicking on the authors' names (explained in example 2).

4.2.2 Example 2

Here, we concentrate on method-related details provided by CoCoMac-Online for one selected article. We choose the article "Connections between anterior inferior temporal cortex and superior temporal sulcus in the macaque monkey" by Saleem et al. (2000).

In the detailed view (Fig. 2) of the selected article the status fields give information on the *status of data entry* and *proofreading*. In terms of our example, the status of data entry is "started but not completed" for experimental data and "completed" for mapping data. The status of proofreading for both data types is "not started". In the line below, the *BrainSites* that have been specified in this article are presented by their abbreviations according to the nomenclature used in the specific article, here "TEO" for "posterior inferior temporal cortex".

The next part of the table contains important details about the labelling procedures, the animals, and injection sites. The Precision of Data representation Code *(PDC)* is listed for the injected brain area, the extent of the injection, affected neighbours, and the injected laminae (compare Figure 2).

In our example, the tracer substance Phaseolus vulgaris leucoagglutinin ("PHA-L") was applied iontophoretically ("I") at different injection sites of an unknown number of male and female Macaca fuscata, weighing between 3.3 and 6.9 kg. The animals survived 16-18 hours until their brains were removed and cut in sections of 35-40 μm for further investigation.

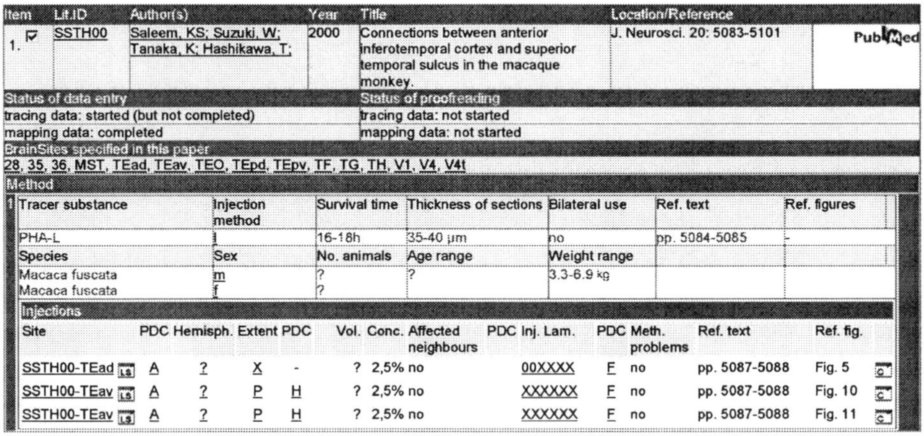

Item	Lit.ID	Author(s)	Year	Title	Location/Reference	
1.	SSTH00	Saleem, KS; Suzuki, W; Tanaka, K; Hashikawa, T;	2000	Connections between anterior inferotemporal cortex and superior temporal sulcus in the macaque monkey.	J. Neurosci. 20: 5083-5101	PubMed

Status of data entry	Status of proofreading
tracing data: started (but not completed)	tracing data: not started
mapping data: completed	mapping data: not started

BrainSites specified in this paper

28, 35, 36, MST, TEad, TEav, TEO, TEpd, TEpv, TF, TG, TH, V1, V4, V4t

Method

Tracer substance	Injection method	Survival time	Thickness of sections	Bilateral use	Ref. text	Ref. figures
PHA-L	i	16-18h	35-40 µm	no	pp. 5084-5085	-
Species	**Sex**	**No. animals**	**Age range**	**Weight range**		
Macaca fuscata	m	?	?	3.3-6.9 kg		
Macaca fuscata	f	?				

Injections

Site	PDC	Hemisph.	Extent	PDC	Vol.	Conc.	Affected neighbours	PDC	Inj. Lam.	PDC	Meth. problems	Ref. text	Ref. fig.
SSTH00-TEad	A	?	X	-	?	2,5%	no		00XXXX	F	no	pp. 5087-5088	Fig. 5
SSTH00-TEav	A	?	P	H	?	2,5%	no		XXXXXX	F	no	pp. 5087-5088	Fig. 10
SSTH00-TEav	A	?	P	H	?	2,5%	no		XXXXXX	F	no	pp. 5087-5088	Fig. 11

Figure 2. Example for the representation of the methodological information, here for Saleem et al. (2000).

The *BrainSite-ID* of the first injection (here: SSTH00-TEad) is composed of the first letters of the authors' surnames plus two digits for the year of publication (here: SSTH00 for Saleem,KS; Suzuki,W; Tanaka,K; Haschikawa,T and the year 2000) and TEad for the "dorso-anterior part of the inferotemporal area". The PDC for the injection site (SSTH00-TEad) is coded with "A", which means that this area was named explicitly in the text and was additionally shown in one or more figures. Since no information is given about the hemisphere, in which the injection was placed, the column *hemisphere* contains a question mark rather than "L" (left) or "R" (right). The spatial extent of the injected tracer was not characterized. It has not been further characterized whether the area was injected completely ("C") or just partially ("P"), which is reflected by the Extension Code of the injection as "X" (existing label of unknown extent). Neighbouring areas were not affected by the injection (see field *Affected Neighbours*). For this injection, only the concentration, but not the volume of the applied tracer substance is known. Additional information is given about the cortical layers (I – VI), in which the tracer was injected (e.g. "00XXXX": tracer was found in layers III to VI, but not in layers I or II). The corresponding PDC is "F", since only the text, but no figure contains information about the laminar injection pattern.

The second injection site (here: SSTH00-TEav), which will not be explained in detail, occurs twice because the author described injections using the same method and into the same brain sites but in two macaque monkeys of different species.

The experimental results of the respective injection sites can be obtained by clicking the *LabelledSites* (LS) button (shown in Fig. 2) In addition the information is displayed by changing the selection in the option bar, which is shown in the following example.

4.2.3 Example 3

In this example, we focus on the *LabelledSites* of a single injection because we are interested in connectivity data of the auditory system. We choose the article "Connectional analysis of the ipsilateral and contralateral afferent neurons of the superior temporal region in the rhesus monkey" by Cippolloni and Pandya (1989) from the *Literature Output List* gained by a search with the authors names as keywords. After selecting *Methods, Injections, Labelled Sites* in the option bar at the top of the screen we activate this view by clicking the *Update*-Button (shown in Fig. 1).

The experimental details including *Methods, Injections* and *LabelledSites* follow the general bibliographical information and *status of data entry* (Fig. 3a). In this example, three different methods are listed, each followed by a list of *Injections*. These injections led to the presence or absence of label in various *BrainSites*, which are presented as lists following the injections and are called *LabelledSites* (Fig. 3c).

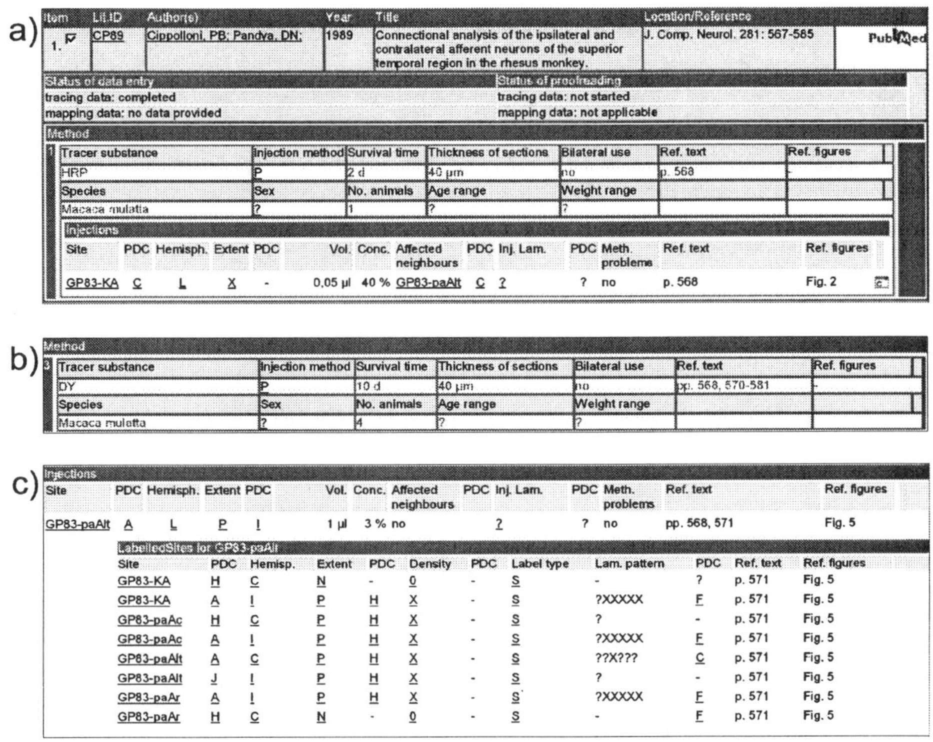

a)

Item	Lit.ID	Author(s)	Year	Title	Location/Reference	
1.	CP89	Cippolloni, PB; Pandya, DN;	1989	Connectional analysis of the ipsilateral and contralateral afferent neurons of the superior temporal region in the rhesus monkey.	J. Comp. Neurol. 281: 567-585	PubMed

Status of data entry	Status of proofreading
tracing data: completed	tracing data: not started
mapping data: no data provided	mapping data: not applicable

Method

Tracer substance	Injection method	Survival time	Thickness of sections	Bilateral use	Ref. text	Ref. figures
HRP	P	2 d	40 µm	no	p. 568	-
Species	**Sex**	**No. animals**	**Age range**	**Weight range**		
Macaca mulatta	?	1	?	?		

Injections

Site	PDC	Hemisph.	Extent	PDC	Vol.	Conc.	Affected neighbours	PDC	Inj. Lam.	PDC	Meth. problems	Ref. text	Ref. figures
GP83-KA	C	L	X	-	0,05 µl	40 %	GP83-paAlt	C	?	?	no	p. 568	Fig. 2

b)

Method

Tracer substance	Injection method	Survival time	Thickness of sections	Bilateral use	Ref. text	Ref. figures
DY	P	10 d	40 µm	no	pp. 568, 570-581	-
Species	**Sex**	**No. animals**	**Age range**	**Weight range**		
Macaca mulatta	?	4	?	?		

c)

Injections

Site	PDC	Hemisph.	Extent	PDC	Vol.	Conc.	Affected neighbours	PDC	Inj. Lam.	PDC	Meth. problems	Ref. text	Ref. figures
GP83-paAlt	A	L	P	I	1 µl	3 %	no		?	?	no	pp. 568, 571	Fig. 5

LabelledSites for GP83-paAlt

Site	PDC	Hemisp.	Extent	PDC	Density	PDC	Label type	Lam. pattern	PDC	Ref. text	Ref. figures
GP83-KA	H	C	N	-	0	-	S	-	?	p. 571	Fig. 5
GP83-KA	A	I	P	H	X	-	S	?XXXXX	F	p. 571	Fig. 5
GP83-paAc	H	C	P	H	X	-	S	?	-	p. 571	Fig. 5
GP83-paAc	A	I	P	H	X	-	S	?XXXXX	F	p. 571	Fig. 5
GP83-paAlt	A	C	P	H	X	-	S	??X???	C	p. 571	Fig. 5
GP83-paAlt	J	I	P	H	X	-	S	?	-	p. 571	Fig. 5
GP83-paAr	A	I	P	H	X	-	S	?XXXXX	F	p. 571	Fig. 5
GP83-paAr	H	C	N	-	0	-	S	-	F	p. 571	Fig. 5

Figure 3. a) Bibliographic information, b) method-related details and c) list of labelled brain areas following the injection in paAlt for the article by Cippolloni and Pandya, (1989).

One injection with the tracer "DY" (1 µl of a 3% solution, compare Fig. 3b) was placed in lateral auditory parakoniocortex ("GP83-paAlt") of the left ("L") hemisphere. Clicking on the acronym for the tracer substance (for example "DY") shows its full name ("Diamidino Yellow") and properties ("retrograde transport"). Regarding the transport direction, the sub-table *LabelledSites* contains all afferents of GP83-paAlt found in this particular experiment (Fig. 3c).

For example, the auditory konicortex (GP83-KA) is examined in both the ipsilateral ("I") and contralateral ("C") hemispheres. The label in the ipsilateral hemisphere extended through part of the area ("P") and was of unknown density ("X"). Clicking on the acronym for the *density* displays a help text for the density coding and shows that CoCoMac has values 0 (absent), 1 (weak), 2 (medium), 3 (strong) or X (unknown). In contrast, no label was found in GP83-KA of the contralateral hemisphere.

The Precision of Description Codes (*PDC*) of *Labelled Sites* correspond to those of *Injection Sites*. The *PDC* for the *density* information refers to the precision, with which density information is described. Generally, in cases of absent label (EC=N, density=0) and existing label of unknown strength (EC=P/X/C, density=X) it is needless to state the precision of density description. Therefore, the *PDC* in this case is represented by a dash.

The *Label Type* in this row codes the part of the neuron where the label was found, the neuron soma ("S") or axon terminals ("T"). Here, the soma was labelled, compatible with the use of a retrograde tracer substance.

The next entry provides information about the laminar distribution of the label. Six characters represent the typical pattern of the six isocortical layers. For the lamina coding, the same rules apply as for the density coding. In this example the label was found in layers II-VI with unknown density (LaminarPattern = "?XXXXX"). For layer I it remained unclear whether it contains label or not; the density in this layer is, therefore, represented by a question mark.

In addition, the label in the contralateral auditory parakonicortex (GP83-paAlt) shows a partial coverage ("P") of unknown density ("X") and was found in layer III ("??X???"). The status of the other layers remained unclear. In terms of connectivity this implies that the projection from the contralateral to the ipsilateral auditory parakonicortex originates mainly from layer III.

CoCoMac provides an exact reference (including page and figure numbers) for each *LabelledSite*, which allows the user to go back to the original publication for cross-checking.

5. OTHER SEARCH CATEGORIES

The literature search described above provides systematic access to detailed information from more than three hundred articles. The next question, of course, concerns the broad picture across many such articles that cover the entire connectivity of a brain region or the connections between several brain regions as can be accessed by the connectivity search category. Another aspect resulting from our collation is information on the relationships between areas in different partitioning schemes. For example, do the areas named 25 by Brodmann in 1905 and 1909, and by Walker in 1940 refer to the same piece of cortex? Since we do not have the space here to describe these fascinating aspects we must refer the interested reader to forthcoming descriptions and to our documentation at www.cocomac.org.

6. FUTURE PERSPECTIVES

In addition to the original data as collated from the published literature, we plan to provide our computationally interpreted data (integrated and transformed data for inter-map relations and connections) as well. These data provide the gist of the information scattered among the many partitioning schemes and are generated using Objective Relational Transformation (ORT), an algorithmic method to convert connectivity data in user chosen brain maps based on optimal inter-map relations (for a detailed description see Stephan et al., 2000). Using this approach, it seems feasible to obtain a complete picture of large-scale cortical wiring within the next couple of years. In addition, we plan to extend the scope of CoCoMac to include further data on connectivity between subcortical structures, eventually leading to a complete wiring diagram of the primate brain.

Another important issue is data integration. The use of the existing XML interface is not limited to graphical front-ends, but allows data exchange with other databases (e.g. NeuroScholar, BrainInfo) and will facilitate data mining across several databases and types of data. A simple example is the use of CARET to overlay data from different studies including information on cytoarchitectonics, functional imaging and connectivity, but more sophisticated automated methods need to be developed.

Such data integration would be greatly facilitated through the development of a common data model as a formal dictionary of neuroscientific terms and concepts. Such concepts are currently being developed (see NeuroML.org or BrainML.org). Our concepts of defining brain sites, for example, have proved very useful and are applied in data collations of functional connectivity from strychnine experiments (CoCoMac-Stry, Stephan et al., 2000) and in an extended version to cellular and microcircuitry data (CoCoDat, D.-Johnsen et al., this volume). Common concepts of brain maps are a prerequisite for comparing connectivity data to other data types, such as the distribution of receptors in cortical areas (see Kötter et al., 2001).

Finally, computational modelling of neurons and brain systems would be facilitated if software for neural simulation could contact databases directly and retrieve the data required for specific model implementations. As an example, we have created tools that facilitate the generation of network models using connectivity data from CoCoMac (Kötter et al., 2002; Kamper et al., 2002) and we expect the implementation of more general methods and tools in the near future (see Goddard et al., 2001).

ACKNOWLEDGEMENTS

We thank Stefan Küppers and mon.de for their continuing support to provide optimal online access to CoCoMac. This work is supported by the German Research Council (DFG: LIS 4 – 554 95 (2) Düsseldorf).

REFERENCES

Bozkurt A., Kamper L., Stephan K.E., Kötter R. (2001a) Organization of primate amygdalo-prefrontal projections. *Neurocomputing* 38-40, 1135-1140.

Bozkurt A., Kamper L., Stephan K.E., Kötter R. (2001b) Alzheimer's disease: Relationship between structural connectivity and neuropathological changes, *Soc Neurosci Abstr* 323.13.

Bozkurt A. Kamper L. Stephan K.E., Kötter R. (2002) The structural basis of information transfer from medial temporal lobe to prefrontal cortex in the macaque monkey. *Neurocomputing*; in press.

Burns G.A. (2001) Knowledge management of the neuroscientific literature: the data model and underlying strategy of the NeuroScholar system. *Philos Trans R Soc Lond B Biol Sci* 356, 1187-1208.

Friston K.J. (1999) Schizophrenia and the disconnection hypothesis. *Acta Psychiatr Scand Suppl* 395, 68-79.

Goddard N.H., Hucka M., Howell F., Cornelis H., Shankar K., Beeman D. (2001) Towards NeuroML: model description methods for collaborative modelling in neuroscience. *Philos Trans R Soc Lond B Biol Sci* 356, 1209-1228.

Kamper L., Bozkurt A., Nielsen P., Dyhrfjeld-Johnsen, J., Stephan K.E., Kötter R. (2002) Modelling the influence of thalamo-cortical projections on prefrontal activity. *Neurocomputing;* in press.

Kötter R., Nielsen P., D.-Johnsen J., Sommer F.T., Northoff G. (2002) Multi-level neuron and network modeling in computational neuroanatomy. In *Computational Neuroanatomy: Principles and Methods*, G.A. Ascoli, ed. Totowa: Humana.

Kötter R. (2001) Neuroscience databases: tools for exploring brain structure – function relationships theme issue. *Philos Trans R Soc Lond B Biol Sci* 356, number 1412.

Kötter R., Stephan K.E., Palomero-Gallagher N., Geyer S., Schleicher A., Zilles K. (2001) Multimodal characterisation of cortical areas by multivariate analyses of receptor binding and connectivity data. *Anat Embryol* 204, 333-350.

Northoff et al. (2000) Functional dissociation between medial and lateral prefrontal cortical spatiotemporal activation in negative and positive emotions: a combined fMRI/MEG study. *Cereb Cortex* 10, 93-107.

Stephan K.E., Zilles K., Kötter R. (2000) Coordinate-independent mapping of structural and functional data by objective relational transformation (ORT). *Philos Trans R Soc Lond B Biol Sci* 355, 37-54.

Stephan K.E., Hilgetag C.C., Burns G.A., O'Neill M.A., Young M.P., Kötter R. (2000) Computational analysis of functional connectivity between areas of primate cerebral cortex. *Philos Trans R Soc Lond B Biol Sci* 355, 111-126.

Stephan KE, Kamper L, Bozkurt A, Burns GAPC, Young MP, Kötter R (2001) Advanced database methodology for the collation of connectivity data on the macaque brain (CoCoMac). *Phil. Trans. R. Soc. Lond. B* 356, 1159-1186.

Tononi G., Edelman G.M. (2000) Schizophrenia and the mechanisms of conscious integration. *Brain Res Rev* 31, 391–400.

Chapter 12

Graph Theory Methods for the Analysis of Neural Connectivity Patterns

Olaf Sporns
Department of Psychology, Indiana University, Bloomington, IN 47405,
Correspondence to: osporns@indiana.edu

Abstract: This paper summarizes a set of graph theory methods that are of special relevance to the computational analysis of neural connectivity patterns. Methods characterizing average measures of connectivity, similarity of connection patterns, connectedness and components, paths, walks and cycles, distances, cluster indices, ranges and shortcuts, and node and edge cut sets are introduced and discussed in a neurobiological context. A set of Matlab functions implementing these methods is available for download at http://php.indiana.edu/~osporns/graphmeasures.htm.

Key words: networks, neuroanatomy, cortex, correlation, complexity

1. INTRODUCTION

As is evident from many of the other contributions in this volume, recent advances in neuroscience and neuroinformatics are bringing about fundamental changes in the collection, access, and computational analysis of neuroscience data sets (Kötter, 2001). An increasing number of neuroscience databases are being developed and are made available online. Specifically, significant progress has been made in the collation and representation of neuroanatomical connection patterns. These efforts provide researchers with comprehensive data sets of connectional relationships between individual neurons or brain areas. The availability of such data sets necessitates the development of appropriate computational tools for their principled and comprehensive analysis. One avenue for such an analysis is provided by graph theory, a branch of mathematics and combinatorics with many applications in diverse fields, ranging from physics, communication science and electrical engineering to genetics, linguistics, and sociology.

All graph theory methods are based on a network's connection (adjacency) matrix, which can be derived from several different sources:

- Databases of cortico-cortical and cortico-thalamic pathways yield connection matrices representing the large-scale connectional organization of cortical networks. Several such matrices have been published (e.g. Felleman and Van Essen, 1991; Young, 1993; Scannell et al., 1999) and computationally analysed.
- Studies of individual neurons and local circuits provide statistical estimates of connection densities and indications of spatial patterns (e.g. Braitenberg and Schüz, 1998), which can be used to compile probabilistic connection matrices.
- In computational neuroscience, network models incorporate often highly specific patterns of connections, which can be represented as artificial connection matrices.

Supplementing this chapter is a set of Matlab (The Mathworks, Inc., Natick, MA, http://www.mathworks.com/) functions designed to allow the computational exploration of specific neuroanatomical connection data sets. The functions have been developed and tested with Matlab version 6.0 and are available at: http://php.indiana.edu/~osporns/graphmeasures.htm. This web resource will be continually updated. All Matlab functions contain extensive comments on their usage and limitations. The information in this chapter is designed to provide some mathematical background and a rationale for the potential use of these functions in neurobiology. All corrections, suggestions for expansion or modification of the software, additional neurobiological applications, or additional functions are welcome (contact osporns@indiana.edu).

This chapter is not intended to be a general review of graph theory, or a review of cortical networks or of computational neuroanatomy. Rather, the chapter summarizes a set of graph theory methods of particular relevance for neuroscience applications, specifically the computational analysis of neural connectivity data. Other computational methods, such as multivariate statistical approaches used to display and organize connection data sets, are not covered (instead, see Hilgetag et al., 2002, for a comprehensive overview). The methods reviewed here represent only a small selection of graph theory tools, collected from a variety of textbooks and papers. The main criterion for their inclusion in this chapter is their potential usefulness for computational analyses of neural connection patterns. Individual methods focus on characterizing average measures of connectivity and degree distributions (3.1, 3.2), within-network similarity of connection patterns (3.3), paths and cycles (3.4, 3.5, 3.6), connectedness and components (3.7), distances (3.8), cluster indices (3.9), shortcuts (3.10) and vulnerability measures such as node and edge cut sets (3.11). Other methods, not included in this chapter, may be implemented in the future and will be made available at the website listed above.

2. REPRESENTING NETWORKS AS GRAPHS

To allow mathematical analysis, we represent neuronal connectivity patterns (networks) as graphs. All graphs are composed exclusively of vertices (nodes, points, units) and edges (arcs, connections). While much of classical graph theory (covered in numerous textbooks, e.g. Harary, 1969; Chartrand and Lesniak, 1996) is dealing with un-directed (random) graphs, graphs that represent neural connectivity patterns contain directed edges and are thus called directed graphs, or digraphs. A digraph $G_{n,k}$ consists of a set $\{v_l\}$ of n vertices and a set $\{e_{ij}\}$ of k directed edges, each linking a source vertex v_j to a target vertex v_i. Excluding self-connections, the number of edges ranges between 0 and n^2-n. The structure of the graph is described by the graph's adjacency matrix $A(G)$ (also called the connection matrix), with binary entries a_{ij} ($a_{ij} = 1$ if the connection from j to i is present, and $a_{ij} = 0$ if the connection is absent) and an all-zero main diagonal. An important concept in graph theory is the path. Paths are all ordered sequences of *distinct* edges and vertices, linking a source vertex j to a target vertex i. If $j = i$, the corresponding paths link the source vertex to itself and are called cycles.

The Matlab functions provided with this chapter are designed for graphs with these characteristics:

- directed edges,
- unweighted edges (i.e. all-positive, binary adjacency matrix),
- simple edges (no multiple connections between j and i), and
- edges e_{ii} (self-connections) are not allowed.

For most Matlab functions, there are no constraints (other than CPU time or memory) on size (n, k), sparseness or connectedness of the graph. No distinction is made between connections that are absent or presumed absent (i.e. for which no empirical information is available). This distinction (a measure of "confidence" in the existence or non-existence of individual pathways) is sometimes taken into account in the analysis of large-scale connection matrices assembled from individual neuroanatomical studies.

3. GRAPH THEORY METHODS

3.1 Average Connection Density (*density.m*)

The average connection density k_{den} of an adjacency matrix $A(G)$ is the number of all its non-zero entries, divided by the maximal possible number of connections (n^2-n for a directed graph, excluding self-connections). Thus, we have $0 \leq k_{den} \leq 1$. The sparser a graph, the lower its k_{den}.

Average connection densities can vary widely, depending on the particular neural structure, on the level of analysis (i.e. populations or single cells), and on the spatial extent of the neural network (e.g. entire brain versus local circuit). While the average connection density between cells across the entire cerebral cortex is approximately 10^{-6}-10^{-7}, local connection densities are significantly higher within single cortical columns (10^{-1}-10^{-3}). Matrices of connection pathways linking cortical areas tend to have k_{den} ~ 0.2-0.4. Connection matrices of patches of local cortical circuits comprising multiple columns would likely have very low k_{den} ~ 0.01-0.001.

3.2 Indegree and Outdegree (*degrees.m*)

The indegree $id(v)$ and outdegree $od(v)$ of a vertex is defined as the number of incoming (afferent) or outgoing (efferent) edges, respectively. The degree of a vertex $deg(v)$ is the sum of its $id(v)$ and $od(v)$. Note that the sum of $id(v)$ over all vertices equals the sum of $od(v)$ (and both are equal to k). Individual vertices may show imbalances in their indegree and outdegree. These imbalances are recorded in the joint degree distribution matrix of the graph $J(G)$, whose entries J_{tu} correspond to the number of vertices with an indegree $id(v) = u$ and an outdegree $od(v) = t$. Entries of J_{tu} far away from the main diagonal correspond to vertices with a high imbalance in incoming and outgoing edges. Vertices above the main diagonal have an excess of outgoing edges, while vertices below the main diagonal have an excess of incoming edges.

Indegrees and outdegrees have obvious functional interpretations. A high indegree indicates that a neural unit is influenced by a large number of other units, while a high outdegree indicates a large number of potential functional targets. For most neural structures, indegrees and outdegrees of neural units are subject to constraints due to growth, tissue volume or metabolic limitations. Connections cannot be attached or emitted beyond the limits imposed by these constraints. The distribution of $deg(v)$ (or $id(v)$, $od(v)$) over the entire graph may be inspected for scale-free attributes such as power laws (Albert and Barabási, 2002). The joint degree distribution matrix $J(G)$ may be compared to corresponding matrices obtained from same-sized random graphs (Newman et al., 2001), to identify significant deviations from balanced or symmetrical degree distributions.

3.3 Connectivity Matching Index (*compare.m*)

The connectivity matching index m_{ij} between two vertices i and j ($i \neq j$) can be defined as the amount of overlap in their connection patterns (Hilgetag et al., 2000). For example, $m_{ij,in}$ for afferent connections refers to the proportion of existing identical connections, divided by the total number of connection pairs for which at least one connection exists (excluding connections between i and j). Analogously, m_{ij} can be defined for efferent connections ($m_{ij,out}$) or for all connections ($m_{ij,all}$, afferent and efferent). Note that $m_{ij} = m_{ji}$, and $m_{ii} = 1$. All entries m_{ij} form the connectivity matching matrix $M(G)$. We distinguish between $M_{in}(G)$, $M_{out}(G)$, and $M_{all}(G)$. In some computational schemes (not implemented here, but see Hilgetag et al., 2000; Kötter et al., 2001; Stephan et al., 2001), connections that have been shown to be absent in anatomical studies have been distinguished from connections for which no evidence has been collected.

This measure provides an indication of the extent to which the connectivity patterns (afferent, efferent or both) of two neuronal units coincide or correlate and has been applied in several previous studies of neural connection patterns. High m_{ij} indicates that both units maintain similar anatomical and functional connections with other units within the system. Strong overlap in the structural relationships of two units within a network may be predictive of an overall similarity in their functional contributions. Matching indices have been used in developing optimal clustering schemes for systems of cortical areas (Hilgetag et al., 2000). Interestingly, these anatomical clustering approaches yield results that are very similar to functional clustering methods based on covariance matrices capturing the pattern of statistical deviations between dynamically interacting brain areas (functional connectivity, see Sporns et al., 2000; Sporns and Tononi, 2002). Thus $M(G)$ may be interpreted as a simple representation of expected baseline correlation in functional activity across a network.

3.4 Reciprocal Connections (*reciprocal.m*)

Reciprocal connections are pairs of edges that reciprocally link two vertices. Thus, in $G_{n,k}$, a reciprocal connection is found if both e_{ij} and e_{ji} exist. Edges that participate in reciprocal connections can be labeled and counted; their number divided by k results in $\rho(G)$, the fraction of reciprocal connections for the graph.

In the brain, reciprocal connections are found in many systems and at many levels of scale, from local circuits to pathways between brain areas (where they are very abundant). Evaluating $\rho(G)$ provides a first-order estimate of the extent of reciprocal dynamical coupling present within the entire network. Typical large-scale cortical connection matrices have $\rho(G)$ of around 0.7-0.8 (perhaps higher, given that reciprocity of connection pathways may be underestimated due to missing anatomical information). This is much higher than the $\rho(G)$ of random networks of corresponding size (n,k) (Sporns et al., 2000). We note that the existence of reciprocal connections does not, in general, imply a symmetrical functional relationship between the linked neural units. For example, connection pathways in the cortex show characteristic laminar termination patterns that differ for feedforward and feedback connections (Felleman and Van Essen, 1991) and that may have different functional impact on their neuronal targets.

3.5 Paths and Walks (*findpaths.m, findwalks.m*)

As defined above, paths are ordered sequences of *distinct* edges and vertices, linking a source vertex j to a target vertex i. Distinctness means that no vertex or edge may be visited twice along a path from j to i. If $j = i$, the corresponding path links a source vertex to itself and is called a cycle. The length of a path or a cycle is defined as the number of distinct edges. The number of unique paths of length q from vertex j to i are recorded as entries in the path matrix of the graph, $P_q(G)$. Entries on the main diagonal of $P_q(G)$ indicate the number of cycles of length q. While all paths between j and i are unique sets of edges and vertices, they are usually not edge- or vertex-disjoint (see below). The path length distribution $pl(q)$ plots the total number of paths found within the graph for each path length q and t_{path} indicates the total number of paths of all lengths. Finding and recording all existing paths within even modest sized graphs is a significant computational challenge. The supplied Matlab function 'findpaths.m' uses a breadth-first approach to "grow" paths from source vertices to target vertices. Walks are defined as any sequence of vertices and edges leading from j to i, without the condition of distinctness. Walks are easily obtained from the powers of $A(G)$ (Buckley and Harary, 1990), yielding the walk matrix of the graph, $W_q(G)$. The walk length distribution $wl(q)$ contains the total number of walks of given length q, and t_{walk} records the total number of walks of all lengths for the graph.

The number of paths of different lengths linking two neuronal units is an important measure of the number of ways through which signals can travel between them. A large number of paths from j to i implies a potentially high degree of redundancy in their interconnections and interactions. However, the number of paths alone is not a good measure of the vulnerability of the interaction (see discussion on disjoint paths in 3.11). Paths provide a more selective and specific view of a graphs's structural connectivity and are widely preferred over walks as the principal basis for the evaluation of reachability and distance (see 3.7, 3.8). Most of the discussion in this chapter will therefore focus on paths. Clearly the length of a path (or a walk) has a dramatic influence on its potential impact on real neuronal interactions. We must assume that potential interactions between two neuronal units, be they neurons or brain areas, rapidly decrease as the length of paths between them increases. In other words, neurons will tend not to influence each other through paths that are highly indirect, i.e. more than a few synaptic steps in length. It is unlikely that even large numbers of weak indirect interactions occurring in parallel (through large numbers of long, indirect paths) would have a significant cumulative functional impact in neuronal networks. Thus, we may assume that $P_q(G)$ and $W_q(G)$ with $q > 3$ or 4 are of relatively little functional significance in the brain. If this assumption is correct, then a complete derivation of $P_q(G)$ or $W_q(G)$ for all path lengths may be rather unnecessary and computational analyses should focus on entries of these matrices for small values of q.

3.6 Cycles and Cycle Probability (*cycprob.m*)

Cycles are paths that link a given source vertex j to itself, with otherwise distinct vertices and edges. The main diagonals of $P_q(G)$ record the number of cycles of a given length q. A measure of the overall probability of continuing a given non-cyclic path as a cycle can be defined, called cycle probability (Sporns et al., 2000). This measure, $p_{cyc}(q)$, estimates the probability that a non-cyclic path of length $q-1$ can be continued as a cycle of length q (i.e. by the addition of one edge), obtained from the ratio between the total number of non-cyclic paths of length $q-1$ and the total number of cycles of length q. If all non-cyclic paths of length $q-1$ can be continued as cycles of length q, then $p_{cyc}(q) = 1$, if none of them can be continued as a cycle, $p_{cyc}(q) = 0$. Computationally, $p_{cyc}(q)$ is calculated from the path matrix, as the ratio between the sum of all entries of $P_q(G)$ on the main diagonal and the sum of all off-diagonal entries of $P_{q-1}(G)$. Note that $p_{cyc}(1) = 0$ and $p_{cyc}(2) = \rho(G)$ (see 3.4). Another measure, the cycle frequency $f_{cyc}(q)$, captures the overall abundance of cycles as a fraction of all paths for a given path length q. Both p_{cyc} and f_{cyc} range between 0 and 1.

Frequencies and probabilities of cycles within a graph are important indicators of the graph's local connectivity, and are often associated with the amount of clustered connections within the graph (see 3.9). Just as high proportions of reciprocal connections (high $p_{cyc}(2) = f_{recip}$) indicate high levels of dynamic coupling between pairs of vertices, high values for $p_{cyc}(3)$ indicate an abundance of triads of edges (forming cycles of length 3). Important reference points for comparison are random graphs of equivalent n and k. Distributions of $p_{cyc}(q)$ and $f_{cyc}(q)$ for a given neuronal connection matrix should be compared to averaged distributions for such random graphs, to determine if cycles of varying length are more or less abundant than expected based on a random model.

3.7 Reachability Matrix, Connectedness, and Components (*reachdist.m, components.m*)

The reachability matrix $R(G)$ records whether or not at least one path (of any length) exists between j and i (Harary, 1969). Its entries are $r_{ij} = 1$ if i is reachable from j ($r_{ij} = 0$ otherwise). Note that (deviating from Buckley and Harary, 1990) we set $r_{ii} = 1$ only if at least one cycle linking vertex i to itself exists. Computationally, the reachability matrix can be obtained using a variety of algorithms. Following Buckley and Harary (1990), the Matlab function 'reachdist.m' uses the powers of $A(G)$ to derive $R(G)$, together with the distance matrix (see below). The reachability matrix allows the assessment of connectedness, which, in digraphs, is a subtle concept, with gradations from strong, to unilateral, weak, and trivial. Most importantly, if all entries of $r_{ij} = 1$, then all vertices can be reached from all other vertices within the graph and the graph is strongly connected. If $R(G)$ contains multiple (mutually exclusive) subsets of vertices that are strongly connected within each subset, but are not connected between subsets, then the graph contains multiple (strong) components. If there are vertices whose entries r_{ij} are zero for either an entire row or a column, they cannot be reached from any other vertex or no other vertex can be reached from them (their $id(v)$ and $od(v)$ would also be zero, respectively).

Reachability is a fundamental and insurmountable prerequisite for the transmission of functional effects. If reachability is lost, for example by disconnection of areas after brain injury, functional interactions are abolished. Thus, the reachability matrix (and the distance matrix) is a good candidate for monitoring and measuring the vulnerability of neuronal circuits to functional dissociation (see 3.11). Studies of the connection patterns of extended cortical systems in the primate and the cat have shown that (at the level of segregated areas) the cortex is a strongly connected graph (Sporns et al., 2000).

3.8 Distance Matrix, Eccentricity, Radius, Diameter, and Characteristic Path Length (*reachdist.m, breadthdist.m, charpath.m*)

The distance matrix *D(G)* records the distances between *j* and *i*, defined as the length of the shortest path between them (Harary, 1969). If no path exists, d_{ij} = Inf. Thus, d_{ij} = Inf if $r_{ij} = 0$. Note that (deviating from Buckley and Harary, 1990) d_{ii} = Inf, unless at least one cycle linking vertex *i* to itself exists (in which case d_{ii} is the length of that cycle). Computationally, *D*(*G*) is derived together with *R*(*G*) from the powers of *A*(*G*) (see Buckley and Harary, 1990). Aternatively, distances can be determined using Floyd's algorithm, Dijkstra's algorithm, or through breadth-first search (implemented in 'breadthdist.m'). The eccentricity of a vertex *j* is the maximum of its finite distances to all other vertices, i.e. $ecc(j) = \max(d_{ij})$. Computationally, *ecc(j)* is the maximum of each row of *D(G)*. If all other vertices can be reached from *j*, then *ecc(j)* is the maximum number of steps needed to reach all of them. The radius of a digraph is the minimum eccentricity of all its vertices, $radius(G) = \min(ecc(j))$. The diameter of a digraph is the maximum eccentricity, $diameter(G) = \max(ecc(j))$. Following Watts and Strogatz (1998; see also Watts, 1999), we define the characteristic path length $\lambda(G)$ of a graph as the global mean of the finite entries of its distance matrix. Thus, the characteristic path length constitutes a measure of central tendency of *D(G)*.

The distance matrix provides information about the "directness" with which two units in a network can interact. For example, if $d_{ij} = 2$, then *j* can influence *i* through just one intermediate unit. In general, the length of the shortest path is indicative of the potential strength of functional interactions. Note that another factor determining this strength may be how many different short paths exist. The entries of the path matrix $P_q(G)$ for $q = d_{ij}$ give the total number of paths between *j* and *i* that have shortest possible length. Distance *per se* in graphs makes no reference to the physical distances between neuronal units located in metric space. It is possible for one pair of neuronal units to be separated by $d_{ij} = 3$ and be 200 μm apart, while another pair is directly connected ($d_{ij} = 1$), but separated by a metric distance of 20 mm. It seems reasonable to assume that, in many cases, graph distance is a better predictor of the degree and strength of functional interactions than metric distance. However, as a result of developmental and volumetric factors, metric distance may constrain the probability that a connection between two neuronal units actually exists.

3.9 Cluster Index (*clustind.m*)

Introduced by Watts and Strogatz (1998), the cluster index of a vertex $\gamma_v(v)$ indicates how many connections are maintained between a vertex's neighbours. For digraphs, neighbours are all those vertices that are connected, either through an incoming or an outgoing connection, to the central vertex v. The number of neighbours per vertex is b_v. The vertex's cluster index is defined as the ratio of actually existing connections between the b_v neighbours and the maximal number of such connections possible $(b_v^2 - b_v)$. If $b_v = 0$, $\gamma_v(v) = 0$. The average of the cluster indices for each individual vertex is the cluster index of the graph $\gamma(G)$.

The cluster index for a network expresses the extent to which the units within the network share common neighbours that "talk" among each other, an attribute that has been called the "cliquishness" of the network. A high cluster index $\gamma(G)$ points to a global organizational pattern consisting of groups of units that mutually share structural connections and can thus be surmised to be functionally closely related. However, the cluster index does not provide information about the number or size of these groups and only captures local connectivity patterns involving the direct neighbours of the central vertex. Watts and Strogatz (1998) used the cluster index as well as the characteristic path length as attributes defining "small-world" network architectures, characterized by an abundance of surprisingly short paths linking units within a very large and sparsely connected network. The small-world model has had a tremendous impact for studies of a wide range of networks, from social and communications networks to the brain.

3.10 Ranges and Shortcuts (*range.m*)

The range g_{ij} of an edge e_{ij} is the length of the shortest path from j to i, after the edge e_{ij} has been removed from the graph (Watts, 1999). If removal of e_{ij} results in disconnecting j from i, g_{ij} = Inf. Before removal of e_{ij}, there is a direct link (formed by e_{ij}) from j to i ($d_{ij} = 1$). Thus, g_{ij} must be 2 or greater. The average range $\eta(G)$ for the entire graph is the sum of the ranges of all edges for which $g_{ij} \neq$ Inf, divided by their number. If $g_{ij} > 2$, then the removed edge forms a shortcut from j to i (Watts, 1999). The fraction of shortcuts f_s averaged over the entire graph is the number of edges with $r_{ij} > 2$, divided by k. Note that high values for cycle probabilities and cluster indices will generally be associated with low incidences of shortcuts resulting in $\eta(G)$ just slightly higher than 2.

In a neural context, removal of edges corresponds to cutting or lesioning of connections or pathways. The range of an edge provides information about its role in linking vertices. If the edge is a shortcut, its presence may be crucial in functionally connecting two vertices that would otherwise be separated by a longer (and thus functionally ineffectual) distance. The fraction of shortcuts f_s is related to the prevalence of densely clustered (triadic) local connection patterns throughout the graph. A low value of f_s indicates a high number of short paths linking all pairs of vertices and is found in many graphs with high k_{den}, or in graphs with locally clustered connections (see 3.9).

3.11 Disjoint Paths, Cut-Vertices and Bridges, Vertex Cut Sets and Edge Cut Sets (*vertexcutsets.m*, *edgecutsets.m*)

Two paths linking vertex j to vertex i are vertex- (edge-) disjoint, if they have no vertices (edges) in common (apart from source j and target i). Disjoint paths may be of equal or unequal lengths. The maximum number of vertex- (edge-) disjoint paths (of any length) between j and i is equal to the minimum number of vertices (edges) that need to be removed in order to disconnect j from i (a version of Menger's theorem; see Harary, 1969; Buckley and Harary, 1990). A cut-vertex is a vertex whose removal from the graph increases the number of the graph's components. For example, the number of components of a graph would increase, if removal of a vertex from a connected graph would result in a disconnected graph (with multiple components). Correspondingly, a bridge is an edge whose removal increases the number of components. A vertex cut set (edge cut set) is a set of vertices (edges) whose removal results in a disconnected graph (White and Harary, 2001). The vertex- and edge-connectivity of a graph G are defined as the smallest number of vertices (edges) that, upon removal, result in a disconnected graph, and are denoted as $\kappa_v(G)$ and $\kappa_e(G)$, respectively. For example, if a graph contains at least one cut-vertex or one bridge, its $\kappa_v(G)$ or $\kappa_e(G)$ are equal to one.

The number of disjoint paths between two neuronal units provides a strong indication of the amount of (structural) redundancy in their mutual interconnectivity. For example, if two edge-disjoint paths exist between two units j and i, removal of only one edge cannot disconnect j from i; at least two edges must be removed. The vertex- or edge-connectivity of a neuronal network may be interpreted as a measure of its internal cohesiveness, vulnerability, or resistance to disintegration upon damage. The higher $\kappa_v(G)$ or $\kappa_e(G)$, the more vertices or edges need minimally to be removed in order to bring about the disconnection (disintegration or dissociation) of the original network. Of particular interest are cut-vertices or bridges that result in the disconnection of a strongly connected graph. Such cut-vertices and bridges correspond to single neuronal units or pathways whose removal abolishes functional interactions between two separate components of the network.

4. EXAMPLES

As an example of a structural analysis of a neuronal network using the methods and Matlab functions outlined in this chapter we select a connection matrix of cortico-cortical pathways of the macaque cortex, originally published by Young (1993) (an electronic copy of the matrix was kindly provided by Claus Hilgetag). For the present analysis, the amygdala and area A14 (which has an indegree of 0 in the original matrix) are omitted. In many cases, it is useful to compare actual neurobiological connection data sets to other connection matrices whose main organizational features (i.e. randomness, clusters, defined degree distributions) are known. Here, the macaque connection matrix is compared to equivalent random matrices and highly ordered ring lattice matrices (Watts, 1999), both with $n = 71$, $k = 755$.

The adjacency matrix (with reciprocal connections highlighted), the connectivity matching matrix, the joint degree distribution, and the distance matrix of the macaque cortex are shown in Figure 1. For $n = 71$, $k = 755$ we obtain $k_{den} = 0.152$. Macaque cortex contains a high proportion of reciprocal connections, with $\rho(G) = 0.824$ (random: 0.117; ring lattice: 0.940). The joint degree distribution is balanced, with 27 units having a higher $id(v)$ than $od(v)$ and 25 units *vice versa*. We find $p_{cyc}(2) = 0.824$, $p_{cyc}(3) = 0.386$, and $p_{cyc}(4) = 0.311$ (random: 0.117, 0.152, 0.150; ring lattice: 0.940, 0.674, 0.539). The characteristic path length $\lambda(G)$ is 2.312 (random: 2.022; ring lattice: 3.772), with radius = 3 and diameter = 5 (random: 3, 3; ring lattice: 7, 7). The cluster index $\gamma(G)$ is 0.461 (random: 0.151; ring lattice: 0.656). The average range $\eta(G)$ is 2.019 (random: 2.109; ring lattice: 2.000), with only 14 connections serving as shortcuts (random: 143; ring lattice: 0). The vertex connectivity is at least $\kappa_v = 3$ and the edge connectivity is at least $\kappa_e = 2$, as no cut sets are found up to these levels.

Having derived these structural measures, it is now possible to inspect the adjacency matrix in detail and derive explicit information about the structural contributions and relationships of identified cortical areas. For example, identified shortcuts in the graph can be traced to specific pathways leading to hypotheses about their role in facilitating direct interactions between specific areas. Another possibility is the use of the connectivity matching matrix to derive hypotheses about functional overlap between sets of areas and their cluster structure (Hilgetag et al., 2000). Other, more global comparisons, such as those between macaque cortex and corresponding random and ring lattice matrices are possible (see Sporns et al., 2000) allowing insight into the potential "small-world" architecture of the cortex.

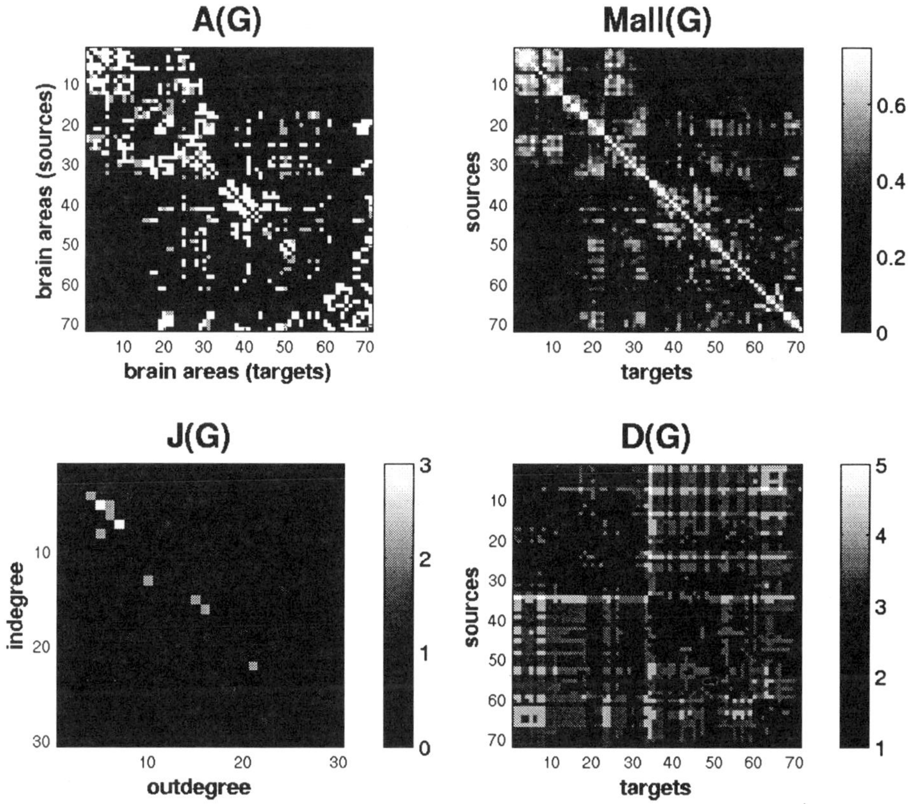

Figure 1: Adjacency matrix *A(G)*, connectivity matching matrix $M_{all}(G)$, joint degree distribution *J(G)* and distance matrix *D(G)* for the connectivity matrix of the macaque cortex (Young, 1993). *A(G)* shows reciprocal connections in white and unidirectional connections in light gray. Brain areas are V1, V2, V3, VP, V3A, V4, VOT, V4T, MT, MSTD, MSTL, FST, PITD, PITV, CITD, CITV, AITD, AITV, STPP, STPA, TF, TH, PO, PIP, LIP, VIP, DP, A7A, FEF, A46, TGV, ER, HIPP, A3A, A3B, A1, A2, A5, R1, S2, A7B, IG, ID, A35, A4, A6, SMA, A3, A23, A24, A9, A32, A25, A10, A45, A12, A11, A13, G, PAAR, PAAL, PAAC, KA, PAL, PROA, REIT, TGD, TS1, TS2, TS3, TPT (1-71); the connection matrix can be found at the website http://php.indiana.edu/~osporns/graphmeasures.htm.

5. DISCUSSION

Sparked by an increasing need in science and technology for understanding complex interwoven systems as diverse as the world wide web, cellular metabolism and human social interactions, there has been an explosion of interest in "network dynamics", the computational analysis of the structure and function of large physical and virtual networks (Strogatz, 2001; Albert and Barabási, 2002). Brain networks are of central interest to neuroscience, as their structural patterns are of fundamental importance for all neuronal information processing and, ultimately, for the emergence of coherent perceptual and cognitive states (Tononi et al., 1998; Sporns and Tononi, 2002).

The methods outlined in this chapter represent only one approach to the computational analysis of brain connectivity patterns. Additional methods are needed, including methods that take into account the sign and weight of connections, the density of projection pathways, or their physiological effects on their targets. The computational analysis of neural connection patterns is still in its infancy and future applications and studies will almost certainly utilize additional methods, likely including some that are not yet developed or even envisioned. Of particular significance are multivariate methods designed to display and extract significant structural features of connection or other matrix representations, such a multidimensional scaling or clustering approaches (reviewed in Hilgetag et al., 2002).

The structure of brain networks is a key for understanding the brain's functional dynamics. Therefore, structural analyses of neuronal connectivity patterns are best interpreted in a functional context. This can be done by making explicit reference to the specific physiological and cognitive contributions of identified sets of neurons or brain areas. Another avenue includes the simultaneous evaluation and analysis of functional connectivity patterns (Sporns and Tononi, 2002), empirically derived through neuroimaging or obtained from computer simulation of a structurally based neuronal model. In the nervous system, structural and functional connectivity are strongly interdependent, with structural connections shaping statistical and informational patterns across large-scale networks, while such dynamical interactions in turn lead to alterations in the structural substrate. These and other considerations lead us to expect that the computational analysis of structural and functional connectivity patterns will be a main contributor to our understanding of brain dynamics and its role in perception and cognition.

REFERENCES

Albert, R, and Barabási, A.-L. (2002) Statistical mechanics of complex networks. *Reviews of Modern Physics*, 74, 47-97.

Braitenberg, V., and Schüz, A. (1998) *Cortex. Statistics and Geometry of Neuronal Connectivity.* Springer, Berlin.

Buckley, F., and Harary, F. (1990) *Distance in Graphs.* Addison-Wesley, Redwood City, CA.

Chartrand, G., and Lesniak, L. (1996) *Graphs and Digraphs.* Chapman and Hall, Boca Raton etc.

Felleman, D.J., and Van Essen, D.C. (1991) Distributed hierarchical processing in the primate cerebral cortex. *Cerebral Cortex* 1, 1-47.

Harary, F. (1969) *Graph Theory.* Addison-Wesley, Reading, MA.

Hilgetag, C.C., Burns, G.A.P.C., O'Neill, M.A., Scannell, J.W., and Young, M.P. (2000) Anatomical connectivity defines the organization of clusters of cortical areas in the macaque monkey and the cat. *Philosophical Transactions of the Royal Society London B* 355, 91-110.

Hilgetag, C.C., Kötter, R., Stephan, K.E. and Sporns, O. (2002) Computational methods for the analysis of brain connectivity. In: Giorgio Ascoli (ed.) *Computational Neuroanatomy: Principles and Methods*, pp. 295-335, Humana Press, Totowa, NJ.

Kötter, R., Stephan, K.E., Palomero-Gallagher, N., Geyer, S., Schleicher, A., and Zilles, K. (2001) Multimodal characterisation of cortical areas by multivariate analyses of receptor binding and connectivity data. *Anatomy and Embryology* 204, 333-350.

Kötter, R. (2001) Neuroscience databases: tools for exploring brain structure-function relationships. *Philosophical Transactions of the Royal Society London B* 356, 1111-1120.

Newman, M.E.J., Strogatz, S.H., and Watts, D.J. (2001) Random graphs with arbitrary degree distributions and their applications. *Physical Review E* 64, 026118.

Scannell, J.W., Burns, G.A.P.C., Hilgetag, C.C., O'Neil, M.A., and Young, M.P. (1999) The connectional organization of the cortico-thalamic system of the cat. *Cerebral Cortex* 9, 277-299.

Sporns, O., Tononi, G., and Edelman, G.M. (2000) Theoretical neuroanatomy: Relating anatomical and functional connectivity in graphs and cortical connection matrices. *Cerebral Cortex* 10, 127-141.

Sporns, O., and Tononi, G. (2002) Classes of network connectivity and dynamics. *Complexity* 7, 28-38.

Stephan, K.E., Kamper, L., Bozkurt, A., Burns, G.A.P.C., Young, M.P., and Kötter, R. (2001) Advanced database methodology for the collation of connectivity data on the macaque brain (CoCoMac). *Philosophical Transactions of the Royal Society London B* 356, 1159-1186.

Strogatz, S.H. (2001) Exploring complex networks. *Nature (London)* 410, 268-277.

Tononi, G., Edelman, G.M., and Sporns, O. (1998) Complexity and coherency: Integrating information in the brain. *Trends in Cognitive Sciences* 2, 474-484.

Watts, D.J., and Strogatz, S.H. (1998) Collective dynamics of 'small-world' networks. *Nature (London)* 393, 440-442.

Watts, D.J. (1999) *Small Worlds.* Princeton University Press, Princeton, NJ.

White, D.R. and Harary, F. (2001) The cohesiveness of blocks in social networks: node connectivity and conditional density. *Sociological Methodology* 2001, 31, 305-359.

Young, M.P. (1993) The organization of neural systems in the primate cerebral cortex. *Proceedings of the Royal Society London B* 252,13-18.

Chapter 13

The PUPS-MOSIX Environment: A Homeostatic Environment for Neuro- and Bio-informatic Applications

Mark A. O'Neill[1], Gully A.P.C Burns[2] and Claus C. Hilgetag[3]
[1]*Newcastle Univ, Dept Psychology, Newcastle upon Tyne, NE2 4HH, UK;*
[2]*USC, Hedco Neurosci Bldng, 3614 Watt Way, Los Angeles, CA90089, USA;*
[3]*Int'l Univ Bremen, Campus Ring 1, D-28759 Bremen, Germany.*
Correspondence to: mao@ermine.ox.ac.uk.

Abstract: Computational analyses of biological data can be formidably complex. This complexity leads to computations which may have a lifetime of days or even weeks. In order to support these computations, a novel computing environment inspired by biological principles, **PUPS**, has been implemented. **PUPS** provides mechanisms for parallelisation of, dynamic interaction with, and homeostatic protection of, such computations. **PUPS** has been used to implement a variety of applications including the **CANTOR** network processor, the **DAISY** automated insect identification system and a number of geographical information systems (**GIS**), and neuro- and bio-informatic tools. Some familiarity with the **UNIX** operating system and its programming tools is required in order to write **PUPS** applications. The **PUPS** environment is freely available in both source and binary forms from the **PUPS** website: http://chasseur.usc.edu/pups.

Key words: homeostasis, persistent computation, organic computing, neuroinformatics, bioinformatics

1. INTRODUCTION

PUPS, The **P**ortable **U**NIX **P**rogramming **S**ystem (O'Neill and Hilgetag, 2001) represents an attempt to produce a standard environment which supports complex and persistent computations, including the CANTOR network processor, neuroinformatic databases such as ***Neuroscholar*** (Burns et al, 2000), the **DAISY** Computer Aided Taxonomy (**CAT**) package (Weeks et al, 1997), and a number of **GIS** applications, for example the **ITG** stereo digital terrain model generation system (O'Neill and Denos, 1995). **PUPS**

supports many mechanisms for homeostatic protection of, and dynamic interaction with, running processes including:

1. Automated process migration via **MOSIX**, an extension to **LINUX**, developed by Hebrew University of Jerusalem (Barak et al, 1993), and natively via the process tunnel daemon, *rtunneld.*
2. Support for checkpointable (e.g., recoverable) processes via CONDOR (Litzkow et al, 1988), a high throughput parallel computer environment developed at the University of Wisconsin and the University of Tennessee Checkpointing Library (Plank et al, 1995) *libckpt*[1].
3. Homeostatic protection of objects (files, **FIFOS**, etc.).
4. Peer-to-peer and user-to-peer dynamic interaction with running processes.
5. Support for dynamic goal and computational payload reassignment and steering parameter update in running applications.
6. Support for manipulation and storage of complex datasets via a practical multihost/ multiprocess implementation of a *persistent object store.*

The **PUPS** system is coded in **ANSI-C**. Although it has been developed under the **LINUX** operating system using the Free Software Foundation **GNU** compiler tools, **PUPS** has been ported to a number of **POSIX.1b**[2] compliant operating systems including **OSF1**, **Solaris, SunOS 4.1** and **BSD 4.4**. With relatively little effort, it could also be ported to the **Microsoft Windows-NT** and **Windows-2000** environments, as both these **PC** operating systems offer **POSIX.1b** compliance[3].

2. STRUCTURE OF THE PUPS SYSTEM

2.1 Core libraries

There are six core libraries in the standard **PUPS** distribution. These are linked with *computational payload code* in order to produce a **PUPS** application binary. The standard libraries are:

[1] Note that both of these libraries require modification so that they can deal with **FIFO**'s and redirected I/O in a sensible manner.

[2] **POSIX** is a standard defined by the IEEE which defines programming and system call interfaces for UNIX like operating systems,

[3] The *Cygwin package* (available from www.cygnus.com) is required for **NT** and its derivatives to support **POSIX.1b**. The built-in Microsoft support is incomplete and should not be used.

1. *utilib*: which contains general purpose utilities including command tail decoding, extended I/O functionality, application process name overloading[4], extended asynchronous event handling[5] and a number of standard homeostats for protecting files and interprocess communication channels.
2. *casino*: Provides pseudo-random number generators and generators for *Poisson*, *Normal*, *Binomial* and other well known distribution functions
3. *netlib*: Provides support for network computing including asynchronous signalling of processes running on remote hosts, remote execution of processes, gathering of status information from remote nodes.
4. *psrplib*: Implements **PSRP** (**P**rocess **S**tatus **R**equest **P**rotocol). **PSRP** is a novel Internet Protocol that enables both users and also peer processes (***PSRP** clients*) to interact asynchronously with running applications on an Internet-wide basis. **PSRP** is a secure protocol. With the provision of appropriate authentication tokens, it is possible for a ***PSRP** client* to access and make use of resources which it does not own[6]. In addition, the **PSRP** protocol is *multithreaded.* A running ***PSRP** server* may be simultaneously accessed by multiple ***PSRP** clients.* The **PSRP** library also provides inbuilt functions which support standard **PSRP** client-server interaction.
5. *mvmlib*: provides support for **PUPS** processes to (optimally) manage their own virtual memory.
6. *tadlib*: provides support for multithreaded **PUPS** applications via the **POSIX** threads libraries.

2.2 PUPS/PSRP daemons

The ***PUPS** daemons* facilitate the transparent networking and parallel computation aspects of **PUPS** including signal relaying to remote hosts, remote process execution, initialising remote ***PSRP** client* sessions and disseminating host node performance statistics to remote nodes. The **PUPS** daemons use a novel protocol in order to propagate authentication information to **PUPS** applications (e.g., *PSRP clients* and *servers*), which are running on remote hosts, and to check and authenticate the identity of clients making requests for *daemon* services. There are nine **PUPS/PSRP** *daemons*:

[4] This enables applications to be tracked by name rather than **PID** within a **POSIX.1b** environment.

[5] Asynchronous events (signals) are used by one process to tell anther that it wants its attention. They are also used by the operating system to get the attention of application processes.

[6] This facility was provided to enable effective resource sharing for multi-person, multi-site projects, as it permits seamless access to an instance of key **PSRP** applications such as *smboltzmann*by many users irrespective of who actually owns that instance.

1. *rttyd:* makes a connection between a remote ***PSRP*** *client* running on a remote host and a ***PSRP*** *server* running on the *daemons localhost* via a *full duplex* terminal interface.
2. *xrshd*: makes a connection between a remote ***PSRP*** *client* running on a remote host and a ***PSRP*** *server* running on the *daemons localhost* via a **half duplex** *BSD socket interface*[7].
3. *rfifod:* permits *FIFOS* and other non regular-file filesystem objects to be exported via network files systems, such as **NFS**.
4. *rpsd,* (5.) *rloadd,*(6.) *rupd*: disseminate information about localhost computational load and any ***PSRP*** *server* processes which it may be running to **PUPS** processes running on remote nodes.
7. *xkilld*: relays signals sent by processes running on remote hosts to target processes which are running on *xkilld*'s localhost.
8. *rtunneld*: is a daemon which permits processes (and associated state context information) to migrate between host nodes. In contrast to the **MOSIX** cluster computing environment, *rtunneld* migrates both the *compute* and *system* contexts of a process. **MOSIX** only migrates the *compute* context.
9. *connected*: is a daemon which periodically checks whether it is possible to connect to the Internet. When Internet connectivity changes, *connected* takes appropriate action. Although *connected* is not part of **PUPS,** it is included as it works around reverse **DNS** problems encountered in some implementations of the **POSIX** standard.

2.3 PUPS service functions

The **PUPS** service functions are tools which facilitate building of user applications within the **PUPS** environment. Many of the functions, for example *fsw, xcat* and *xtee,* are effectively used as *glue modules* in *virtual dataflow machines* (which consist co-operating *pipelines* of application processes). Other functions, for example *nkill* and *xrshd,* extend the functionality of existing **UNIX** tools (e.g. *kill* and *rsh*) in potentially useful ways. The **PUPS** environment currently provides nine service functions:

1. *nkill:* Permits processes (which may be remote) to be signalled using a unique name rather than their **P**rocess **ID**entifier (**PID**).
2. *xrsh*: Uses the *xrshd* **PUPS** daemon to start application processes running on remote hosts. *Xrsh* is intended to be a replacement for the BSD *rsh* command; it is more secure than *rsh* and also incorporates support for *static load balancing* and execution of *multihost process pipelines and virtual dataflow machines (**VDM**'s).*

---

[7] *Sockets* are a form of communication channel which permit reliable, stream based communications between a pair of computers connected using **TCP/IP** protocols.

3. *fsw, xcat, xtee*: permit the efficient implementation of *virtual dataflow machines*. *Fsw* is a *file system watcher* which is able to halt the execution of any **VDM** of which it is a member in the event of the **VDM** output filesystem becoming full[8]. *Xcat* is primarily intended to act a *place marker* within **VDM**'s which can be *overforked* by ephemeral processes (e.g., **X**[9] and other interactive applications). Typically, *xcat* processes provide targets within **VDM**'s which can be overlaid by interactive monitoring and/or visualisation applications.
4. *embryo*: is an uncommitted *PSRP server* application. Typically *it* is used to test new **PUPS/PSRP** installations, but it may also be used to build applications whose *computational payloads* are *dynamic*.
5. *protect*: is a file system object protection homeostat, which provides homeostatic protection for file system objects.
6. *pass:* an *application carrier* which enables non-**PUPS** processes to access **PUPS** and **PSRP** services. *Pass* provides homeostatic protection for application payload pipelines, and acts as an I/O manager which is capable of reading/writing data to and from regular files, **FIFOS** and **SIDS**. In addition, it provides homeostatic protection for the file system objects it is performing I/O on, and can provide its *payload pipeline* with *thread-of-execution* protection.
7. *maggot*: is a service function which cleans the **PUPS** environment, removing stale checkpoint files, **PUPS** shared heaps and other derelict file system objects.
8. *lyosome* is a lightweight utility similar in function to *maggot* which protects files for a specified period of time. When this lifetime has expired, *lyosome* deletes the files and exits.
9. *psrp*: is a text based *client* which allows the user sitting at an interactive terminal to interact with ***PSRP** server* processes.

In addition to the service function described above, the **PUPS** environment also uses an enhanced version of the *Secure Shell* (*ssh*) communication client. The version of *ssh* used by **PUPS** has extensions which permit the *ssh client* to work non-interactively[10] allowing **PUPS** processes to use *ssh* functionality to build encrypted tunnels for data traffic between networked hosts.

---

[8] Alternatively *fsw* can try and move output files to an alternative filesystem, which is not full, and then resume execution of its **VDM**.

[9] **X** is a public domain network capable windowing environment developed at MIT which is a popular graphical fronted for **UNIX**-like Operating Systems.

[10] The standard version of the ssh client expects authentication token etc. to be passed to it via a terminal (e.g. */dev/tty* or */dev/pty* devices). ***PUPS**-ssh* will also accept authentication data from **FIFO** devices. This facilitates the creation of authenticated encrypted tunnels between hosts by *processes* (as opposed to *users*).

3. THE MOSIX ENVIRONMENT

Although it is possible to build stand-alone **PUPS** environments, the full potential of **PUPS** is realised when it is deployed with the **MOSIX** cluster computing environment. **MOSIX** effectively turns a network of commodity computers into a parallel virtual machine (**PVM**) with sophisticated load balancing algorithms which move running computations around the **PVM** in order to optimise the use of resources. **MOSIX** achieves this by dividing every running process into system and computational *contexts*. The *system context* (process-kernel interaction) stays *rooted* on the host which initially spawned the process (*home node*) The *compute context* (*computational payload*) moves under the aegis of the load balancing algorithms in order to optimise cluster-wide resource utilisation. The system and compute contexts communicate via sockets. System calls from the compute context are effectively implemented as **RPC**[11] calls on the process home node. The **MOSIX** load balancing algorithms try to simultaneously maximise processor utilisation across the cluster while minimising communication overhead. In early versions of **MOSIX**, communication overhead was minimised at the expense of maximisation of processor utilisation, leading to an inability to deal effectively with processes which are both *compute* and *I/O bound.* Processes with significant I/O were tied to their home node limiting the advantage gained by running **MOSIX** for this class of computation[12]. The **MFS** filesystem which is shipped with the standard **MOSIX** distribution has overcome this limitation. **MFS** is a network file system which is visible to all nodes in a **MOSIX** cluster. The compute contexts of migrated processes can perform I/O operations directly via **MFS** without having to talk to the system context on their home node, reducing inter-context communication traffic. Consequently the load balancing algorithms will not force the migrated compute context back to its home node. In parallel computational environments, **MFS** has a significant advantage over standard network filing systems such as **NFS**, *it is synchronous*: changes which are made to a file within the **MFS** filesystem by a process are committed to hardware before any other process is allowed to access it. In contrast, **NFS** is *non-synchronous*, which means that co-operating processes which need to atomically access **NFS** mounted files may experience *race conditions.*

3.1 Integration of PUPS and MOSIX - the best of both worlds?

---

[11] **RPC** stands for *remote procedure call.* **RPC** is a mechanism first introduced by Sun Microsystems which allows processes running on one host to call functions on another.

[12] Many (if not all) **PUPS-PSRP** processes fall into this class!

Inter-operation of the **PUPS** system with **MOSIX** confers three advantages:

1. It provides **PUPS** with transparent optimal load balancing across a computer cluster.
2. **MFS** provides a synchronous network file system required for multiple node access to certain classes of **PUPS** filesystem objects such as shared heaps.
3. **MOSIX** automatically elucidates the computational capabilities of host nodes and the speed inter-node communications when load balancing.

Currently, the **MOSIX** project is implementing *network* ***RAM***[13]. The idea behind this is simple. When a typical *fat process* is migrated it is moving its *data context* which causes much of the load on the communication network. This can be reduced dramatically if the data stays in a fixed location and *thin processes* with minimal data contexts move around the network in order to process it. In this scenario, processes which wish to access the data push their compute context onto the data host and then use memory mapping to access it. When the data has been processed, the compute context unmaps it and retreats back to its home node. The ***PUPS shared heap*** is also an ideal vehicle for building multiple access network **RAM** as it supports the memory mapping and the locking mechanisms required for simultaneous atomic access by multiple processes. Information about the **MOSIX** system, and the **MOSIX** software itself is freely available from the **MOSIX** site: http://www.mosix.org.

3.2 The PUPS-MOSIX cluster

The **PUPS-MOSIX** cluster is a **PVM** which consists of a number of host nodes based on networked commodity **PC** hardware running **MOSIX** and **PUPS-PSRP** enabled application processes. Typically the cluster has the following structure:

1. A *fileserver* which exports all cluster-wide **NFS** and **MFS** filesystems. Typically this server also exports a /pups filesystem containing the binaries for **PUPS** service functions and daemons, and the **PUPS** libraries and header files.
2. A *NIS-server* which provides centralised network information services [**NIS**] for the cluster.
3. A *gateserver* which controls access to the rest of the cluster from The Internet. Typically the gateserver is a relatively low powered machine:

[13] Ironically **PUPS** grew out of the **MSPS** system, an early attempt to implement network-RAM on the BBC Microcomputer over EcoNet!

processes started on the gateserver rapidly migrate onto compute nodes within the cluster.

PUPS-MOSIX clusters need not work in isolation. The daemon *rtunneld* may be used to migrate processes between clusters should the need arise. Of course, this will require modifications to be made to the **MOSIX** load balancing algorithm, so processes are handed to *rtunneld* for intercluster migration when the mean loading on the outbound cluster exceeds some defined maximum[14].

4. EXAMPLE PUPS APPLICATIONS

To date, the **PUPS** environment has been used to build a number of applications including:

1. The **ITG** stereo ranging system (O'Neill and Denos, 1992; 1996).
2. Neuroinformatic analysis applications, for example, the clustering and hierarchical analysis tool **CANTOR** (Hilgetag et al, 1996; 2000a; 2000b).
3. Bioinformatic applications, for example, the **DAISY** automated insect identification system (Weeks et al, 1998).

Experience with these applications has led to a number of extensions to the **PUPS** environment. These include:

1. The use of *sensitive I/O directories* (**SID**'s) as repositories for input data: when data arrives in the **SID**, the application process associated with it is signalled and can take appropriate action.
2. The use of *data tagging* techniques to facilitate the processing of data. Data placed into **SID**'s which are associated with multiple applications may be tagged with a *unique key*. Only those processes which possess a *lock* which matches the key are allowed to process the data. This mechanism, which is crudely based on cellular protein transport allows groups of co-operating processes of arbitrary complexity to be constructed whose global topology is *emergent*[15].

---

[14] Complex **MOSIX** network topologies may be defined which may also be able to deal with inter-cluster process migration.

[15] This means the implementer only need define a set of *input keys* which processes require to read input data from a **SID** and a set of *output keys* which processes use to stamp data they place in the **SID**. The keys then implicitly define the spatial and temporal topology of the process network.

5. INSTALLING PUPS MOSIX

5.1 System requirements

In order to install the **PUPS** system, the following basic hardware requirements must be met:

1. All the nodes in the **PUPS** cluster must be connected by a **LAN** which supports **TCP/IP** protocols under POSIX.
2. If **MOSIX** is installed, the CPU installed in all nodes must be Intel compatible (e.g., *Intel Pentium*, *AMD K6/K7*, *Cyrix* etc.).

Installation and programming of the **PUPS-MOSIX** environment is fully documented on the **PUPS** web site, http://chasseur.usc.edu, which also has a number of examples of how to write **PUPS** applications, libraries and dynamic link libraries.

5.2 Installation from binary RPM

The **PUPS** system may be installed either from a *binary distribution* or *source distribution.* The binary distribution (which will be available for Intel compatible processors in the near future) is in *RedHat* ***RPM*** format. In this case, the basic **PUPS** environment is installed by obtaining the **RPM** (from the **PUPS** web site) and then typing:

```
rpm -iv pups.rpm
```

assuming that the name of the retrieved **RPM** is pups.rpm. The RedHat Package Manager, `rpm,` will then take care of installing all the components of the **PUPS** system in the correct locations.

5.3 Building PUPS from source tarball

In order to build and install **PUPS** from source, the following operations are required:

1. Uninstall current version of secure shell (ssh).
2. Build **Linux-MOSIX** (and install **MOSIX** services).
3. Make default shell (and default root shell) *csh* or *tcsh.*
4. Build **PUPS** services and then (*as root*) install **PUPS** and **PUPS** compliant version of secure shell (*ssh*).

5. Build **NIS**[16] on **PUPS-MOSIX** cluster *master* and *slave* nodes.
6. Add timed daemon start-up line to */etc/rc.local.*
7. Add connected daemon startup line to */etc/rc.local.*
8. Add **PUPS-MOSIX** local network i.d. to */etc/networks.*
9. Export **PUPS-MOSIX NFS** mounted filesystems from *master* node (and mount them on the *slave* nodes).
10. Connect manually to each new system (in order to set up *key* information for secure shell clients on each system).

Fortunately, there are tools available which automate much of the work needed to build and install **PUPS** from source. Only step 10 needs to be done manually.

Although installing the **PUPS** package from source is a little more complex than a binary installation, it allows much more control to be excercised over the capabilities of the system installed. In addition, the system can be built from source for *any* system which supports **POSIX.1b.** In addition to **LINUX, PUPS** has been built for[17] *SunOS, Solaris, BSD4.4, IriX* and *OSF1/TRU64 UNIX*. Installation of **PUPS** from source needs three shell scripts to be run:

```
build_pups_services
install_pups_services
project
```

As these scripts are *csh* scripts, the default shell (for the installers' User IDentifier) must be changed to *csh* or *tcsh* prior to installation. After unpacking the distribution, go to the **PUPS** source directory:

```
cd dist.supups/pupscore.src
```

Then run the **PUPS** build script:

```
build_pups_services cluster
```

This compiles the **PUPS** libraries and service functions, leaving **PUPS** libraries in *dist.supups/lib* and corresponding **PUPS** service functions and daemons in *dist.supups/bin. There* are a number of build options in addition

---

[16] For some systems this needs to be done manually. An excellent account of **NIS**, **NFS** and other advanced **POSIX** subsystems utilised by the **PUPS** environment is given in *RedHat LINUX-7 Unleashed* (2001).

[17] The **MOSIX** component of **PUPS-MOSIX** is currently restricted to *Intel*, *Cyrix*, *AMD* and similar architectures (although this will change when **U-MOSIX** is released in the near future).

to `cluster` which may be specified as parameters to the `build_pups_services` *csh* script. However, for practical purposes only two will be required when installing *production* versions of **PUPS**:

1. `cluster`: which builds a full version of **PUPS** (with all extensions e.g. *shared heaps, networking, dynamic functions* etc. enabled).
2. `vanilla`: which builds a very basic version of **PUPS** (with no extensions) for a standalone machine. This option is particularly useful when building **PUPS** for primitive versions of **UNIX,** for example **MINIX** (Tanenbaum, 1987).

The **PUPS** system can also be built (and periodically updated) from a nominated ***PUPS*** *source distribution server* using the *pups_bootstrap csh* script and the *pups_isys* and *pups_iapp* distribution and application update tools. Documentation explaining how this is accomplished is available at the **PUPS** web site, http://chasseur.usc.edu/pups.

5.4 Installing PUPS

In order to install **PUPS** on a node (so that it can be utilised by any user of that node), become root (type `su` *not* `su -`) and then run the ***PUPS*** *installation script,* typing:

```
install_pups_services cluster master
install_pups_services cluster slave
install_pups_services cluster gateserver
```

for *master node*, *slave nodes* and *gateserver node*, respectively. The installation script does a number of things:

1. It creates a directory, */home/pups* (in which the **PUPS** binaries, libraries, header files and documentation is stored). This directory is *symbolically linked* to */pups.*
2. It creates **PUPS** compliant versions of a number of system configuration files (in */etc*). The original files are of course backed up with a *.save* extension.
3. It builds a **PUPS** compliant version of secure shell (ssh) and its utilities and installs them in the default locations (usually */usr/local/bin* and */usr/local/sbin*).
4. It build a version of *top*, *mtop* which has **MOSIX** extensions and installs it in */usr/local/bin.*
5. In the case of a *master node* it builds */etc/exports* (**NFS** exported directories) and **NIS** maps for the cluster. It also arranged for a *master*

timed daemon to run on the node (which synchronises all *slave* and *gateserver* nodes with the clock on this node).

6. In the case of a **slave node**, it modifies */etc/fstab* (so that **NFS** volumes exported by master node are mounted by the *slave node*). It also arranges for a *slaved timed daemon* to run on the *slave node* (which synchronises the *slave nodes clock* with the *master node*).

A third parameter to *install_pups_services*, *kernel* may also be specified, in which case the *Linux-**MOSIX** distribution* is unpacked into /usr/src/linux in preparation for a kernel build. When this option is selected, the installation script expects *lilo* rather than *grub* to be the default **LINUX** *boot loader*. If this is not the case, **this script could cause damage to the system** when it tries to automatically update the boot block. Specifying *kernel* causes the following to happen:

1. The *install_mosix* script is run (within */usr/src/linux*). This is an interactive script maintained by the **MOSIX** Project, which patches the standard LINUX kernel sources with **MOSIX** extensions. After patching the kernel, this script also builds and installs a number of **MOSIX** tools and modifies configuration files in */etc*.
2. The LINUX *xconfig* utility is then run which allows the **LINUX-MOSIX** kernel to be customised. The distribution comes with a sensible set of options but in order to optimise performance the CPU type should be set appropriately. The distribution assumes **IDE** disks and CD-ROM: If a given target has **SCSI** devices, or needs **SCSI** emulation over **ATAPI** protocols (e.g. **SCSI** CD-Writer driver), these options must be configured at this stage. The only other items which will need specific configuration are drivers for non-standard hardware, and the driver for the network device (an ***NE2000** compatible device* is assumed).
3. The kernel is built and installed in */boot*. This kernel is labelled **linux-mosix** and becomes the *default boot kernel*.
4. The system reboots automatically at the end of the **PUPS** installation.

5.5 The PUPS System Tree

The *install_pups_services* script creates a *system tree* consisting of the following files and directories:

1. */pups*: root of the PUPS distribution.
2. */pups/bin.<arch>*: PUPS services functions for (CPU, OS and **PUPS**) architecture *<arch>*. For example for a **PUPS** cluster running under **LINUX** on an Intel compatible this would be */pups/bin.IX86.linux.cluster*.

3. */pups/lib.<arch>*: **PUPS** libraries (.o, .a and .so formats).
4. */pups/include.libs*: **PUPS** header files (for development of application code which is linked against the **PUPS** libraries).
5. */pups/examples*: contains a number of program and makefile templates which are used by the **PUPS** build tools to generate specific templates for new user application (using the **PUPS** application generator build tool, *application*).

The **PUPS** build tools and build scripts are located in the top level directory */pups*.

5.6 Building PUPS applications

Once the **PUPS** environment has been built on a target node, the **PUPS** libraries and the binaries for the **PUPS** service tools, daemons and build tools will be accessible by all users of that node. In order to develop **PUPS** applications, the user must first create a **PUPS** Project Directory (**PPD**) in their home directory. To do this type:

```
project ppd IX86.linux.cluster.
```

This creates a **PPD** directory called *ppd* in the users current working directory and preloads it with binaries and libraries for Intel compatible CPU's running the **LINUX** operating system in a networked environment. The **PPD** directory has following structure:

ppd (top level directory): build *csh* scripts.
ppd/include.libs: library headers
ppd/src: source code and makefile templates .
ppd/lib (link to the *lib.IX86.linux.cluster*): **PUPS** .o, .a and .so libraries.
ppd/bin: build tools

Once the user has defined a **PPD** directory, an *application template* forming the basis of a user application, may be generated by typing:

```
application
```
[18]

The application build tool will prompt the user interactively for:

1. An application name.
2. A string describing the purpose of the application.

[18] There are similar template generators available for **PUPS** library and **DLL** code. These are documentted on the **PUPS** website.

3. A string describing the application.
4. The author (of the application).
5. The author's e-mail (if any).
6. The authors institution (if any).
7. Date (year).

The application builder then takes the files *skelpapp.c* and *Make_skelpapp.in* and generates application specific templates for the main function of the application and its makefile. Once these application specific templates are generated, application specific code is added to the template, and the development of a **PUPS** application proceeds in the same manner as any standard **C** or **C++** application in the **POSIX** environment, using standard tools such as:

1. Your favourite editor, for example *vi* or *emacs*[19].
2. Debugging tools such as *gdb* (or windowing tools based on gdb. e.g. *ddd*).
3. Additional code integrity tools such as *electric fence* or *checker* which help to locate and correct runtime errors such as buffer overruns.

A good account of **POSIX** application development is given in Johnson and Troan (1998). The account of **PUPS** application development given here is necessarily brief: further, up-to-date information on the **PUPS** environment and **PUPS** application development is available at the **PUPS** web site, http://chasseur.usc.edu/pups together with the sources and binary **RPM**'s and tarballs. In order to develop high quality **PUPS** applications, the programmer must have significant experience of application development under **POSIX**. In particular, knowledge of interactive debugging, ideally for multithreaded applications, is required in order to write clean, high quality code. Furthermore, adherence to a *high software engineering standard* is to be encouraged. There are a number of (open source) utilities (e.g., *man2html* and *c2man*) which will be integrated within the **PUPS** environment to encourage this.

6. DISCUSSION

There is a single theme which unifies the rather disparate set of applications which have been implemented under **PUPS**. In order to complete their *payload computations* these applications have long lifetimes compared to typical computer programs. Being long lived gives rise to its own unique

[19] One can in fact set up a complete **CASE** environment for systems like **PUPS** using the *emacs* environment.

set of problems: *persistent computations* must expect, and be able to deal with:

1. Crashing and malfunctioning host computers.
2. Damage to an application data or its process address space.
3. Acute fluctuations of critical system resources such as disk space, CPU time and available memory.

This means that *persistent applications* and their support environments must embrace homeostatic mechanisms at many levels leading to many similarities with biological systems[20]. **PUPS** freely appropriates many biological mechanisms including:

1. Non-centralised control of computation.
2. Use of *cumulative evolution* in order to optimise critical aspects of computations.
3. Provision for *self-assembling applications* (whose properties are emergent).
4. Redundancy: applications must *fail gracefully* when they sustain damage.
5. *Irritability*: applications must be *aware* of their environment so they can react to changes in it.
6. *Intelligence*: persistent systems *must be able to learn about their environment* and then use that knowledge to *their advantage*.

At the present time, the process migration and checkpointing facilities offered by **PUPS-MOSIX** are limited to clusters composed of machines with identical hardware architectures. Functionality of the system could be significantly enhanced if process migrations between heterogeneous architectures were possible. This could be accomplished in two ways:

1. Writing **JAVA** wrapper functions for **C** function call within the **PUPS** libraries, thus embedding **PUPS** style functionality within the ***JAVA*** *sandbox.*
2. Writing a backend for an appropriate compiler, e.g., *gcc,* which produces object code for a *virtual RISC architecture*. The virtual object code is then either interpreted on the target architecture using ***JIT*** *translation techniques* or alternatively a *batch translator* simply translates the virtual object code to that of the target dynamically.

---

[20] In fact, a biological system such as a bacteria, an oak tree or an elephant may be viewed as a naturally occurring example of a *persistent computation*. Conversely, a persistent computation is an example of an *artificial lifeform*.

Although the first approach is an attractive short term solution to architecture independent process migration the second approach is superior as in principle *any* high level language (e.g., **C**, **C++**, **FORTRAN**, **Pascal**, etc.) can be compiled to *shrink wrapped* (e.g., hardware architecture independent) object code. Thus, the virtual processor approach allow applications to run efficiently in a heterogeneous cluster environment irrespective of the language that an application is written in. Proof of concept systems already exist for both the **JIT** approach (Microsoft's *.net*) and for batch translation (the **TAOS** system, see http://www.taos.com), but both of these systems are *static*. They do not provide mechanisms for moving a running process between different hardware architectures without loss of state.

REFERENCES

Ball, B., and Pitts, D. (2001) *RedHat LINUX-7 Unleashed.* SAMS Publishing.

Barak, A., Guday, S., and Wheeler, R.G. (1993) *The MOSIX Distributed Operating System: Load Balancing for UNIX.* In: *Lecture Notes in Computer Science,* pp.672, Springer-Verlag.

Burns, G.A.P.C. (2000) Knowledge management of the neuroscientific literature: the data model and underlying strategy of the NeuroScholar System. *Phil. Trans. R. Soc. Lond. B* 356, 1187-1208.

Burns, G.A.P.C, Khan, A.M., Ghandeharizadeh, S., O'Neill, M.A., and Chen, Y.-S.(2001) Neuroinformatics and the literature: can the depth of our understanding scale with the depth of our knowledge? *Neuroinformatics* 1,

Felleman, D.J. & Van Essen, D.C. (1991) Distributed hierarchical processing in the primate cerebral cortex. *Cereb. Cortex* 1, 1-47.

Hilgetag C.C., Burns, G.A., O'Neill, M.A., and Young, M.P. (2000a) Anatomical connectivity defines the organization of clusters of cortical areas in the macaque monkey and the cat. *Philos Trans R Soc Lond B Biol Sci* 355, 91-110.

Hilgetag C.C., O'Neill, M.A., and Young, M.P. (1996) Indeterminate organization of the visual system, *Science* 271, 776-777.

Hilgetag C.C., O'Neill, M.A. and Young, M.P. (2000b) Hierarchical organization of macaque and cat cortical sensory systems explored with a novel network processor. *Philos Trans R Soc Lond B Biol Sci* 355, 71-89.

Johnson M.K., and Troan, E.W. (1998) *LINUX Applications Development*, Addison Wesley.

Litzkow, M, Livny, M., and Mutka, M.W. (1988) *Condor - A Hunter of Idle Workstations*, Proc. 8th Int. Conf. Distrib. Comp. Sys. pp.104-111.

O'Neill, M.A. (2001) *Soft network architectures: how to raise the efficiency of persistent and parallel computations.* Morpho technical discussion document 02/2002.

O'Neill, M.A., and Hilgetag, C.C. (2001) PUPS and CANTOR - A computational environment for dynamical representation and analysis of complex neurobiological data, *Philos Trans R Soc Lond B Biol Sci* 356, 1259-1276.

Plank, J.S, Beck, M., and Kingsley, G. (1995), *Libckpt: transparent checkpointing under UNIX,* In: Proc. USENIX Winter Tech. Conf. New Orleans Louisiana.

Tannenbaum, A. (1987) *Operating Systems: design and implementation,* Prentice Hall, Englewood Cliffs, NJ.

Weeks, P.J.D., Gauld, I.D., Gaston, K.J., and O'Neill, M.A. (1997) Automating the identification of insects: A new solution to an old problem. *Bull. Ent. Res.* 87, 203-211.

Chapter 14

The NeuroHomology Database: An Online KMS for Handling and Evaluation of the Neurobiological Information

Mihail Bota and Michael A. Arbib
Dept. of Neurobiology, University of Southern California, Los Angeles, CA 90088, USA.
Correspondence to: mbota@rana.usc.edu, arbib@pollux.usc.edu

Abstract: The NeuroHomology Database (NHDB) is an online knowledge management system designed for handling, systematization and processing of the neurobiological information about brain structures, neuroanatomical connections and similarities between brain nuclei from different species, as found or inferred from the literature. The NHDB has a modular structure, with a knowledge-base designed for storing the data and a set of inference engines for processing of the stored information. In this chapter we briefly describe the structure of the NHDB and discuss the important aspects of the process of searching for data inside of the system.

Key words: online database system, inference engines, expert systems, homology

1. INTRODUCTION

The high volume of scientific information which is available in any direction of research in the neurosciences requires a tremendous effort from the researcher's side for gathering the relevant data, and for the interpretation of the information. This implies that the design of summary databases (Arbib and Bischoff-Grethe, 2001; Burns, 2001) and of knowledge management systems (KMS), which can be used by researchers for investigation, comparison and processing of data, can help the researchers in their activity.

An important aspect of the nature of neurobiological information which has to be considered when a KMS is designed is the heterogeneity of data across different levels of the organization of the nervous system, and also within a single level. Each of those is characterized by specific types of data, and the translation and integration of information specific to a level to another level is not a trivial task in many cases. The design of a KMS for handling and processing of the neurobiological information should be general enough to allow the insertion and processing of data from many levels of organization of the nervous system, but it has also to include a series of expert systems, in order to provide solutions to the heterogeneous nature of the neurobiological data.

In order to address these problems, we have designed different KMS for systematization, evaluation and processing of the neurobiological information. Since these database systems were primarily designed for processing of information related to similarities of brain structures from different species, we have called those the NeuroHomology Database systems (NHDB). However, the functionality of the NHDB systems is not restricted to the evaluation of the neural homologies across species, but includes modules which handle brain structures information or connectivity data, as well as inference engines for evaluation of the reliability of the connectivity information as reflected in the literature, and for translation of the neurobiological information in different neuroanatomical atlases.

The KMS NHDB has at present two online versions. NHDB-I contains a knowledge-base that allows the insertion of neurobiological data characteristic from the cellular to the structural level of the nervous system and two inference engines for evaluation of the connectivity information and of similarities between brain structures from different species. Despite of its capabilities, the NHDB-I does not address the problem of translation of the neurobiological data in different parcellation schemes, nor it contains a scheme for grasping the essential aspects of the cellular level of the nervous system. Since the problem of translation of data pertaining to brain structures in different atlases is an important aspect of neuroanatomy and constitutes a subject of debate across different research groups, and that of capturing the cellular descriptions of brain structures in Golgi stain is crucial for understanding the functionality of the nervous system, we have developed the second version of the system, NHDB-II, which has a partially different structure and contains two inference engines designed for translation of connectivity data in different parcellation schemes. Our final goal is to transfer all the information that exists at the moment from the NHDB-I to the NHDB-II, but since the system NHDB-I has a series of modules that have not been replicated yet in the NHDB-II, we consider these KMS as two separate entities, having partially overlapping functionalities.

In this chapter we discuss the main aspects of searching and retrieving information from both versions of NHDB and the type of information which can be retrieved. We only briefly describe the structure of the NHDB in the following section and the interested reader can find details of the KMS and of the inference engines in Bota, 2001; Bota and Arbib, 2001; Bota and Arbib, 2002.

2. NHDB SYSTEM: GENERAL DESCRIPTION

The object-relationship schema (OR) of NHDB is centered on the object "Brain Structure", uniquely defined by three attributes: name, species were it was identified, and atlas used for identification. The OR schema of the NHDB system is presented in Figure 1.

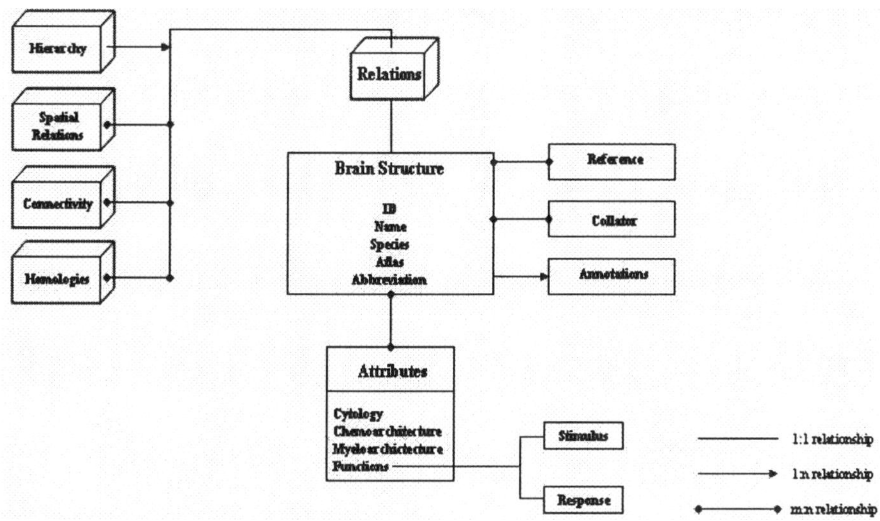

Figure 1. The OR schema of the NHDB systems.

Each of the objects and relations shown in Figure 1 are usually captured in more than a single table. Variables that are allowed to be inserted in the knowledge-base of the NHDB can be Boolean or text.

The object "Brain Structure" has as attributes cell types (cytology), chemoarchitecture, myeloarchitecture and functionality. The relations of all these attributes with the "Brain Structures" are of the type m:n.

The chemoarchitecture refers to that set of chemicals which are specific to the object "Brain Structure". The main attributes of chemoarchitecture are the intensity and pattern of staining of a given chemical, and the localization of it within the structure, as described in the associated references. As for chemoarchitecture, the myeloarchitecture of a brain structure has as properties the intensity of staining and the pattern of staining.

Since the cytology of a brain structure constitutes one of the most important criteria for describing it, we have designed a specific database structure to capture the characteristics of neural cells, as revealed by Golgi staining. The database schema designed to capture the features of neural cell types is general enough to allow the insertion of data pertaining to any class or subclass of neurons. Thus, we have considered those morphological characteristics which can be used to define a generic neural cell as it is described in the literature. The full discussion of the knowledge-base schema for cytology can be found in Bota, 2001 and Bota and Arbib, 2002.

The attribute "Functions" refers to that set of neurophysiological responses of the cellular components, or the behavioral associates of the associated brain nucleus. The functionality of a brain structure is considered to have as properties the fields of "Stimulus" and "Response". By "Stimulus" we refer to any type of procedure followed (from neurophysiological stimulations of single neurons, to inactivation of brain structures, by using specific techniques), and by "Response" we refer to any type of change of activity recorded from the individual cells, or behavioral alterations due to lesions of brain structures. The relationships between "Functions" and "Stimulus" and "Response" are of the type 1:1, since any given function is revealed by using a specific experimental paradigm.

We consider "Annotations" those parts of the inserted reference(s) that describe brain structures or any of the associated attributes, as well as statements associated to reports of brain nuclei and inserted online by users. The relationship between "Brain Structures" and "Annotations" is of the type 1:n. That is, for any unique brain structure, a set of different annotations may be associated. The relationships between "Brain Structures" and "Collator", and "References", respectively, are of the type m:n. A collator can insert information about many brain nuclei and data referring to a single brain structure can be inserted by different collators. The same is true for references: the information about a brain structure can be found in different references and a report can contain data referring to many structures.

We define four types of relationships inside of NHDB: Hierarchy, Spatial Relations, Connectivity and Homologies. The relationship "Hierarchy" is of type 1:n with the object "Brain Structure", since any nucleus has a unique hierarchy, but any brain structure can participate to many hierarchies of other nuclei recorded in the system. This relationship has two attributes: hierarchy path and hierarchy level. The hierarchy level is the number of structures that contain that brain nucleus and the hierarchy path is the string made out of the names of the superstructures. The hierarchy path for each brain structure is established on the basis of the reference that describes it, or inferred from a commonly used frame of reference (Bota and Arbib, 2001). The relationship

"Spatial Relations" refers to the topological and directional relations between different brain structures as found, or inferred from the literature, or established from unrelated information by running the topological inference engine. "Connectivity" refers to the knowledge-base that contains details about the neuroanatomical connections as found or inferred from the literature, and to the inference engines for evaluation of connections strengths and for translation of connectivity information in different atlases. Finally, the relationship "Homologies" refers to the knowledge-base and the inference engine for evaluating the similarities and the degree of homology between the brain structures from different species, by taking into account eight different criteria. The descriptions of the inference engines and of the implemented algorithms can be found in Bota and Arbib, 2000; Bota, 2001; Bota and Arbib, 2002.

In terms of the user interface, the NHDB system contains three interconnected modules, **Brain Structures**, **Connections** and **Homologies**, which can be accessed independently. We have designed the web interface in independent parts to allow queries from a larger category of users. A user who wants to find if there is any homology between two structures, X and Y, from different species, can also inspect the definitions of X and Y found in different sources, as well as the pattern of connectivity of these two structures. Additionally, users can create their own profiles and manage the information for personal use.

NHDB-I is designed in Microsoft Access and uses the WebMerger CGI parser engine as a web interface and can be accessed online at the URL: http://brancusi.usc.edu/scripts/webmerger.exe?/database/homologies-main.html both in terms of search of neurobiological information and insertion of new data. The second version of the KMS designed by us, NHDB-II, is designed in Informix 4.0 and uses the Illustra parser engine as a web interface and it can be accessed online at the URL: http://java.usc.edu/neurohomologies/apb/webdriver?MIval=homologies-main.html. Each of the versions of the NHDB system contains links to the other. NHDB-I contains to date about 500 reports of brain structures, more than 1000 reports of neuroanatomical connections and about 100 established similarities between brain structures from rats, macaques and humans. NHDB-II contains about 100 reports of brain structures and 200 reports of neuroanatomical connections.

3. SEARCHING THE NHDB

In this section we describe the main aspects of searching for information and setting up users accounts in NHDB. Any of the modules of NHDB is available for search from the main page, as shown in Figure 2.

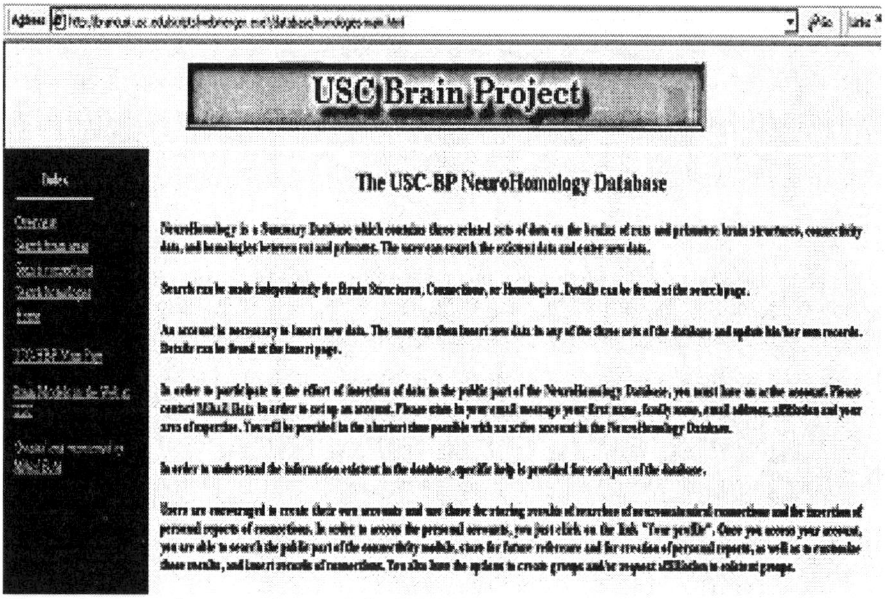

Figure 2. The main page of any of the NHDB systems, containing links to the modules described above.

3.1 Searching Brain Structures

The search of the module Brain Structures can be extended, or can be made specific. One type of extended search can be made using a word or phrase from the description of entered brain structures. Another possible extended search can be made by the name of an author of an article inserted in the database. The search of brain structures can be more specific, by using any combination of three possibilities of search: search by abbreviations of brain structures, by superstructure and by species.

A typical result of search of brain structures is as shown in Figure 3. The information which is retrieved by querying the system contains the ID's of the returned entries, the associated references, the names and abbreviations of the retrieved brain structures, the hierarchy levels (which lead to a separate page for the hierarchy paths) and the species, as well as the descriptions of the brain structures, the collators and the dates of entry of the returned records.

All brain areas for species(genus) Human

Click on *ID* to see all the associated annotations
Click on *Hierarchy* to see the hierarchy path
Click on *Author* to see details
Click on *Collator* to see details

ID	Author	Name of brain structure	Species (strain)	Abbreviation structure	Hierarchy	Description	Collator	Inserted on:
504	Bowden & Martin, 1995	Precentral sulcus	Human	prs	4	Landmark structure in the human frontal lobe.	Mihail Bota	17 July 2000 12 38
493	Bowden & Martin, 1995	Inferior temporal sulcus	Human	its	4	A superficial structure within the temporal lobe.	Mihail Bota	2 July 2000 14 40
505	Petrides & Pandya, 1994	Area 44	Human	44	5	This area occupies the pars opercularis of the inferior frontal gyrus, and is thought to constitute the larger part of Broca's area	Mihail Bota	17 July 2000 13 03

Figure 3. A typical result of search in the Brain Structures module of the NHDB. Users can access additional information about the retrieved records by clicking on the links available in each result of search.

Users can also access the associated annotations or insert new ones to any retrieved entries by clicking on the associated ID.

The options of search of records of brain structures in the NHDB-II are identical to those in NHDB-I. The differences that are specific to the Brain Structures module in the NHDB-II are related to the mode of displaying of information and to the type of data. Figure 4 shows the result of a query performed in the Brain Structures module of the NHDB-II.

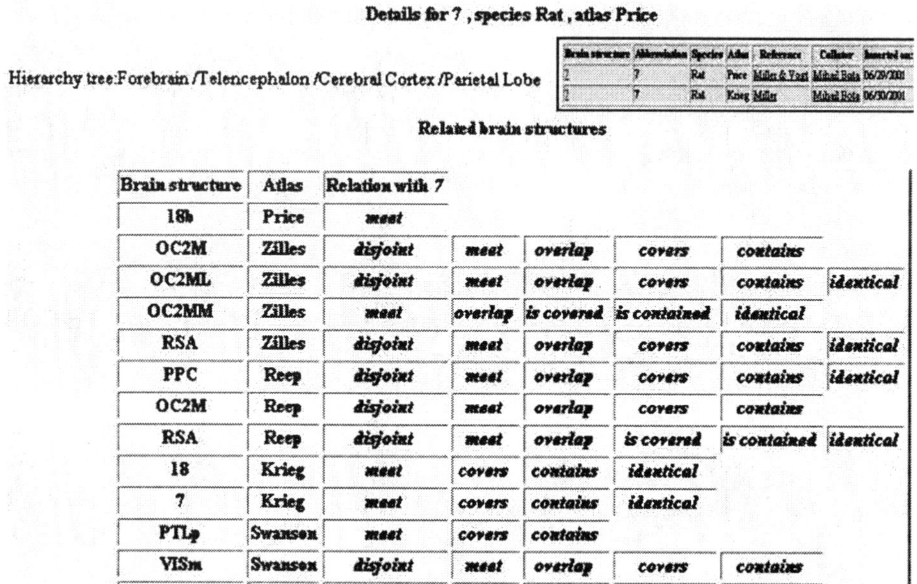

Details for 7 , species Rat , atlas Price

Hierarchy tree:Forebrain /Telencephalon /Cerebral Cortex /Parietal Lobe

Brain structure	Abbreviation	Species	Atlas	Reference	Collator	Inserted on:
7	7	Rat	Price	Miller & Vogt	Mihai Bota	06/29/2001
7	7	Rat	Krieg	Miller	Mihai Bota	06/30/2001

Related brain structures

Brain structure	Atlas	Relation with 7					
18b	Price	*meet*					
OC2M	Zilles	*disjoint*	*meet*	*overlap*	*covers*	*contains*	
OC2ML	Zilles	*disjoint*	*meet*	*overlap*	*covers*	*contains*	*identical*
OC2MM	Zilles	*meet*	*overlap*	*is covered*	*is contained*	*identical*	
RSA	Zilles	*disjoint*	*meet*	*overlap*	*covers*	*contains*	*identical*
PPC	Reep	*disjoint*	*meet*	*overlap*	*covers*	*contains*	*identical*
OC2M	Reep	*disjoint*	*meet*	*overlap*	*covers*	*contains*	
RSA	Reep	*disjoint*	*meet*	*overlap*	*is covered*	*is contained*	*identical*
18	Krieg	*meet*	*covers*	*contains*	*identical*		
7	Krieg	*meet*	*covers*	*contains*	*identical*		
PTLp	Swanson	*meet*	*covers*	*contains*			
VISm	Swanson	*disjoint*	*meet*	*overlap*	*covers*	*contains*	

Figure 4. A typical result of search in the Brain Structures module of the NHDB-II. The result of the query which was performed is shown in the inset. Users can access additional information, such as the established set of topological relations between the retrieved structure and other nuclei.

As shown in Figure 4, users can inspect in the NHDB-II different attributes of brain structures, as well as the possible topological relations with other nuclei from the same, or different atlases, as found, or inferred from the investigated references.

3.2 Searching Neuroanatomical Connections

The search of the Connections module of NHDB-I is similar to that of the Brain Structures. The search of reports of connections can be broad, or can have different degrees of specificity. The broad search can be made by a word or phrase from the description of a connection, or by name of author. A search can be narrowed by using a combination of three additional possibilities: afferent structure, efferent structure, and species. The narrowest query allows users to evaluate the confidence level of the retrieved connections by taking into account the number of reports referring to the same tract, the strengths of staining and the techniques used in each of the retrieved cases. An example of a result of such query is shown in Figure 5.

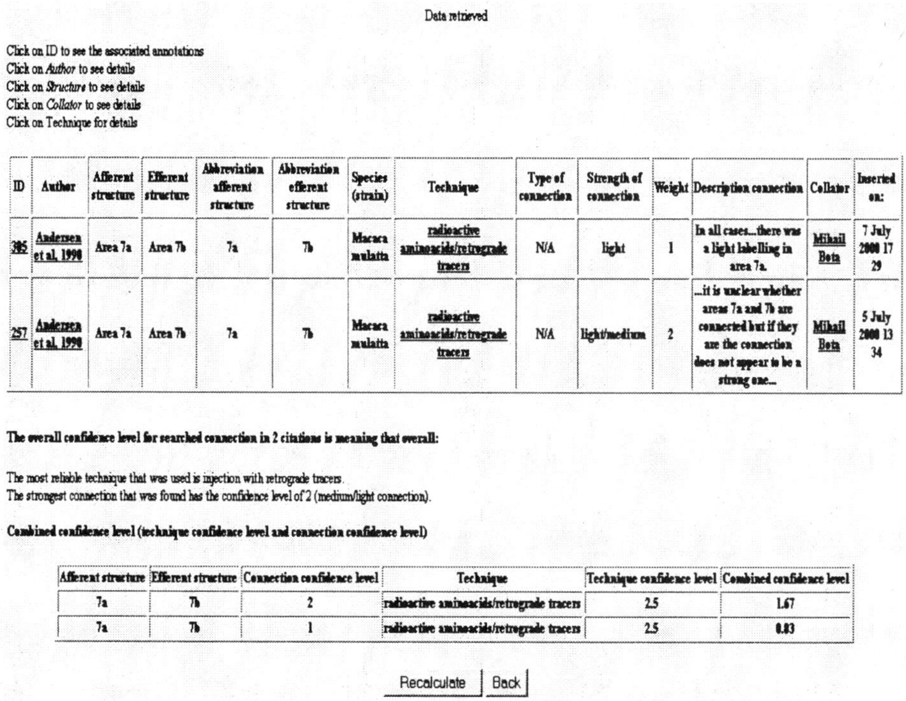

Data retrieved

Click on ID to see the associated annotations
Click on *Author* to see details
Click on *Structure* to see details
Click on *Collator* to see details
Click on Technique for details

ID	Author	Afferent structure	Efferent structure	Abbreviation afferent structure	Abbreviation efferent structure	Species (strain)	Technique	Type of connection	Strength of connection	Weight	Description connection	Collator	Inserted on:
385	Andersen et al, 1990	Area 7a	Area 7b	7a	7b	Macaca mulatta	radioactive aminoacids/retrograde tracers	N/A	light	1	In all cases...there was a light labelling in area 7a.	Mihail Bota	7 July 2000 17 29
257	Andersen et al, 1990	Area 7a	Area 7b	7a	7b	Macaca mulatta	radioactive aminoacids/retrograde tracers	N/A	light/medium	2	...it is unclear whether areas 7a and 7b are connected but if they are the connection does not appear to be a strong one...	Mihail Bota	5 July 2000 13 34

The overall confidence level for searched connection in 2 citations is meaning that overall:

The most reliable technique that was used is injection with retrograde tracers.
The strongest connection that was found has the confidence level of 2 (medium/light connection).

Combined confidence level (technique confidence level and connection confidence level)

Afferent structure	Efferent structure	Connection confidence level	Technique	Technique confidence level	Combined confidence level
7a	7b	2	radioactive aminoacids/retrograde tracers	2.5	1.67
7a	7b	1	radioactive aminoacids/retrograde tracers	2.5	0.83

Recalculate | Back

Figure 5. Users can access connectivity information in the NHDB as well as evaluate the connections confidence levels, or customize the connections inference engine.

The interested reader can find the complete description of the inference engine for evaluation of neuroanatomical connections from the neurobiological literature in Bota and Arbib 2002 and Bota 2001.

The search of Connections module in the NHDB-II is similar to that in NHDB-I. Users can additionally evaluate the probability of existence of the retrieved connections in different parcellation schemes, as shown in Figure 6. The inference engine for translation of the connectivity information evaluates the spatial relations of the injection site and of the terminal field in the new parcellation scheme and computes the probability of existence of the connection in the new atlas by taking into account the sets of inferred topological relations for the injection site and terminal field. All this information is available to the user to be inspected. The full description of the inference engine of translation of neuroanatomical connections in different atlases can be found in Bota 2001.

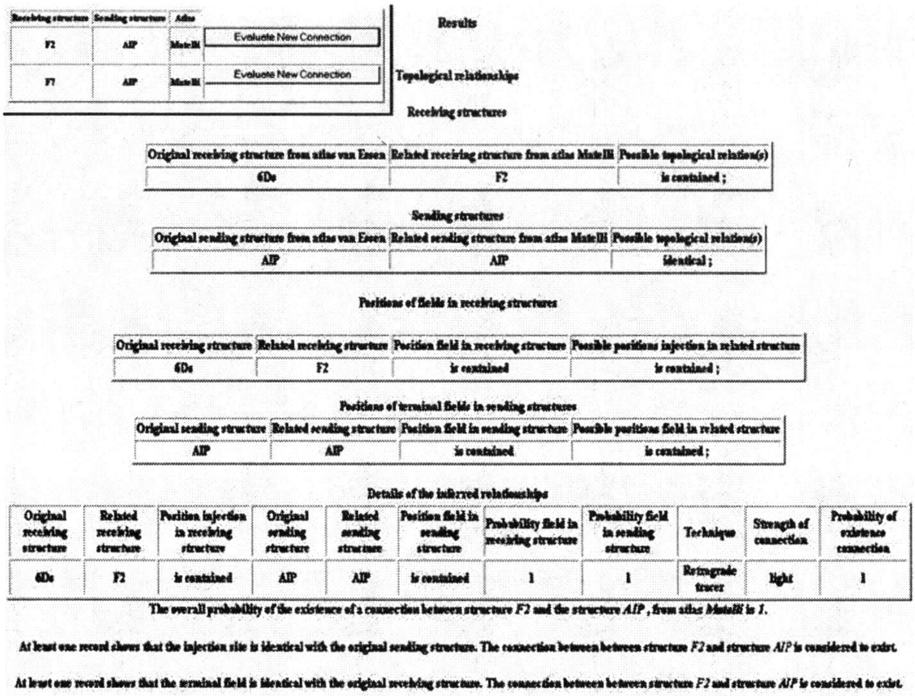

Figure 6. A typical result of search in the Connections module of the NHDB-II. The query which was performed is shown in the inset. Additional to the information shown in Figure 5, users can evaluate the probabilities of existence of retrieved connections in other parcellation schemes. In this case, the connections between areas F2 and AIP (Rizzollatti and Luppino, 1998) are evaluated in the atlas defined by Lewis and van Essen, 2000.

3.3 Searching Homologies

As described earlier, we consider eight criteria to evaluate the homology between two brain structures in different species. We have argued (Bota and Arbib, 2001, Bota, 2001) that a better approach to compare brain structures from different species is to evaluate the degree of homology (DG) as a function of the number of fulfilled similarity criteria and of the retrieved references.

The information existent in the Homologies part of the NHDB-I can be inspected in two ways: browsing all the similarities existent in the knowledge-base at a given moment, or searching the system for homologies by abbreviations of brain structures and species.

The information which is retrieved when browsing the similarities existent in the database includes the abbreviations of the compared nuclei, the species, the common features, the reference, and the collator, as well as the previously calculated DG. As for the other modules of the NHDB, users can access details of the associated references and of the collator, and insert annotations to each of the retrieved entries.

The second option of viewing the comparative data allows the user to evaluate the degree of homology from the information related to the searched structures and species, and recorded in the NHDB-I. An example of search of similarities and evaluation of DG for area 8A in the macaque and the precentral medial cortex (PrCM) in the rat is shown in Figure 7.

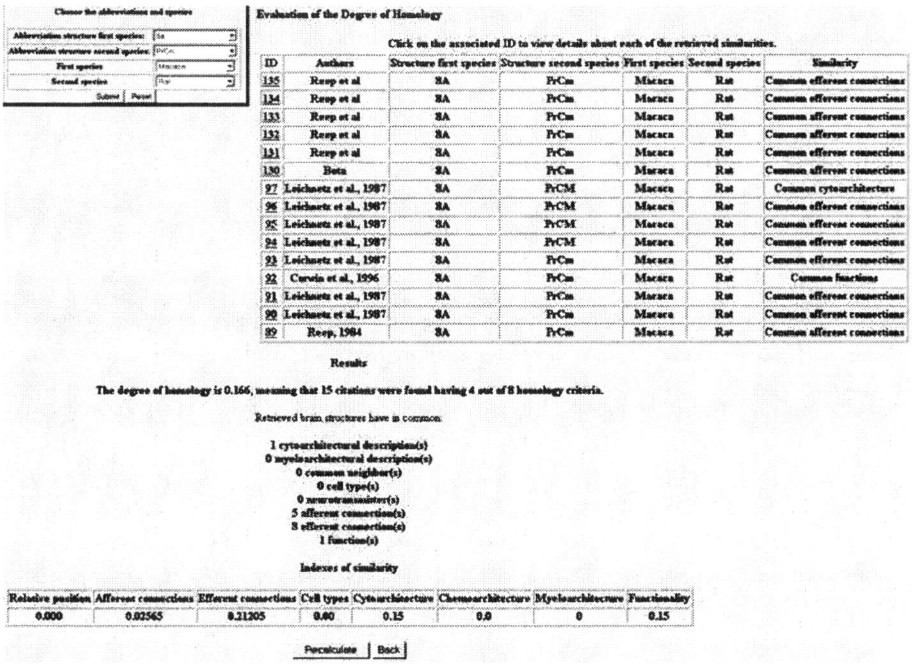

Figure 7. The result of searching for homologies between area 8A in the macaque and PrCM in the rat, in the NHDB-I. See text for details.

The result of the query shown in the inset of Figure 7 returns all the records existent in the knowledge-base of the NHDB and related to similarities between areas 8A and PrCM. The retrieved information includes the summary of the fulfilled criteria, the values of indexes of similarity associated to each criterion, and the calculated DG. As in the case of evaluation of connections confidence levels, users have the possibility of customizing the homology inference engine, by changing the relative importance of each of similarity criteria. The complete description of the inference algorithm for evaluation of DG can be found in Bota and Arbib, 2002 and Bota, 2001.

3.4 Creating User Profiles in the NHDB

We have created an additional module in the NHDB-I system which allows users to create their personal accounts, and populate those with records of neuroanatomical connections.

This module also allows users to participate in different groups inside of NHDB-I. Users can access the menu of the personal account from the main page, and this is shown in Figure 8.

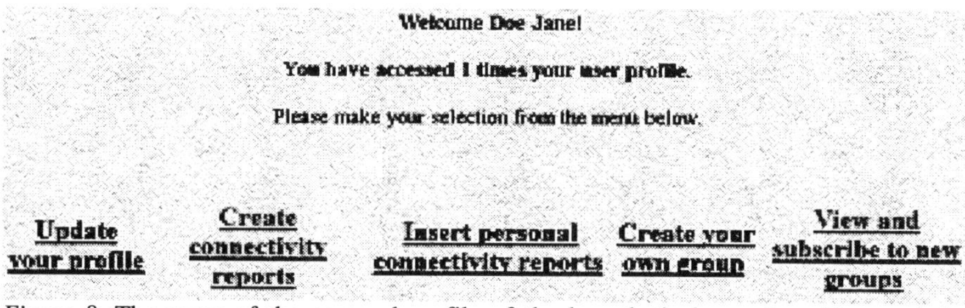

Figure 8. The menu of the personal profile of the hypothetical user Jane Doe, as it is accessed for the first time. Users have a menu of five options to manage and populate their personal accounts.

The option "Update your profile" can be chosen whenever the user wishes to update the personal profile. Options "Create your own group" or "View and subscribe to new groups" (not shown in Figure 8) manage the groups of users. Any individual user has the possibility to create his or her personal group and unsubscribe from any groups. The option “Create connectivity reports” allows users to search the public part of the database and the inserted connections in the personal account, or to create new reports or connections.

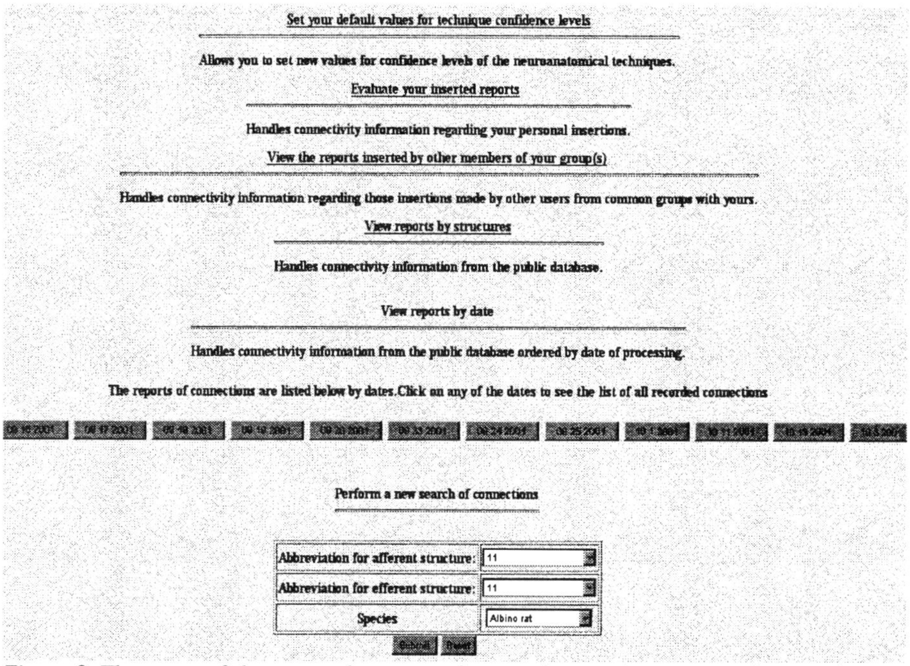

Figure 9. The menu of the personal account of a user for creating reports of connections and for searching both the public part of the NHDB-I and the personal reports.

The menu which is specific to the option of creating connectivity reports includes the possibility of searching and evaluating connections from the public part of the system, viewing the results of queries by date or by structure, evaluating the connectivity reports inserted by the user in his or her profile and viewing the connectivity reports inserted by other users from the same group(s). Users also have the possibility of setting their own values of the technique confidence levels and customize the evaluation of the connections confidence levels. Users have also the possibility of inserting simple personal reports of neuroanatomical connections. This can be performed by choosing the option "Insert your personal connectivity reports", shown in Figure 9. The part of inserting new personal reports has two additional features: if the user has also the privileges of a collator, he or she is allowed to make available the newly inserted information to the public part of the system. The second feature is related to updating and deletion of personal reports. If a user has already inserted connectivity information in his or her account, then he or she will be able to delete or update any of the inserted reports.

4. CONCLUSIONS

The structure of NHDB is general enough and sufficiently detailed to support insertion of neurobiological information pertaining to any mammalian species at different levels of organization of the central nervous system.

The interfaces of the system developed by us allow users to search it in different ways and also to insert new data either in the public part of the system, or in personalized accounts. Since the evaluation of the neurobiological information depends also on the expertise of the person who performs it, we designed the user interface of NHDB in such a way that it allows users to process the data according to their experience. Thus, users have the possibility of customizing the parameters of the inference engines for neuroanatomical tracts and homologies.

A partial list of the research directions in neurosciences that may benefit from using the NHDB systems includes neuroanatomy, neurophysiology and computational neuroscience. Researchers from fields of neuroanatomy and neurophysiology can inspect data about brain structures and fiber tracts in different species as found in collated references and also the similarities between cortical structures, as reflected from the literature or inferred by collators. Neuroanatomists have the possibility to establish spatial relations between cortical structures from different atlases by using the topological inference engine in the NHDB-II and evaluate neuroanatomical tracts in alternative atlases.

Computational neuroscientists can use NHDB as tools for validation of the formal models they develop. The knowledge-base and the inference engines of NHDB allow the computational neuroscientists to inspect information about the brain structures that are modeled, details of the connections between different structures and similarities between brain structures in different species. Moreover, computational neuroscientists can use the NHDB-I to evaluate the degree of reliability to model neural systems of a species by using data from other species.

The NHDB can be considered as a part of the ongoing effort specific to neuroinformatics and is one of the KMS that have been proposed to handle and process heterogeneous neurobiological information as found in the literature. However, a number of features of NHDB make it unique. One of those is that it contains the Homologies module. If the modules of Brain Structures and Connections can be found in different forms in systems developed by other neuroinformatics groups (Bowden and Martin, 1997; Burns, 2001; Burns et al. 2001, Marenco et al. 1999, Stephan et al. 2001), then the module of Homology is specific only to NHDB. Finally, another important advantage of NHDB is that it is an open system to any member of the neuroscientific community. The NHDB system can be inspected online and enriched in information by any member of the neuroscientific community who has the privileges of a collator. Moreover, the option of creating the personal account enables users to share information within a research group or across different laboratories.

Therefore, the NHDB system can be considered as a proper informatics environment for searching, sharing and processing of the neurobiological information as collated or inferred from the literature. We invite the members of the neuroscientific community to use both versions of NHDB and actively participate to enrich any of those.

REFERENCES

Arbib M.A. (2001) The NeuroInformatics: The Issues. In Arbib M.A., Grethe J. (eds) *Computing the Brain: A Guide to Neuroinformatics*, Academic Press, 3-28.

Arbib M.A., Bischoff-Grethe, A. (2001) Summary databases and model repositories. In Arbib M.A., Grethe J. (eds) *Computing the Brain: A Guide to Neuroinformatics*, Academic Press, 287-297.

Bota, M. (2001) *Neural Homologies: Principles, Databases and Modeling*. University of Southern California, Ph.D. Thesis.

Bota M., Arbib M.A. (2000) The NeuroHomology Database. In Arbib M.A., Grethe J. (eds) *Computing the Brain: A Guide to Neuroinformatics*, Academic Press, 337-351.

Bota, M., Arbib, M.A. (2001) The NeuroHomology Database. *Neurocomputing*, 38-40, 1627-31.

Bota, M., Arbib, M.A. (2002) Integrating Databases and Expert Systems for the Analysis of Brain Structures, Connections and Homologies. Submitted for publication.

Bowden, D.M, Martin R.F. (1997) A digital Rosetta stone for primate brain terminology. In: Bloom, F.E, Bjorklund, A., Hokfelt, T. (eds), *Handbook of Chemical Neuroanatomy vol. 13: The Primate Nervous System, part I*, 1-37.

Burns G.A.P.C. (2001) Knowledge mechanics and the NeuroScholar project: a new approach to neuroscientific theory. In Arbib M.A., Grethe J. (eds) *Computing the Brain: A Guide to Neuroinformatics*, Academic Press, 319-336.

Burns G.A.P.C., Stephan K.E., Ludäscher B., Gupta A., Kötter R. (2001) Towards a federated neuroscientific knowledge management system using brain atlases. *Neurocomputing* 38-40, 1633-1641.

Butler, A.B., Hodos, W. (1996) *Comparative vertebrate neuroanatomy: evolution and adaptation*. New York: Wiley-Liss, 7-13.

Cavada C, Goldman-Rakic PS. (1989) Posterior parietal cortex in rhesus monkey: I. Parcellation of areas based on distinctive limbic and sensory corticocortical connections. *Journal of Comparative Neurology* 287(4), 393-421.

Egenhofer, M. and Franzosa, R. (1991) Point-Set Topological Spatial Relations. *International Journal of Geographical Information Systems*, 5(2), 161-174.

Felleman DJ, Van Essen DC. (1991) Distributed hierarchical processing in the primate cerebral cortex. *Cerebral Cortex*. 1(1), 1-47.

Hilgetag CC, O'Neill MA, Young MP. (2000) Hierarchical organization of macaque and cat cortical sensory systems explored with a novel network processor. *Philos Trans R Soc Lond B Biol Sci*. 355(1393), 71-89.

Jones, E.G. (1990) History of cortical cytology. In Peters, A. and Jones, E.G (eds): *Cerebral Cortex. Volume I: Cellular components of the cerebral cortex*, Plenum Press, 1-34.

Kötter R., Hilgetag C.C., Stephan K.E. (2001) Connectional characteristics of areas in Walker's map of primate prefrontal cortex. *Neurocomputing*, 38-40, 741-46.

Krieg, W. J. S. (1947) Connections of the cerebral cortex. I. The albino rat. A topography of the cortical areas. *Journal of Comparative Neurology,* 84, 221-275.

Lewis JW, Van Essen DC. (2000) Mapping of architectonic subdivisions in the macaque monkey, with emphasis on parieto-occipital cortex. *Journal of Comparative Neurology* 428(1), 79-111.

Marenco L., Nadkarni P., Skoufos E., Shepherd G., Miller P. (1999) Neuronal database integration: the Senselab EAV data model. *Proceedings AMIA Symposium*, 102-106.

Nieuwenhuys R. (1998) Comparative Neuroanatomy: place, principles and programme. In Nieuwenhuys, R., ten Donkelaar, H.C., Nicholson, C. (eds) *The central nervous system of vertebrates, Volume 1*. Springer, 273-326.

Papadias, D. and Sellis, T. (1994) Qualitative Representation of Spatial Knowledge in Two Dimensional Space. *Very Large Data Bases Journal*, 3(4), 479-516.

Paxinos, G. and Watson, C. (1986) *The rat brain in stereotaxic coordinates.* San Diego: Academic Press.

Preuss T.M., Goldman-Rakic P.S. (1991) Myelo- and cytoarchitecture of the granular frontal cortex and surrounding regions in the strepsirhine primate Galago and the anthropoid primate Macaca. *Journal of Comparative Neurology* 310(4), 429-474.

Rizzolatti G., Luppino G., Matelli M. (1998) The organization of the cortical motor system: new concepts. *Electroencephalogrical Clinical Neurophysiology* 106(4), 283-296.

Seltzer B., Pandya D.N. (1984) Further observations on parieto-temporal connections in the rhesus monkey. *Experimental Brain Research* 55(2), 301-312.

Stephan K.E., Kamper L., Bozkurt A., Burns G.A.P.C., Young M.P. & Kötter R. (2001) Advanced database methodology for the Collation of Connectivity data on the Macaque brain (CoCoMac) *Philosophical Transactions of the Royal Society London B: Biological Science B* 356, 1159-1186.

Swanson L.W. (1992) *Brain Maps: Structure of the Rat Brain.* Elsevier.

Swanson L.W. (2000) Interactive brain maps and atlases. In Arbib M.A., Grethe J. (eds) *Computing the Brain: A Guide to Neuroinformatics*, Academic Press, 167-177.

Walker A.E. A cytoarchitectural study of the prefrontal area of the macaque monkey. *Journal of Comparative Neurology* 1940; 73, 59-86.

Zilles, K., Wree, A. (1985) *Cortex: Areal and laminar structure.* In Paxinos G. (Eds.) *The rat nervous system*, Academic Press, 375-415.

Chapter 15

The fMRI Data Center: Software Tools for Neuroimaging Data Management, Inspection, and Sharing

John Darrell Van Horn[1,2], Jeffrey B. Woodward[1], Geoffrey Simonds[1], Bennet Vance[1], Jeffrey S. Grethe[1], Mark Montague[1,3], Javed Aslam[3], Daniela Rus[3], Daniel Rockmore[3], and Michael S. Gazzaniga[1,2]
[1]The fMRI Data Center, [2]Center for Cognitive Neuroscience, [3]Computer Science Department, Dartmouth College, 6162 Moore Hall, Hanover, New Hampshire 03755
Correspondence to: John.D.Van.Horn@dartmouth.edu

Abstract: Efforts for the sharing of neuroscientific data depend on the ability of researchers to intelligently organize and manage their study information prior to it being shared. This involves the need for software tools for investigators to use in visualization, navigation, and exploration of their study data in an easy and accessible manner in which they may quickly become familiar. Additionally, the method of information storage should be one that permits not only the individual investigator to manage their data locally but also permits ease of sharing that information with their colleagues at remote locations or with centralized data archives. In this chapter, we describe a suite of data-management tools for use in the organization, viewing, and sharing of functional magnetic resonance imaging data. Designed to be flexible and extensible, these software tools should help researchers organize their data more carefully, inspect their neuroimaging experimental data more closely, and share their data more readily.

Key words: Neuroinformatics; fMRI; Databasing; JAVA™; XML

1. INTRODUCTION

In many ways, functional neuroimaging of the human brain is one of the most visually-oriented disciplines within neuroscience. As a digital representation of a three-dimensional object, brain image data is best examined by means of graphical image display, from inspection of neurological anomalies in patient groups to the examination of cognitively related patterns of activation in studies of in vivo brain function. It is now customary for the results of increasingly sophisticated analytic methods to be displayed as rendered overlays or superimposed on flatten representations of the cortical sheet (Van Essen, Drury, Dickson, et al., 2001). New visualization methods are appearing with increasing frequency, many of which detail abstract representations of brain morphological information, for instance, as it occurs during normal patterns of human aging (Toga and Thompson, 2002).

Despite the increasing mathematical and computational sophistication being brought to bear on the analysis of functional magnetic resonance imaging (fMRI) experiments, little attention has been paid to how such large data sets might be easily managed, manipulated, visualized, and shared via computer networks with other neuroimaging colleagues as well as with centralized study archives. Often, study information pertaining to subject demographics, experimental time-course information, and MR scanner parameters are kept in hand-written log books or at best in spreadsheet software files. However, even if records are kept using a computer, the level of detail or organization is certainly of variable quality even for a single investigator. This makes the examination of study metadata cumbersome, incomplete, and often hinders the ability of others to make sense of the study architecture. Therefore, a need exists for computer-based tools that assist investigators in managing their study data, and facilitate the sharing of that information in an organized fashion.

Parallel to the growth of neuroimaging as a dominant research tool for examination of the living brain has been the development of the Internet as a tool for information exchange. Internet browsers enable news and information to be viewed as never before with image graphics, hyperlinked text, and multimedia capabilities. Over the last decade, the Internet has also transformed scientific communication with the ability to send large quantities of data between collaborators across the globe as well as the proliferation of online journals wherein not only are the results of studies presented but also the raw data is available for other researchers to download or request for their own inspection.

Recent years have also witnessed increased sophistication of database technology for the management of large data archives. Researchers from many disciplines have realized the enormous benefits of having large-scale data archives and have argued convincingly that the preservation of the raw data is essential to progress in their field. For example, The Human Genome Project set a historic and scientific milestone by working to sequence the human genome and then making its contents publicly available via the Internet (Collins and Mansoura, 2001; Collins and McKusick, 2001). Sophisticated searches can now be performed for homologues to human genetic sequences across the genomes of different species (Ma, Tromp and Li, 2002). Information taken from cell proteins is now being catalogued and subjected to X-ray crystallographic decomposition and entered into large-scale proteomic databases with the promise of helping to guide the development of improved pharmaceuticals (Ezzell, 2002). To overcome some of the resistance to sharing primary research data, several leading scientific periodicals (e.g. The Journal of Cognitive Neuroscience) now make the contribution of data to publicly available data repositories a prerequisite for publication.

2. THE DATA-SHARING EFFORT OF THE fMRIDC

The fMRI Data Center (fMRIDC; http://www.fmridc.org), based at Dartmouth College, is making steady progress in the archiving and sharing of functional brain imaging data. The central purpose of the fMRIDC has been to make fMRI data more easily available to all researchers for continued study (Van Horn, Grethe, Kostelec, et al., 2001). The first step in creating a large-scale data-sharing database is acquiring the data. From the beginning, the fMRIDC has been fortunate to have support from the Journal of Cognitive Neuroscience (JOCN) that made submission of the primary research data for human fMRI studies a condition of publication. However, acquiring the data is only the beginning of a much larger effort that requires interpreting, organizing, and packaging data for data sharing. As a result, the Data Center currently archives raw, processed, and results image data from nearly 30 complete studies, totaling more than 500 Gigabytes (GB) of storage space with the holdings expected to double every nine months. Several larger fMRI studies are almost 15GB compressed, and when uncompressed are close in size to the entire Human Genome database (~45GB). Researchers from around the world may request the studies, and the Data Center provides compact disks of the data at no cost (Van Horn and Gazzaniga, 2002).

2.1 Early Tools for Contributing Study Data

In its first generation of tools, the fMRIDC developed a series of common gateway interface (CGI) web-forms to collect the study parameters from authors. The forms focused on the core information common to most fMRI studies: information about subjects, experimental protocols, scanner protocols, etc. Unfortunately, the variation of study data is not well suited to such a rigidly structured system. In many cases authors have found it difficult or impossible to adequately describe their study using only the fixed structure of the web-forms. First, subject assessments beyond just the normal handedness information can not be readily entered into the web-forms in a consistent manner, nor can the web-forms be easily extended to accommodate such variation. Second, experimental paradigm time-course information is laborious to enter - the web-forms offer no provisions for importing the ASCII text files often used by researchers (e.g. stimulus regressor files). Finally, a more subtle, but equally problematic, issue is that authors supply the correct information but in the wrong units of measure. For example, it is common for an author, measuring in milliseconds, to report a value of 2,000 in the TR field of an EPI scanner protocol overlooking the fact that the field may be clearly labeled as expecting seconds. As a result, the fMRIDC handles a large amount of supplemental information that the web-forms don't address. Ultimately, the fMRIDC staff works with the authors to ensure that the data they provide is correctly interpreted and compiled into a document detailing the dataset. Final packaging couples the vast amount of imaging data with the detailed document.

This lack of automation is clearly an indication of deficient tools. There are a number of ways that the current CGI based web-forms could be enhanced to address the issues outlined above; however, an important obstacle would remain; researchers would still be contributing study data retrospectively. This retrospective process is laborious, time consuming, and error-prone for the researcher.

2.2 New Directions for Managing Experimental Data

In light of the above-mentioned limitations of our first generation tools, the fMRIDC is developing a new suite of tools to transition the process of data sharing from retrospective to prospective via integrated tools that enable researchers to manage their own data locally. The tools provide a large degree of organization and flexibility to 1) represent all of the experimental variables and parameters under study; 2) reflect the context of the experimental question being examined; and 3) enable researchers to visualize their experimental data and thereby permit others to understand the data in the manner in which the original authors intended.

Figure 1 shows the main user interface of the Data Management Tool. The tool breaks the data up into a collection of hierarchically organized entities. As an example, subject assessments are entities; their relation in the data hierarchy can be seen in the tree structure within the upper left-hand pane, and the data values contained in the entity are shown in the right-hand pane.

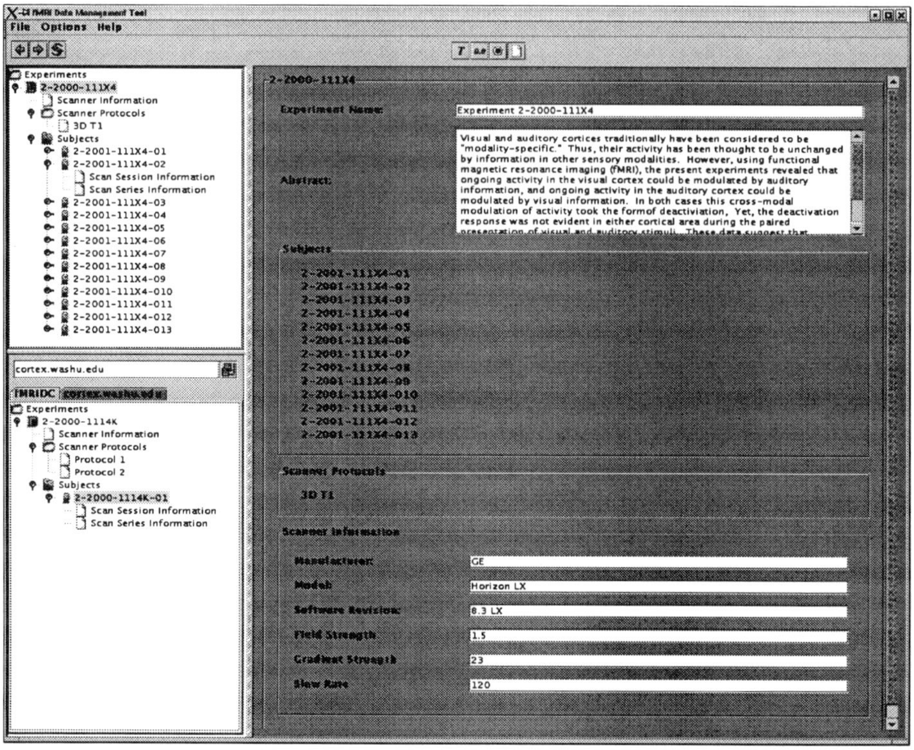

Figure 1. Experiment Management Tool displaying fMRI study data. In the upper-left hand pane, the experimental data hierarchy is displayed where users may expand a heading, select an item, and have the enclosed data displayed in the main viewer pane on the right. Data contained in the right-most browser pane includes experiment name and abstract information, subject lists, scanner protocol information, etc. Each of these may be expanded to view further information about that portion of the experiment. The lower-left hand pane permits the display of studies that are contained on remote sites. This enables authenticated users in the same laboratory to securely view data in a common repository within the lab, allows researchers at different institutions to collaborate on studies, and enables users to interact and share data with centralized study repositories like the fMRIDC.

In Figure 2, a subject designated as 2-2001-111X4 has an Edinburgh Handedness assessment selected, as indicated by highlighting; the detailed assessment information is displayed and accessible for modification within the right-hand pane. In this figure, the data input template for the Edinburgh

Handedness Inventory is predefined; however, researchers may save new templates as well as extend existing ones. To add a new assessment to a subject, the researcher would right-click on the assessment folder of the corresponding subject and a list of known assessments would be provided to choose from. Likewise, right-clicking on the scanner protocol folder will guide the user through adding a new scanner protocol from a template. Variation of a particular instance of an entity is also possible. This can be done without modifying the underlying template and without creating a new template. For instance, a researcher could annotate a particular Edinburgh Handedness Inventory assessment by right-clicking on the assessment and adding custom fields; this provides a solution to the scenario in which a researcher wishes to note that a particular subject lost the use of his left hand in a childhood accident. This feature is not restricted to assessments; rather, it is made possible by the entity framework of the tool.

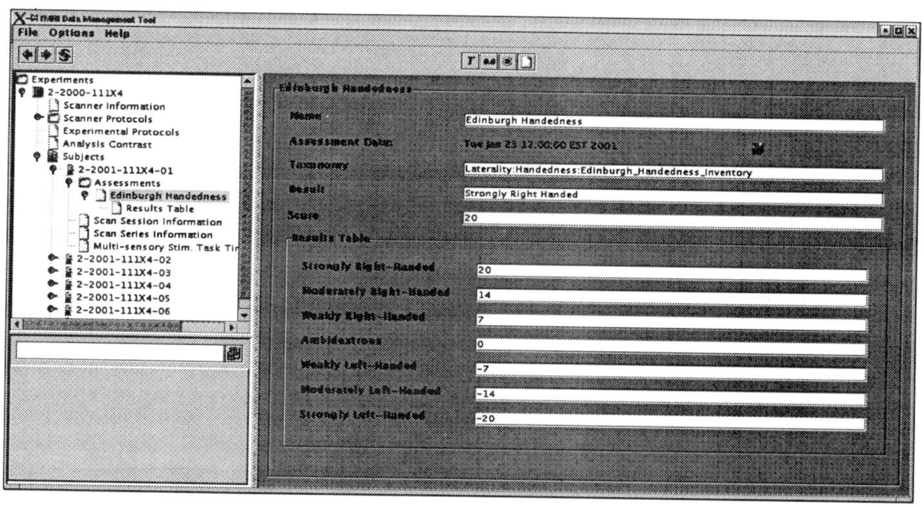

Figure 2. The Data Management Tool displaying subject data pertaining to hand preference as measured using the Edinburgh Handedness Inventory. This demonstrates how a great degree of specificity can be achieved for describing subject-level information which is also the case for scan series, experimental protocol, and other information. Assessment templates, such as the one in this example, may be predefined or may be created as desired by the user for their specific needs.

Within the entity framework, entities are composed of properties of predetermined types. For example, a subject has the properties of a name, age, gender, etc, which can be one of several types. Types exist for representing data values such as integers, strings, long text (plaintext or HTML), dates, and measurements to name just a few. Having rich data types enables a context for a value that would otherwise be lost if all values were simply treated as text. This is of particular importance when entering and storing measurements such as the TR of a scanner protocol or size of image voxels. Having a field that explicitly requires a number with units of the appropriate type overcomes the issue of a value being ambiguous. Further, it makes the conversion from compatible units automatic, such as the conversion from milliseconds to seconds.

Fully populating the properties of an entity such as the paradigm time-course can be laborious. The entity framework allows users to customize the display and edit the plug-ins associated with entities in such cases. Figure 3 shows the plug-in for viewing, constructing, and editing paradigm time-course entities. This particular plug-in allows researchers to import and export task timing information from various common representations of task time-courses, often stored in ASCII text files (for instance a single column time-course regressor or a multi-column format).

A tool suite for neuroimaging data wouldn't be complete without a tool for viewing brain image data. The fMRIDC has developed an Image Data Browser to fulfill this need (Figure 4). The image browser may be used not only as a stand-alone application but also to display fMRI time-series image data from within the Data Management Tool. Note, however, that this tool is not meant to be a fully-featured image analysis package as there are several other statistical packages which are highly optimized for the inferential analysis of fMRI data (e.g. Statistical Parametric Mapping, AFNI, Brain Voyager, etc.). Rather, this image browser is to enable the rapid viewing and inspection of image data that is contained in an experiment. The image viewer is modular in that new file readers can be developed to handle the multitude of neuroimaging file formats that exist. Furthermore, a novel wavelet decomposition streaming module has been developed that permits near-real time "voxel surfing" over a network allowing image data from remote sites to be displayed locally.

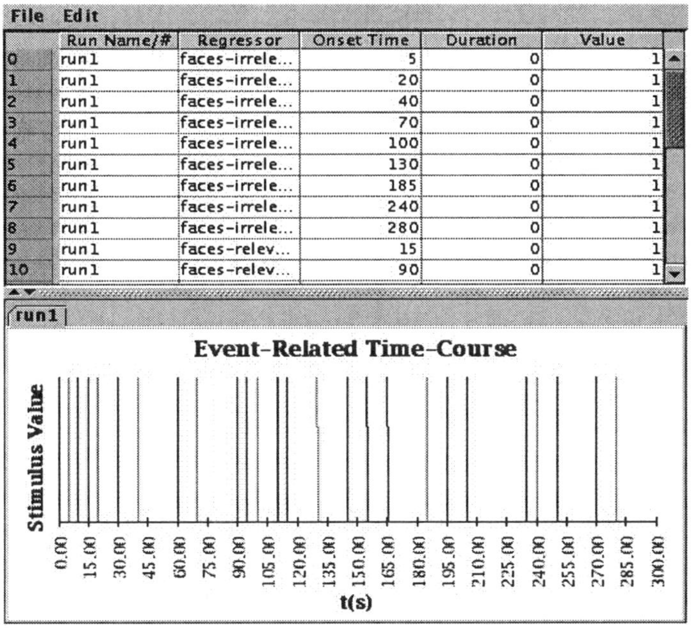

Figure 3. fMRI Stimulus Time-Course Viewer. In the upper pane, data collection run-specific time-course stimulus information is listed by stimulus type, onset time, duration, and stimulus value. In the lower pane, this information is displayed as a time course, color-coded by stimulus type. The example presented here shows an event-related experimental run of the presentation of context-relevant pictures of human faces, context-irrelevant faces, and images of stimulus-neutral noise patterns. Events may also be specified by using the computer mouse to click at a location on the time axis. The plot is updated with the new event at that temporal location with the indicated duration (in the units specified by the user) and value information, and the table in the upper panel is automatically updated. The events displayed in this example are shown here prior to convolution with a model of the hemodynamic response function.

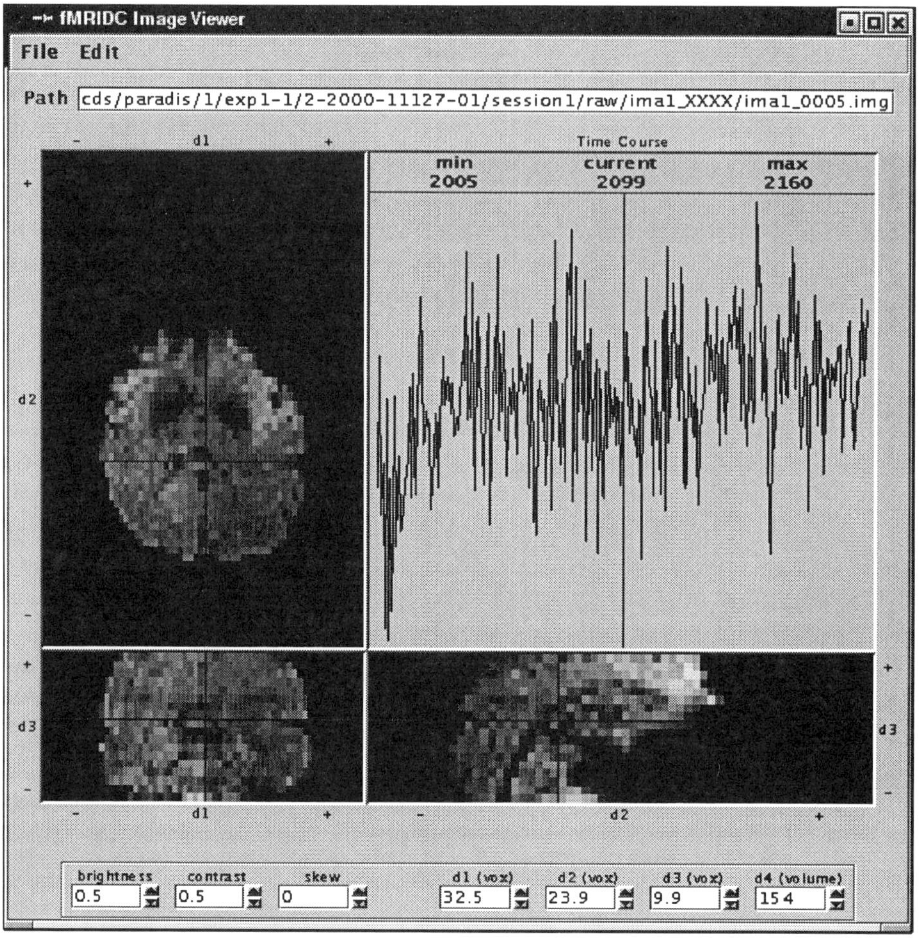

Figure 4. Image Data Browser. Brain image time data may be viewed in each of three orthogonal planes and the voxel data at the intersection of the cross-hairs displayed as a time course. Movement of the cross-hairs in any of the images updates the data in the time course pane. Movement of the indicator in the time course pane across the duration of the image sequence updates the image panes with the data from the image volume at that time point.

2.3 *The Technology Guiding Software Tool Development*

Our technology choices in implementing the fMRIDC tools have been driven by these considerations: 1) the tools need to be able to run on different kinds of PCs, as well as on workstations and servers running different operating systems - in short, on the full variety of platforms that neuroscience researchers might use; 2) the software should run "as is" - it should not require complex installation or configuration procedures, nor should it depend on availability of third-party packages and tools; 3) the fMRIDC tools should be easily adaptable to future needs, including interoperability with other tools.

To address the first two points, the fMRIDC tools are being developed entirely in the Javatm programming language, which is intrinsically platform-independent. (These tools do make use of open-source third-party packages, but these packages are also Javatm-based and are compiled into the fMRIDC tools so that users need not deal with them explicitly.) Moreover, tools such as Javatm Web Start fully automate installation and maintenance of Javatm-based applications. Consequently, researchers will be able to use the fMRIDC tools without relying on any system-administration expertise.

To address the third objective—adaptability and interoperability—the fMRIDC tools use semi-structured data representation techniques. As we have seen above, neuroscience entities such as subject assessments do not always fit exactly the same mold. Though different assessment instances typically have some fields in common (for example, an assessment name, date, and summary score), some assessments might also include idiosyncratic information such as special annotations. Traditional relational (i.e., tabular) database representations can accommodate optional fields through the use of "null" values, but make no provision for fields that were not anticipated at schema-design time; relational databases also limit flexibility in the representation of hierarchies. At the other extreme, free-format data representations (e.g., text files) provide infinite flexibility, but fail to capture common structure. Semi-structured databases (Abiteboul, Buneman and Suciu, 2000) offer a middle ground. To store experiment metadata, the fMRIDC tools use XML, which has become an industry-wide standard for semi-structured data storage. The tools also build indexes supporting structured queries over the experiment metadata. For example, it would be possible to query an fMRIDC database to find all experiments involving juvenile, left-handed subjects.

The XML representation provides versatility beyond the fMRIDC tools themselves. So-called Extensible Style-sheet Language Transformations (XSLT) can be constructed to transform the XML-based experiment data into HTML documents, which are then in a format that can be posted on the web. In a similar fashion, the XML document can be transformed into Adobe PDF to obtain a neatly formatted, printable document. Indeed, any of the many general-purpose XML-based tools now available can be applied to fMRIDC experiment metadata. Additionally, essential to the utility of these representations will also be their interoperability with other knowledge-based systems embedded applications, such as Protégé-2000 (http://protege.stanford.edu) from the Stanford Medical Informatics group (Musen, 1998; Noy, Sintek, Decker, et al., 2001). Finally, the schema will have the ability to cross-reference neuroanatomical information with the NeuroNames hierarchical anatomy model (Bowden and Martin, 1995; Martin, Mejino, Bowden, et al., 2001).

2.4 Summary

The fMRIDC project to develop a second-generation suite of neuroimaging data-management tools began with the realization that our first-generation data-acquisition tools were too limited. The new tools have a broader scope, and can be used for multiple purposes. First, quite apart from data-sharing considerations, researchers can benefit from using the tools to record, organize, search, and view their own experiment data and metadata. Second, researchers can transmit metadata recorded using these tools to the fMRIDC for archiving without having to reenter it, much less reorganize it, as was required with the first-generation tools. Third, the fMRIDC can incorporate new study data into its archives more rapidly and more accurately by bypassing the human intervention that was previously involved. Fourth, when researchers request experiment data from the archive, they will receive machine-readable XML metadata that can be directly manipulated by the fMRIDC tools. Finally, researchers may use the fMRIDC tools and the corresponding data formats for authenticated peer-to-peer exchange of experiment data (whether or not it has been published in the peer-reviewed literature). This provides more structured information exchange across the continuum of data sharing efforts (Van Horn, et al., 2001).

Integral to achieving these multiple goals is our use of platform-independent, easily installable Javatm technology in the implementation of the second-generation tools. This engineering choice, combined with the use of interoperable, XML-based data representations, allows the fMRIDC tool suite to manage neuroimaging experiment data and metadata through its full life-cycle: from acquisition through archiving to end-user viewing.

3. DISCUSSION

The size of some functional neuroimaging studies already rivals that of the working version of the human genome database. With increasing technological advances in image acquisition and with ever more complex experimental paradigms being employed, we can readily anticipate that the size of future studies will only be greater still. The ability to organize the often enormous amount of study information (study descriptions, demographic information, and image time-course data) will be essential for making sense of increasingly complex experimental designs. What is more, the exchange of these data between researchers as well as depositing complete study data in functional imaging archives like the fMRIDC will require thoughtful consideration of logical hierarchical data frameworks using extensible file formats.

Initiatives for the sharing of primary neuroscientific data, like the fMRIDC, rely on the ability of researchers to logically organize and manage their study information so that once it is contributed to the archive it may easily be incorporated with other studies in the study database. This has necessitated software tools for investigators to use in the visualization and exploration of their study descriptive and meta-data in addition to their fMRI results images. Further, study management tools must be easy to use, and must allow the user to interact with locally, as well as remotely, stored data.

In this chapter, we have presented a suite of data management tools for use in the organization of functional magnetic resonance imaging data. Designed to be flexible and extensible, these software tools should help researchers organize their large neuroimaging data sets more carefully, inspect their studies more closely, and share the information more readily. We believe that these tools will help usher in an era where functional neuroimaging study data will be universally shared. As the field of cognitive neuroscience conducted with fMRI grows in scope and scale, such tools take on increasing importance for data management. For additional information, the reader may visit the fMRIDC web site (http://www.fmridc.org).

ACKNOWLEDGEMENTS

The authors wish to recognize Anne Taggart and Wendy Starr for their valuable efforts on the fMRI Data Center project. The fMRIDC receives support from the National Science Foundation, The William M. Keck Foundation, The National Institute of Mental Health, and Sun Microsystems, Inc. The authors would also like to thank Dr. Abigail Baird for her helpful comments on an early draft of this chapter.

REFERENCES

Abiteboul, S., Buneman, P. and Suciu, D. (2000) *Data on the Web*. San Francisco: Morgan-Kauffman.

Bowden, D. M. and Martin, R. F. (1995) NeuroNames Brain Hierarchy. *Neuroimage*, 2, 63-83.

Collins, F. S. and Mansoura, M. K. (2001) The human genome project. *Cancer*, 91, 221-225.

Collins, F. S. and McKusick, V. A. (2001) Implications of the Human Genome Project for medical science. *Journal of the American Medical Association*, 285, 540-544.

Ma, B., Tromp, J. and Li, M. (2002) PatternHunter: faster and more sensitive homology search. *Bioinformatics*, 18, 440-445.
Martin, R. F., Mejino, J. L., Bowden, D. M., Brinkley, J. F. and Rosse, C. (2001) Foundational model of neuroanatomy: implications for the human brain project. *Proc AMIA Symp*, 438-442.
Musen, M. A. (1998) Domain Ontologies in Software Engineering: Use of Protege withthe EON Architecture. *Methods of Information in Medicine*, 37, 540-550
Noy, N. F., Sintek, M., Decker, S., Crubezy, M., Fergerson, R. W. and Musen, M. A. (2001) Creating Semantic Web Contents with Protege-2000. *IEEE Intelligent Systems*, 16, 60-71
Toga, A. W. and Thompson, P. M. (2002) New approaches in brain morphometry. *Am J Geriatr Psychiatry*, 10, 13-23.
Van Essen, D. C., Drury, H. A., Dickson, J., Harwell, J., Hanlon, D. and Anderson, C. H. (2001) An integrated software suite for surface-based analyses of cerebral cortex. *J Am Med Inform Assoc*, 8, 443-459.
Van Horn, J. D. and Gazzaniga, M. S. (2002) Databasing fMRI Studies - Toward a 'Discovery Science' of Brain Function. *Nature Reviews Neuroscience*, 3, 314-318
Van Horn, J. D., Grethe, J. S., Kostelec, P., Woodward, J. B., Aslam, J. A., Rus, D., Rockmore, D. and Gazzaniga, M. S. (2001) The Functional Magnetic Resonance Imaging Data Center (fMRIDC): The Challenges and Rewards of Large-Scale Databasing of Neuroimaging Studies. *Philos Trans R Soc Lond B Biol Sci*, 356, 1323-1339

Chapter 16

Statistical Parametric Mapping

Karl J. Friston
Wellcome Dept. of Imaging Neuroscience, UCL, UK. k.friston@fil.ion.ucl.ac.uk

Abstract: This chapter deals with the experimental design and analysis of functional brain imaging studies. It considers the neurobiological motivations for different designs and describes some standard approaches, developed to analyse the ensuing data. Functional neuroimaging (positron emission tomography - PET and functional magnetic resonance imaging fMRI) are generally used to make inferences about functional anatomy on the basis of evoked patterns of cortical activity. Functional anatomy refers to the anatomical organisation of the brain, as it relates to function. It involves an understanding of what each part of the brain does, and how different brain systems interact to support various sensori-motor and cognitive functions. This chapter discusses how this understanding is attained from the perspective of design and analysis of neuroimaging data.

Key words: Functional Neuroimaging, SPM, PET and fMRI

1. INTRODUCTION

Large-scale brain organisation can be inferred from techniques that image the hemodynamic and metabolic sequelae of evoked neuronal responses. PET measures regional cerebral blood flow (rCBF) and fMRI measures oxygenation changes. Their spatial resolution is in the order of a few millimetres. Because PET uses radio-tracers, its temporal resolution is limited by the half-life of the tracers employed (usually a minute or so). fMRI is limited only by the biophysical time-constants of hemodynamic responses themselves (a few seconds). However, these are quite long in relation to neuronal dynamics as measured by EEG and MEG. An important issue, in the design and analysis of functional anatomic experiments, is the relationship between conceptual models of brain organisation, neurophysiological responses, and how this relationship is realised in terms of the statistical models used in analysis. This chapter begins by reviewing the distinction between *functional specialisation* and *integration* and how these principles serve as the motivation for most analyses.

Statistical parametric mapping (SPM) is used to identify functionally specialised brain regions. Specialised regions respond selectively to experimental cognitive or sensori-motor changes. SPM is used to detect these responses anywhere they occur, irrespective of changes elsewhere. The alternative perspective, namely that provided by functional integration, requires a different set of [multivariate] approaches that examine the relationship between changes in one area and another. SPM is a voxel-based approach (a 'voxel' is a volume element of an image), employing standard inferential statistics, to make some comment about regionally specific responses to experimental factors at each voxel. In contradistinction to the alternative approach, based on pre-specified *regions of interest* (ROI), SPM examines every part of the brain at the highest resolution afforded by the data. SPM has now almost universally superseded ROI approaches because it avoids preconceptions about evoked responses and eschews many forms of bias (induced by ROI selection). Technically, SPM is a *mass-univariate* approach, in the sense that each data sequence, from every voxel, is treated as a univariate response. The *mass*ive number of voxels are analysed in parallel and dependencies among them are dealt with using Random Field Theory, at the point of inference. The alternative, to a mass-univariate approach, is a *multivariate* analysis that considers images as a multi-dimensional observation with as many components as there are voxels. The general ideas behind statistical parametric mapping are described and illustrated with the different sorts of inferences that can be made with different experimental designs.

This chapter concludes by reviewing multivariate analyses and effective connectivity. The software package implementing SPM (SPM99) is available from http://www.fil.ion.ucl.ac.uk/spm

2. FUNCTIONAL SPECIALISATION AND INTEGRATION

The brain appears to adhere to two fundamental principles of functional organisation; functional integration and functional specialisation, where the integration within and among specialised areas is mediated by *effective connectivity*. This distinction relates to that between *localisationism* and *[dis]connectionism* that dominated thinking about cortical function in the nineteenth century. Since the early anatomic theories of Gall, the identification of a particular brain region with a specific function has become a central theme in neuroscience. However, functional localisation *per se* was not easy to demonstrate. For example, a meeting that took place on August 4th 1881 addressed the difficulties of attributing function to a cortical area, given the dependence of cerebral activity on underlying connections (Phillips *et al* 1984). This meeting was entitled "Localisation of function in the cortex cerebri". Goltz, although accepting the results of electrical cortical stimulation, considered that the excitation method was inconclusive, in that movements elicited might have originated in related pathways, or current could have spread to distant centres. In short, the excitation method could not be used to infer functional localisation because localisationism discounted interactions, or functional integration among different brain areas. It was proposed that lesion studies could supplement excitation experiments; ironically, it was observations on patients with brain lesions some years later that led to the concept of *disconnection syndromes* and the refutation of localisationism as a complete or sufficient explanation of cortical organisation. Functional localisation implies that a function can be localised in a cortical area, whereas specialisation suggests that a cortical area is specialised for some aspect of perceptual, motor or cognitive processing, and that this specialisation is anatomically segregated within the brain. The neuronal infrastructure supporting a single function may then involve many specialised areas whose union is mediated by the functional integration among them. In this view functional specialisation is only meaningful in the context of functional integration and *vice versa*. Certain patterns of cortical projections are so common that they could amount to rules of cortical connectivity. "These rules revolve around one, apparently, overriding strategy that the cerebral cortex uses - that of functional segregation" (Zeki 1990). This is the model upon which the search for regionally-specific effects is based.

The analysis of functional neuroimaging data involves many steps that can be broadly divided into; (i) spatial pre-processing, (ii) estimating the parameters of a statistical model and (iii) making inferences about modelled effects using the parameter estimates and their associated statistics (see Figure 1). We will focus on model estimation and inference.

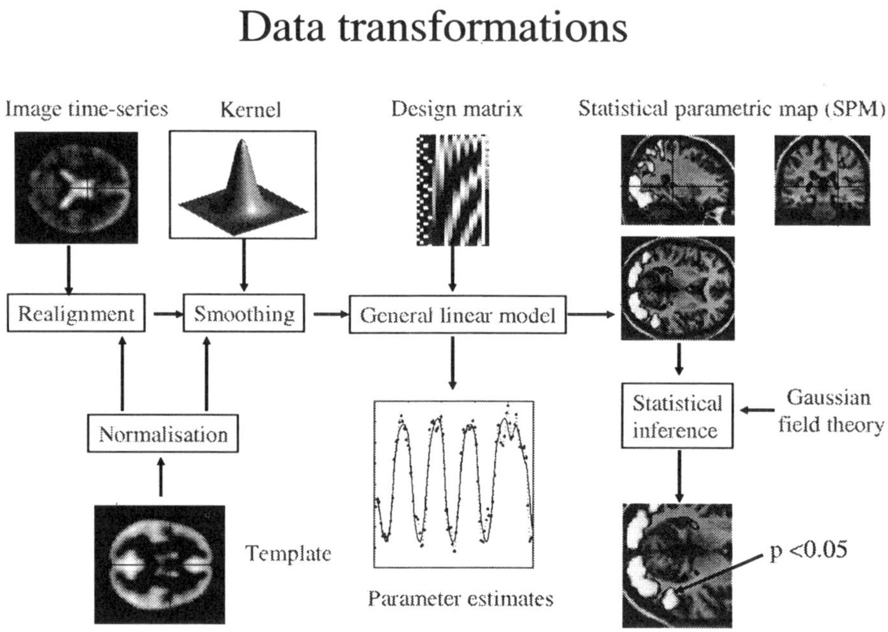

Figure 1.

This schematic depicts the transformations that start with the imaging time-series and end with a statistical parametric map (SPM). SPMs that can be thought of as 'X-rays' of the significance of an effect. Voxel-based analyses require the data to be in the same anatomical space: Realigning the data to remove the effects of subject movement effects this. After realignment the images are subject to non-linear warping so that they match a template that already conforms to a standard anatomical space. After smoothing, the general linear model is employed to (i) estimate the parameters of the model and (ii) derive the appropriate univariate test statistic at each voxel. The test statistics that ensue (usually T or F statistics) constitute the SPM. The final stage is to make statistical inferences on the basis of the SPM and characterise the responses observed using the fitted responses or parameter estimates. The display format in this figure is a maximum intensity projection (MIP) of the SPM{T} from three views of the brain. The anatomical space conforms to that of the Talairach and Tournoux atlas. This MIP display format is often replaced with a thresholded SPM, superimposed on slices through the brain depicting anatomical structures. This is sometimes preferred when the structure-function relationship is important and can be summarised by reporting a single slice of the SPM.

3. STATISTICAL PARAMETRIC MAPPING

Functional mapping studies are usually analysed with some form of statistical parametric mapping. Statistical parametric mapping refers to the construction of statistical maps to test hypotheses about regionally-specific effects like 'activations' (Friston *et al* 1995). Statistical parametric maps (SPMs) are image processes with voxel values that are, under the null hypothesis, distributed according to a known probability density function (usually Student's T or F distributions). The success of statistical parametric mapping is largely due to the simplicity of the idea. Namely, one analyses each and every voxel using any standard (univariate) statistical test. The resulting statistical parameters are assembled into an image - the SPM. SPM{T} refers to a SPM comprising T statistics, similarly SPM{F} denotes a SPM of F statistics. SPMs are interpreted as spatially extended statistical processes by referring to the probabilistic behaviour of stationary Gaussian fields (Friston *et al* 1995, Worsley *et al* 1996). Stationary fields model both the univariate probabilistic characteristics of a SPM and any spatial covariance structure. 'Unlikely' excursions of the SPM are interpreted as regionally-specific effects, attributable to the sensorimotor or cognitive process that has been manipulated experimentally. Statistical analysis corresponds to modelling the data in order to partition observed neurophysiological responses into components of interest, confounds and error. Almost universally, the models employed are general linear models.

3.1 The general linear model

The general linear model is an equation $y = X\beta + \varepsilon$ that expresses the observed *response variable* y in terms of a linear combination of explanatory variables X plus a well-behaved error term (see Friston *et al* 1995). Examples of the general linear model are variously known as 'analyses of covariance' or 'multiple regression analyses' and subsume simpler variants, like the 'two-sample T test' for a difference in means, to more elaborate linear convolution models such as finite impulse response (FIR) models for time series. The matrix X that contains the explanatory variables (*e.g.* designed effects or confounds) is called the 'design matrix'. Each column of the design matrix corresponds to some effect one has built into the experiment or that may confound the results. These are variously referred to as *explanatory variables*, covariates, regressors or, in fMRI, stimulus functions. The example in Figure 1 relates to an fMRI study of visual stimulation under four conditions. The effects on the response variable are modelled in terms of functions of the presence of these conditions (*i.e.* boxcars smoothed with a hemodynamic response function) and constitute the first four columns of the design matrix. There then follows a series of terms that are designed to remove or model low frequency variations in signal due to artefacts such as aliased biorhythms. The final column is whole brain activity. The relative contribution of each of these columns is assessed using generalised least square estimates. These are the estimates of β the parameters of the model. Inferences about parameter estimates are made using T or F statistics, depending upon whether one is looking at a particular linear combination, or several (see below).

The error or noise in neuroimaging has a number of sources and is generally about the same size as the evoked responses themselves. Scan to scan error is due to technical measurement noise and neurophysiological noise, inherent in ongoing brain activity. In multi-subject designs, between-subject variability in responses is another important error variance component, which can sometimes be greater than within-subject or scan to scan variability. In fMRI, aliased biorhythms and drifts in the measurement device induce correlations among the errors. This is known as *non-sphericity* and is accommodated in *generalised* linear models (*e.g.* Worsley and Friston 1995). The non-sphericity corrections, used to deal with temporal correlations in fMRI, are same as those adopted in conventional analysis of variance (*i.e.* those based on the Satterthwaite approximation such as Greenhouse-Geisser).

Generalised linear models can be used to implement a vast range of statistical analyses. The issue is therefore not so much the mathematics but the formulation of a design matrix X appropriate to the study design and inferences that are sought. The design matrix can contain both covariates and indicator variables. Each column of X has an associated unknown parameter. Some of these parameters will be of interest (*e.g.* the effect of particular sensorimotor or cognitive condition or the regression coefficient of hemodynamic responses on reaction time). The remaining parameters will be of no interest and pertain to nuisance or confounding effects (*e.g.* the effect of being a particular subject or the regression slope of voxel activity on global activity). Inferences about the parameter estimates are made using their standard error. This allows one to test the null hypothesis that all the estimates are zero using the F statistic to give an SPM{F} or that some particular linear combination (*e.g.* a subtraction) of the estimates is zero using a SPM{T}. The T statistic obtains by dividing a *contrast* or linear combination (specified by contrast weights) of the ensuing parameter estimates by its standard error. The latter is estimated using the variance of the residuals about the least-squares fit. An example of contrast weights would be [-1 1 0 0.....] to compare the differential responses evoked by the two conditions, modelled by the first two condition-specific regressors in the design matrix. In summary, the T statistic can be regarded as the size of a response, or activation estimate, divided by its expected variability or standard error. The square of this statistic is the ratio of the sum of squares or variance of interest divided by error variance. This is the F statistic (for a single contrast), which can also be computed for several linear combinations or *contrasts* of estimated responses at once. This is useful when several parameter estimates are potentially interesting (*e.g.* using polynomial expansions (Büchel *et al* 1996) or basis functions of some parameter of interest).

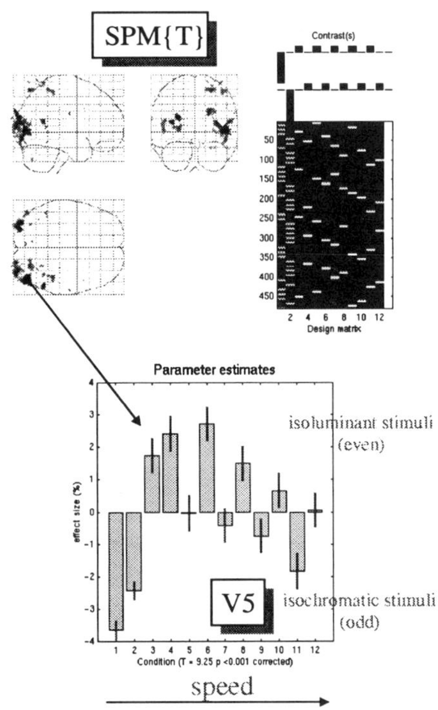

Figure 2. Top right: Design matrix: This is an image representation of the design matrix. Contrasts: These are the vectors of contrast weights defining the linear compounds of parameters tested. The contrast weights are displayed over the column of the design matrix that corresponds to the effects in question. The design matrix here includes condition-specific effects (boxcars convolved with a hemodynamic response function). Odd columns correspond to stimuli shown under isochromatic conditions and even columns model responses to isoluminant stimuli. The first two columns are for stationary stimuli and the remaining columns are for conditions of increasing speed. The final column is a constant term. Top left: SPM{T}: This is a maximum intensity projection of the SPM{T}

conforming to the standard anatomical space of Talairach and Tournoux. The T values here are the minimum T values from both contrasts, thresholded at p = 0.001 uncorrected. The most significant conjunction is seen in left V5. Lower panel: Plot of the condition-specific parameter estimates for this voxel. The T value was 9.25 ($p<0.001$ corrected).

3.2 Statistical Inference and Gaussian fields

Inferences using SPMs can be of two sorts depending on whether one knows where to look in advance: With an anatomically constrained hypothesis, about effects in a particular brain region, the *uncorrected* p value associated with the height or extent of that region in the SPM can be used to test the hypothesis. With an anatomically open hypothesis (*i.e.* a null hypothesis that there is no effect anywhere in the brain) a correction for multiple dependent comparisons is necessary. The theory of Gaussian fields provides a way of computing this *corrected* p-value that takes into account the fact that neighbouring voxels are not independent by virtue of smoothness in the original data. This *Gaussian Field Correction* (GFC) plays the same role for continuous, multidimensional statistical processes like SPM{T} as the *Bonferroni correction* plays in multiple discrete T tests. Provided the data are sufficiently smooth the correction based on Gaussian field theory is much less severe (*i.e.* is more sensitive) than a Bonferroni correction for the number of voxels. This is simply because the GFC ensures the rate of false positive 'clusters' is maintained as a nominal level (usually 0.05). The rate of false positive voxels can therefore be much larger, *e.g.* 0.05 times the expected number of voxels per cluster. The difference between a Gaussian field and a Bonferroni correction increases with smoothness, or expected voxels per cluster.

4. EXPERIMENTAL DESIGN

4.1 Categorical designs, cognitive subtraction and conjunctions

The tenet of cognitive subtraction is that the difference between two tasks can be formulated as a separable cognitive or sensorimotor component and that regionally-specific differences in hemodynamic responses identify the corresponding functionally specialised area. Early applications of subtraction range from the functional anatomy of word processing (Petersen *et al* 1989) to functional specialisation in extrastriate cortex (Lueck *et al*

1989). The latter studies involved presenting visual stimuli with and without some sensory attribute (e.g. colour, motion *etc*). The areas highlighted by subtraction were identified with homologous areas in monkeys that showed selective electrophysiological responses to equivalent visual stimuli.

Cognitive conjunctions (Price and Friston 1997) can be thought of as an extension of the subtraction technique, in the sense that they combine a series of subtractions. In subtraction one tests a hypothesis pertaining to the activation in one task relative to another. In conjunction analyses several hypotheses are tested conjointly, asking whether all the activations, in a series of task pairs, are jointly significant. Conjunction analyses allow one to demonstrate any context-insensitive component of regional responses.

4.2 Parametric designs

The premise behind parametric designs is that regional physiology will vary systematically with the degree of cognitive or sensorimotor processing or deficits thereof. Examples of this approach include the PET experiments of Grafton *et al* (1992) that demonstrated significant correlations between hemodynamic responses and the performance of a visually guided motor tracking task. These relationships or 'neurometric functions' may be linear or nonlinear. Using polynomial regression, in the context of the general linear model, one can identify nonlinear relationships between stimulus parameters (*e.g.* stimulus duration or presentation rate) and evoked responses. To do this one usually uses a SPM{F} (Büchel *et al* 1996).

The example provided in Figure 2 illustrates subtraction, conjunction and parametric approaches to design and analysis. These data were obtained from an fMRI study of visual motion processing using radially moving dots. The stimuli were presented over a range of speeds using isoluminant and isochromatic stimuli. To identify areas involved in visual motion a stationary dots condition was subtracted from the moving dots conditions (see contrast weights on the upper right). To ensure significant motion-sensitive responses, using both colour and luminance cues, a conjunction of the equivalent subtractions was assessed under both viewing contexts. Areas V5 and V3a are seen in the SPM{T}. The responses in left V5 (lower panel) speak to a compelling inverted 'U' relationship between speed and evoked response that peaks at around 8 degrees per second. It is this sort of relationship that parametric designs try to characterise. Interestingly the form of these speed-dependent responses was similar using both stimulus types, although luminance cues are seen to elicit a greater response. From the point of view of a factorial design there is a *main effect* of cue (isoluminant *vs.* isochromatic), a main [nonlinear] effect of speed but no speed by cue *interaction.*

4.3 Factorial designs

At its simplest an interaction represents a change in a change. Interactions are associated with factorial designs where two or more factors are combined in the same experiment. The effect of one factor, on the effect of the other, is assessed by the interaction term. Factorial designs have a wide range of applications (*e.g.* assessing physiological adaptation with time by condition interactions). Psychopharmacological activation studies are examples of factorial designs. In these studies cognitively evoked responses are assessed before and after being given a drug. The interaction term reflects the modulatory drug effect on the task-dependent activation. Factorial designs have an important role in the context of cognitive subtraction and additive factors logic by testing for context-sensitive activations, to demonstrate the fallacy of 'pure insertion'. Pure insertion refers to the assumption that the difference between one task and another is completely specified by the difference in task demands. In other words, the difference is inserted without *interacting* with the context established by the extant task components. See Friston *et al* 1996a. These interaction effects can sometimes be interpreted as (i) the integration of the two or more [cognitive] processes or (ii) the modulation of one process by another.

5. EFFECTIVE CONNECTIVITY AND FUNCTIONAL INTEGRATION

There is a necessary relationship between functional integration and multivariate analyses because the latter model interactions among brain regions. Multivariate approaches can be divided into those that are inferential in nature and those that are data-led or exploratory. Most are based on the singular value or eigen decomposition of the between-voxel covariances in a neuroimaging time-series.

5.1 Eigenimage analysis and related approaches

Friston *et al* (1993) introduced voxel-based principal component analysis (PCA) of neuroimaging data-sequences to characterise distributed brain systems implicated in sensorimotor, perceptual or cognitive processes. These distributed systems are identified with principal components or eigenimages that correspond to spatial modes of coherent brain activity. This approach represents one of the simplest multivariate characterisations of functional neuroimaging time-series and falls into the class of exploratory analyses. Eigenimage analysis generally uses singular value decomposition (SVD) to identify a set of orthogonal spatial modes that capture the greatest amount of variance, expressed over time. As such the ensuing modes embody the most prominent aspects of the variance-covariance structure of a given time-series. Noting that the covariances among brain regions is equivalent to functional connectivity (see below) renders eigenimage analysis particularly interesting because it was among the first ways of addressing functional integration in the human brain.

5.2 Functional and Effective connectivity

Imaging neuroscience has firmly established functional specialisation as a principle of brain organisation in man. The functional integration of specialised areas has proven more difficult to assess. Functional integration is usually inferred on the basis of correlations among measurements of neuronal activity. Functional connectivity has been defined as *correlations between remote neurophysiological events*. However, correlations can arise in a variety of ways: For example in multi-unit electrode recordings they can result from stimulus-locked transients evoked by a common input or reflect stimulus-induced oscillations mediated by synaptic connections (Gerstein and Perkel 1969). Integration within a distributed system is usually better understood in terms of effective connectivity: Effective connectivity refers explicitly to *the influence that one neural system exerts over another*, either at a synaptic (*i.e.* synaptic efficacy) or population level. Effective connectivity is dynamic, *i.e.* activity- and time-dependent and it depends upon a model of the interactions. The key thing here is that effective connectivity may vary from task to task (see below) even though anatomical connections remain the same. The models employed in functional neuroimaging can be classified as (i) those based on regression models (see next section) or (ii) structural equation modelling (McIntosh *et al* 1994).

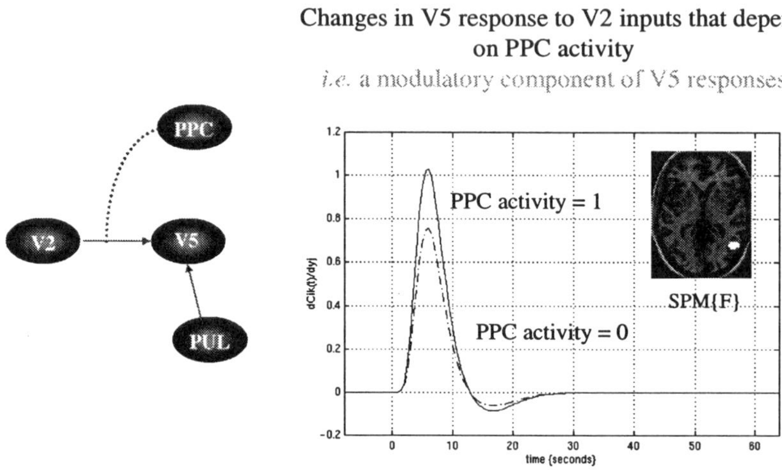

Figure 3. Characterisation of effects of V2 inputs on V5 and the modulation of these effects by posterior parietal cortex (PPC). The broken lines represent estimates of V5 responses when PPC activity = 0 according to a second order Volterra model with inputs based on the activity in V2, PPC and the pulvinar (all normalised to zero mean and unit standard deviation). The input, from V2, corresponded to a square wave of 500ms duration convolved with a hemodynamic response function. The solid curves represent the same response when PPC activity is one. It is evident that V2 has an activating effect on V5 and that PPC increases the responsiveness of V5 to these inputs. The insert shows all the voxels in V5 that evidenced a modulatory effect ($p < 0.05$ uncorrected). These voxels were identified by thresholding SPMs of the F statistic testing for the contribution of second order effects involving V2 and PPC while treating all other components as confounds. Results for one subject are shown. Subjects were studied with fMRI under identical stimulus conditions (visual motion subtended by radially moving dots) whilst manipulating the attentional component of the task (detection of velocity changes).

5.3 Characterising nonlinear coupling among brain areas

Linear models of effective connectivity assume that the multiple inputs to a region are linearly separable. This precludes activity-dependent connections that are expressed in one sensorimotor or cognitive context and not in another. The resolution of this problem lies in adopting nonlinear models that include interactions among inputs. These interactions can be construed as a context- or activity-dependent modulation of the influence that one region exerts over another, where that context is instantiated by activity in other brain regions exerting modulatory effects. These nonlinearities are important in modelling and understanding task-specific changes in effective connectivity mentioned above. From the point of view of regression models, modulatory effects can be modelled with nonlinear input-output models and in particular a Volterra-series formulation (this formulation can be thought of as a polynomial regression on the activity of areas providing inputs, both now and in the recent past). Because the kernels are high-order they embody interactions over time and among inputs. The influence of one region j on another i can therefore be divided into two components; (i) the direct influence of j on i, irrespective of activity elsewhere and (ii) an activity-dependent component that represents an interaction with inputs from the remaining regions. The example provided in Figure 3 addresses the modulation of visual cortical responses by attentional mechanisms and the mediating role of activity- or context-sensitive changes in effective connectivity. Figure 3 shows a characterisation of this modulatory effect in terms of the increase in V5 responses, to V2 input, when posterior parietal activity is zero (broken line) and when it is high (solid lines). This sort of result suggests that parietal activity may be a sufficient explanation for attentional modulation of visually evoked extrastriate responses.

6. DISCUSSION

In summary, functionally specialised brain areas can be identified by some form of statistical parametric mapping. The nature of the specialisation and the interactions between various cognitive or sensorimotor factors in engendering a brain response can be established using simple subtraction, parametric or multifactorial designs. Functional integration is usually assessed using some form of multivariate analysis framed in terms of effective connectivity. Recent advances in modelling effective connectivity emphasise nonlinear interactions among inputs to an area in 'causing' its response.

REFERENCES

Büchel, C., Wise, R.J.S., Mummery, CJ, Poline, J-B. and Friston, KJ, (1996) Nonlinear regression in parametric activation studies. *NeuroImage*, 4, 60-66.

Frackowiak RSJ, Friston KJ, Frith CD, Dolan RJ and Mazziotta JC, (1997) *Human Brain Function.* Academic Press, USA.

Friston, KJ, Frith, C., Liddle, P., and Frackowiak, R.S.J., (1993) Functional Connectivity: The principal component analysis of large (PET) data sets. *J Cereb. Blood Flow Metab.* 13, 5-14.

Friston, KJ, Holmes, AP, Worsley, KJ, Poline, J-B. Frith, C.D., and Frackowiak, R.. SJ. 1995. Statistical Parametric Maps in functional imaging: A general linear approach. *Hum. Brain Mapp.* 2, 189-210.

Friston, KJ, Price, CJ, Fletcher, P., Moore, C., Frackowiak, R.S.J., and Dolan, RJ. (1996) The trouble with cognitive subtraction. *NeuroImage* 4, 97-104.

Gerstein, GL, and Perkel, DH, (1969) Simultaneously recorded trains of action potentials: Analysis and functional interpretation. *Science,* 164, 828-830.

Grafton, S., Mazziotta, J., Presty, S., Friston, KJ, Frackowiak, R.S.J., and Phelps, M., (1992) Functional anatomy of human procedural learning determined with regional cerebral blood flow and PET. *J. Neuroscience,* 12, 2542-2548.

Lueck, CJ, Zeki. S., Friston, KJ, Deiber, MP, Cope, N.O., Cunningham, V.J., Lammertsma, AA, Kennard, C., and Frackowiak, R.S.J., (1989) The colour centre in the cerebral cortex of man. *Nature,* 340, 386-389.

McIntosh, A.R., and Gonzalez-Lima, F., (1994) Structural equation modelling and its application to network analysis in functional brain imaging. *Hum. Brain Mapp.* 2, 2-22.

Petersen, SE, Fox, PT, Posner, MI, Mintun, M., and Raichle, M.E., (1989) Positron emission tomographic studies of the processing of single words. *J. Cog. Neurosci.* 1, 153-170.

Phillips, C.G., Zeki, S., and Barlow, H.B., (1984) Localisation of function in the cerebral cortex: Past present and future. *Brain,* 107, 327-361.

Price, CJ, and Friston, KJ, (1997) Cognitive Conjunction: A new approach to brain activation experiments. *NeuroImage,* 5, 261-270.

Worsley KJ and Friston KJ. (1995) Analysis of fMRI time-series revisited - again. *NeuroImage* 2, 173-181.

Worsley, KJ, Marrett, S., Neelin, P., Vandal, AC, Friston, KJ, and Evans, AC, (1996) A unified statistical approach or determining significant signals in images of cerebral activation. *Hum. Brain Mapp.* 4, 58-73.

Zeki, S. (1990) The motion pathways of the visual cortex. In Vision: coding and efficiency (C Blakemore ed.) Cambridge University Press UK. p321-345.

Chapter 17

The Brain Positioning Software

Volker Schmitt[1], Andrzej Wichert[2], Jo Grothe[2] and Friedrich T. Sommer[1]
[1]Department of Neural Information Processing, University of Ulm, D-89069 Ulm, Germany
[2]Department of Psychiatry III, University of Ulm, D-89075 Ulm, Germany
Correspondence to: Volker.Schmitt@medizin.uni-ulm.de

Abstract: In the assessment and analysis of neuroimaging data often the automated reference to a brain atlas is required. We describe a little software tool, the brain positioning software (BPS), that relies on computerized brain atlases (in Talairach coordinates, by default) and has interfaces to the SPM data analysis software (freely available on http://www.fil.ion.ucl.ac.uk/spm/distrib.html). BPS allows the transformation in two directions: It can project and visualize areas of atlas regions in SPM. When called from SPM, it shows the name of the region under the current cursor position. The default version of the BPS tool uses the Talairach atlas made available on the "Talairach Daemon" website (http://ric.uthscsa.edu/projects/). Other atlases can also be used. For fast performance, the atlas database is converted and label descriptions are managed in a hash-table. The BPS can be downloaded from our website on http://www.informatik.uni-ulm.de/ni/staff/FSommer/BPS.html.

Key words: brain, Talairach, atlas, Brodmann area, SPM, MNI, template

1. INTRODUCTION

In neuroimaging the automated reference to brain atlases is becoming more and more important for reporting results like the spatial distribution of brain lesions or functional activity. Every individual brain has its own size, shape and even a unique configuration of gyri and sulci. Therefore, to locate a site in the individual brain requires a match to the brain on which the atlas is defined. Brodmann classified brain regions on the basis of cytoarchitecture over hundred years ago (Brodmann, 1909). The Brodmann atlas is still

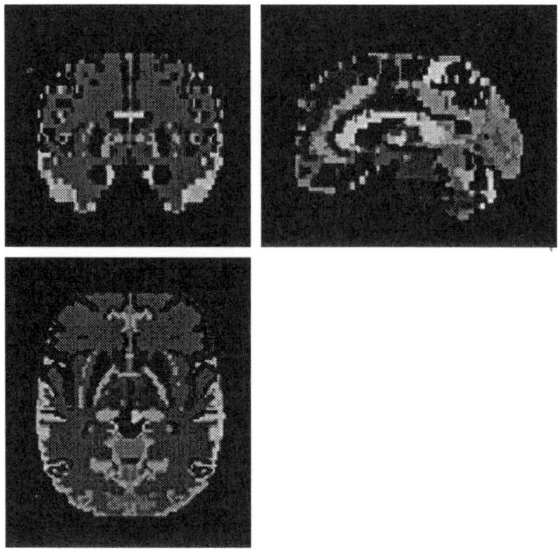

Figure 1. Frontal, sagital and transversal view of the Talairach brain.

commonly referred. Detailed annotation in terms of Brodmann areas is available in the brain coordinate system introduced by Talairach and Tournoux, in the so-called Talairach atlas (Talairach and Tournoux, 1988). SPM, the most commonly used software for data analysis in neuroimaging can normalize individual subject anatomy to the MNI standard template (Montreal Neurological Institute). In the coordinate system of the MNI template, however, the Brodmann atlas is still unavailable. The brain positioning software described below can be called directly from the SPM data analysis software. It gives access to the Brodmann atlas. The atlas data used by default in BPS are retrieved from the so-called Talairach Daemon available under the URL http://ric.uthscsa.edu/projects/ (Lancester et al., 1997). However, the BPS tool can also be used in combination with other brain atlases.

2. MATERIAL

The Talairach and Tournoux atlas has been defined on the brain from an individual subject. The brain was dissected and photographed. The atlas was digitized and divided into anatomical regions that constitute the Talairach atlas. The atlas partitions the brain volume at five different levels of detail: 1) hemispheres, 2) lobes, 3) gyri, 4) white/gray matter and 5) Brodmann areas (level order is not given hierarchically). Brodmann areas were assigned by morphological comparison of the atlas of Brodmann and pictures of the Talairach brain. All Brodmann areas of the cerebral cortex were labeled and expanded into three dimensional space. The Talairach atlas uses stereotaxic coordinates. The origin of Talairach space is located in the anterior commissure. The line from anterior commissure to the center of the posterior commissure is called the AC-PC line. This line defines the plane where Z=0. X-, Y-, and Z-coordinates refer to left-right, posterior-anterior and ventral-dorsal directions. Figure 1 shows a frontal, sagittal and transversal view of the Talairach brain visualized in SPM.

We retrieved the Talairach atlas data as a 140*160*105 voxel volume from the University of Texas Health Science Center at San Antonio database server. In this data set X-, Y- and Z-coordinates are resolved to 1*1*1 mm volume elements within a stereotaxic space. Atlas data consists of an ASCII text file in tabular form. Each line contains 8 columns with coordinates (X, Y, Z) and 5 descriptors corresponding to atlas information. Descriptors are text strings of region names. The following table shows two typical lines of the text.

X	Y	Z	level1	level2	level3	Level4	level5
-2	-54	39	Left Cerebrum	Parietal Lobe	Precuneus	Gray Matter	Brodmann area 7
-1	-54	39	Left Cerebrum	Parietal Lobe	Precuneus	*	*

If there is no information available for a voxel, "level1" description is "none" and following columns are kept empty. A "*" marks no further information.

The SPM software uses standard templates from the Montreal Neurological Institute. The aim of the MNI was to define a brain that is more representative for an average subject than the Talairach brain. 241 brains were scanned and various landmarks were manually defined to identify an origin and the edges of the brain. Thereafter each brain was transformed linearly (translation, rotation, zooms, no warping) to match the landmarks to equivalent positions on the Talairach atlas. In a second stage MNI took 305 normal MRI scans and matched them to the average of the previously transformed scans. This template is called MNI305. SPM96 uses colin27 as its standard template. This is a very highly resolved MRI brain dataset of Colin Holmes, a MNI lab member, who was scanned 27 times. All scans of Colin Holmes were coregistered and matched to MNI305. The current standard MNI template used in SPM99 is ICBM152. It is the average of 152 normal MRI scans that have been matched to the MNI305 by an affine transform. Some regions of the MNI standard brains are slightly larger than the same regions of the Talairach brain. The further one moves from the origin of the coordinate system, the larger the deviations become.

3. METHODS

To compensate for the difference between Talairach atlas coordinates and coordinates of scans coregistered to SPM's standard template we use an approach proposed by Matthew Brett (Brett, 2002) for the conversion of MNI coordinates into Talairach coordinates. This piecewise-linear transform uses different zooms in the Z direction to get a good match between MNI and Talairach spaces. The following linear transformations provide estimate of Talairach coordinates for given MNI coordinates above the AC-PC line:

$X' = 0.9900*X$

$Y' = 0.9688*Y + 0.0460*Z$

$Z' = -0.0485*Y + 0.9189*Z$

Below the AC-PC line transformation for X is the same whereas the Y and Z transforms differ:

$X' = 0.9900*X$

$Y' = 0.9688*Y + 0.0420*Z$

$Z' = -0.0485*Y + 0.8390*Z$

To access the brain atlas from within SPM we converted the Talairach atlas into 3-dimensional SPM image files using the piecewise-linear transform. Color values were used for coding label descriptors of each voxel in MNI space. Color values and descriptors are stored in a table to assign corresponding region names for given colors and vice versa. We generated 5 image files, one for each level of detail. The images and the described table are used to look up Talairach data out of SPM in MNI/SPM voxel coordinates.

The data set retrieved from the Talairach Daemon database server is a tabular text file containing more than 116 MB of data with explicit coordinates. BPS contains two tools to reduce the data file site to a degree that it can be held at once in computer memory: “makeLevelHash” and “TCtxt2TCarr”. The tool “makeTCStd” finally generates the required SPM images.

3.1 Tool 1: makeLevelHash

The main factor of data compression is to manage the relations between descriptors and color values in a hash-table. Descriptors are hashed into the table with the tool “makeLevelHash”. Thereafter each descriptor is assigned to a color value by editing the hash-table manually.

3.2 Tool 2: TCtxt2TCarr

The tool "TCtxt2TCarr" converts the raw atlas data into a smaller file using the edited hash-table. In this file coordinates are no longer given explicitly. They are stored implicitly in the range from X=-70, Y=-100, Z=-40 to X=70, Y=60, Z=65. Voxels can belong to 5 different levels at maximum depending on the information available. Therefore lines of the converted Talairach file contain 5 bytes coding descriptions for each level. If there is no information available a "0" is assigned.

3.3 Tool 3: makeTCStd

For generating Tailairach image files that match the standard template of SPM (as mentioned previously the match is not perfect, but sufficient), there is a final tool called "makeTCStd". This program sets colors for all voxels or voxels of a certain atlas region in SPM image space (X*Y*Z=91*109*91) by transforming coordinates to Talairach space and searching for description/color codes in our less space consuming Talairach database.

The interface of BPS to the SPM software is written in Matlab®, all other programms are written in C.

4. USING THE TOOLS OF THE BPS PACKAGE

The BPS package consists of several tools to convert a Talairach database into an efficient format, to create a hash-table with Talairach region names, and to create SPM image files and query Talairach region names for current SPM's cursor position.

4.1 Usage of BPS from within SPM

BPS can create image files to visualize areas of atlas regions in SPM and to project them on imaging data. The Talairach atlas is used by the BPS software by default.

When called from within SPM, BPS accesses the brain atlas information of the location marked by the current cursor position. The tool used for this purpose is "tal", which requires atlas image files, the hash-table and several supplementary programs. We expected the data from the Talairach Daemon (http://ric.uthscsa.edu/projects/) to be completely symmetric, with respect to the X=Z=0 plane. However, there are some little deviations in the data. Therefore we decided to write two variants of "tal" - "talp" and "taln" - projecting one hemisphere onto the other. The tools can be invoked by typing the program names in the Matlab® window of SPM. The output is written to this window. All levels of detail are displayed. The "tal", "talp" and "taln" programs use five image files containing region descriptors of one level of atlas information each. These were created with "makeTCStd" as described in the following section. A hash-table created with "makeLevelHash" is used to decode color values into readable region descriptions.

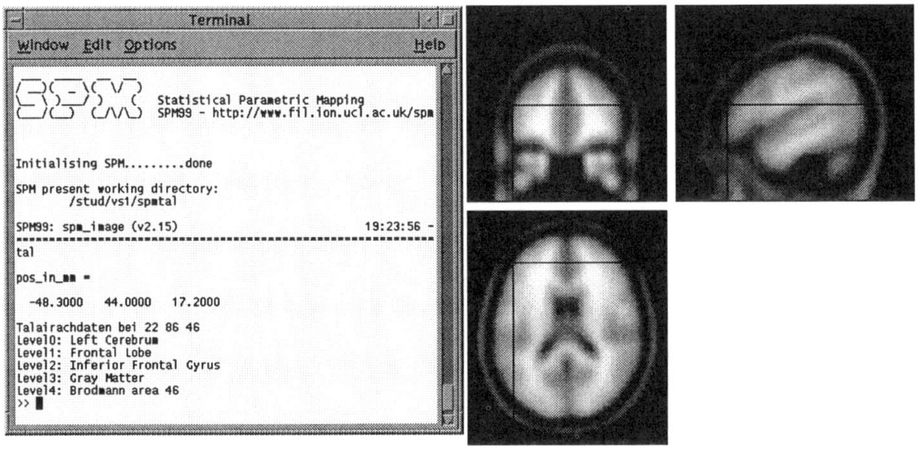

Figure 2. Talairach brain atlas information of a voxel of the SPM t1.img.

4.2 Using brain atlases other than Talairach

By default, BPS uses the Talairach atlas, the atlas information is already provided with the software. Other atlases can also be used after the appropriate SPM images have been generated with the atlas information. The raw information of the atlas has to be provided in a text file of the format described in section 2. The routine "makeLevelHash" creates a hash-table containing region names. The syntax is: makeLevelHash atlas hash. The input file is "atlas" and the output is written to "hash". Thereafter a value representing a color has to be assigned to each region manually using a standard text editor. The field separator is "#", followed by an integer. One is free to choose arbitrary integers in the range from 1 to 255. The atlas file is converted using "TCtxt2TCarr". Syntax: TCtxt2TCarr atlas hash atlas_conv. Two input files are needed - the original brain atlas and the hash-table. The output is written to "atlas_conv" which is used to finally create SPM images with "makeTCStd". Syntax: makeTCStd atlas_conv image. "makeTCStd" asks for further information. The dimensions of a standard SPM image are x=91, y=109, z=91 and the origin is placed in X=46, Y=64 and Z=37. In one image file only one description per voxel is stored. You specify which level should be stored by specifying the minimal and maximal level used. The tool saves descriptors of the highest level specified. If there is no information of this level available then the highest level containing information in level range is used. For generating images containing just one level, minimum and maximum values have to be the same. One can create images containing all or only specified atlas regions with "makeTCStd" (see figure 3 for an image of a specified atlas region).

SPM needs a header file for the image generated. There will be no image header created. You have to copy a suitable image header of another SPM image and rename it or a header can be created (and can be edited) using the 'Display' facility. For more information see "The SPM/Analyze data format" on http://www.mrc-cbu.cam.ac.uk/Imaging/spm_format.html (Ashburner, 2002).

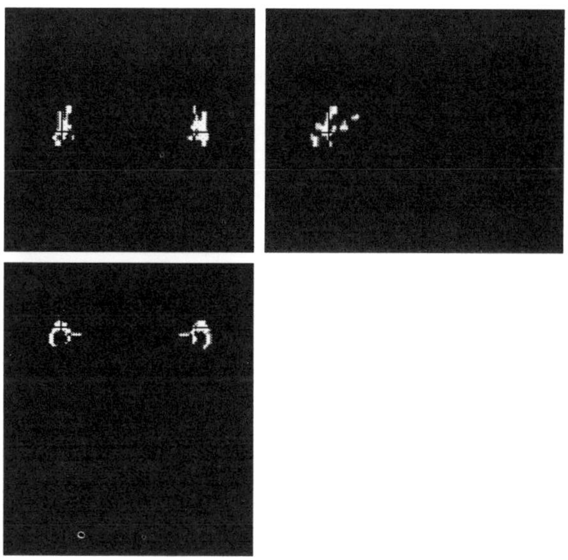

Figure 3. Image of the Brodmann area 46 created with makeTCStd (cursor position is X=-48.3, Y=44.0, Z=17.2).

4.3 Limitations

BPS was written using Matlab® and the GNU C-Compiler on Linux machines. Some of our C programs allocate large arrays of memory to handle atlas data. On operating systems other than Linux (e. g. Solaris) there might be problems with memory allocation.

5. CONCLUSION

We described a software package to visualize and retrieve automated references to brain atlas regions by SPM. The package is freely available on http://www.informatik.uni-ulm.de/ni/staff/FSommer/BPS.html. In the software the Talairach atlas is provided by default. Other computerized atlases can be used with little effort. The described tools can be used to identify brain regions and to define anatomical regions of interest.

REFERENCES

Ashburner, J. (28/01/2002) The SPM/Analyze data format. http://www.mrc-cbu.cam.ac.uk/Imaging/spm_format.html.

Brett, M. (updated 14/02/2002) The MNI Brain and the Talairach atlas. http://www.mrc-cbu.cam.ac.uk/Imaging/mnispace.html.

Brodmann, K. (1909) Vergleichende Lokalisationslehre der Großhirnrinde in ihren Prinzipien dargestellt aufgrund des Zellbaus. Johann Ambrosius Bart, Leipzig.

Lancaster, J. L., Summerlin, J. L., Rainey, L., Freitas C. S. and Fox, P. T. (1997) The Talairach Daemon, a database server for Talairach Atlas Labels. *Neuroimage* 5(4), S633.

Talairach, J. and Tournoux, P. (1988) Co-planar Stereotaxic Atlas of the Human Brain: 3-Dimensional Proportional System - an Approach to Cerebral Imaging. Thieme Medical Publishers, New York.

Chapter 18

BrainInfo

An Online Interactive Brain Atlas and Nomenclature

Douglas M. Bowden and Mark Dubach
Department of Psychiatry and Behavioral Sciences and Washington National Primate Research Center, University of Washington, Seattle WA, USA.
Correspondence to: dmbowden@u.washington.edu

Abstract: BrainInfo is a website that provides neuroinformatic tools and a brain knowledge base useful to students and teachers of neuroanatomy, neuroscientists, and clinicians concerned with brain disorders. For students of neuroanatomy it provides the web-equivalent of an illustrated dictionary of neuroanatomical terminology for human and nonhuman primates that is accessible by common structure names in several languages. For teachers and textbook authors it incorporates the NeuroNames Brain Hierarchy[1] of structures grouped by proximity in the style of Nomina Anatomica, named in English and extended to include all primary structures of the brain; in addition it provides 25-year PubMed citation counts for all English and Latin terms. For neuroscientists it provides access to stereotaxic brain atlases of several primate species, provides data overlays to a template atlas of the macaque brain showing the cortical areas as defined by a number of authors, indicates topologically homologous structures in the human, macaque and rat brain and provides a standard system of abbreviations for labeling brain structures in figures and web displays. It enables clinicians to compose a query to PubMed on disorders of a particular structure that incorporates all synonyms for the structure into a single query. The website is designed to lead users to the information they need by the shortest possible intuitive navigational route.

Key words: neuroinformatic tools, neuroanatomy, brain hierarchy, dictionary, stereotaxic brain atlases, brain templates, cortical areas, homology

[1] The authors make NeuroNames available for incorporation into computerized databases of information about the brain. NeuroNames is distributed in the form of Microsoft Excel spreadsheets or Microsoft Access database tables on request. All rights are reserved; incorporation of NeuroNames in whole or in part requires written permission of the University of Washington. For more information contact dmbowden@u.washington.edu.

1. TO FIND INFORMATION ABOUT A SPECIFIC BRAIN STRUCTURE

BrainInfo currently enables one to find information on more than 2000 brain structures and combinations of structures using any of more than 11,000 terms in several languages. One searches in either of two ways: 1) by entering a name of the structure, or 2) by clicking it in a brain atlas. The first step is to contact BrainInfo (Figure 1) at its web address: **http://braininfo.rprc.washington.edu**. BrainInfo works best with Internet Explorer on a PC. Most of its features work also with Netscape on a PC; some interactive images fail in Netscape on the Macintosh. BrainInfo requires Java Plug-in 1.3, which, if not already installed, will be offered as a free download from Sun Microsystems when BrainInfo first requires it.

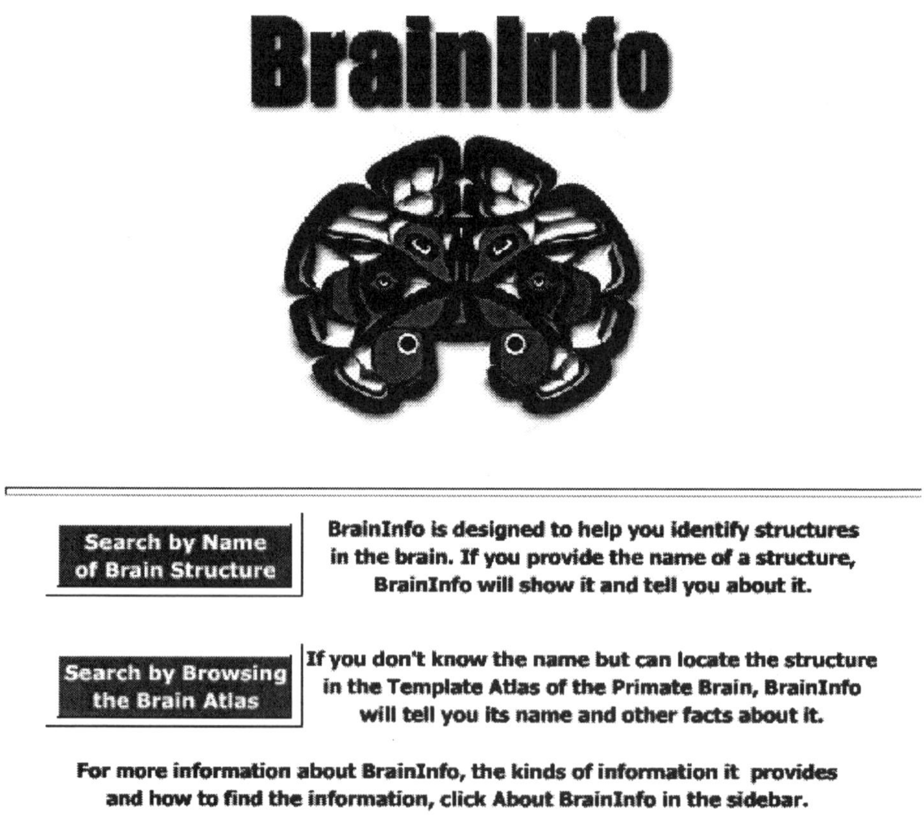

Figure 1. The Home Page of BrainInfo.

1.1 To Search for a Structure by Name

- Click *Search by Name of Brain Structure.*
- If the name is in English or Latin, e.g., nucleus of basis pedunculi, key it into the text box. Click *Submit Query*. If BrainInfo finds an exact match it will tell you BrainInfo's default English name, in this case substantia nigra. Click it. (BrainInfo will also tell you every other neuroanatomical term that contains the character string that you entered; if one of those looks more appropriate, click the default term to its right.)
- If the name is in one of the languages appearing in boxes below the text box, click the box, scroll to the name and click the standard (default) English term for it.

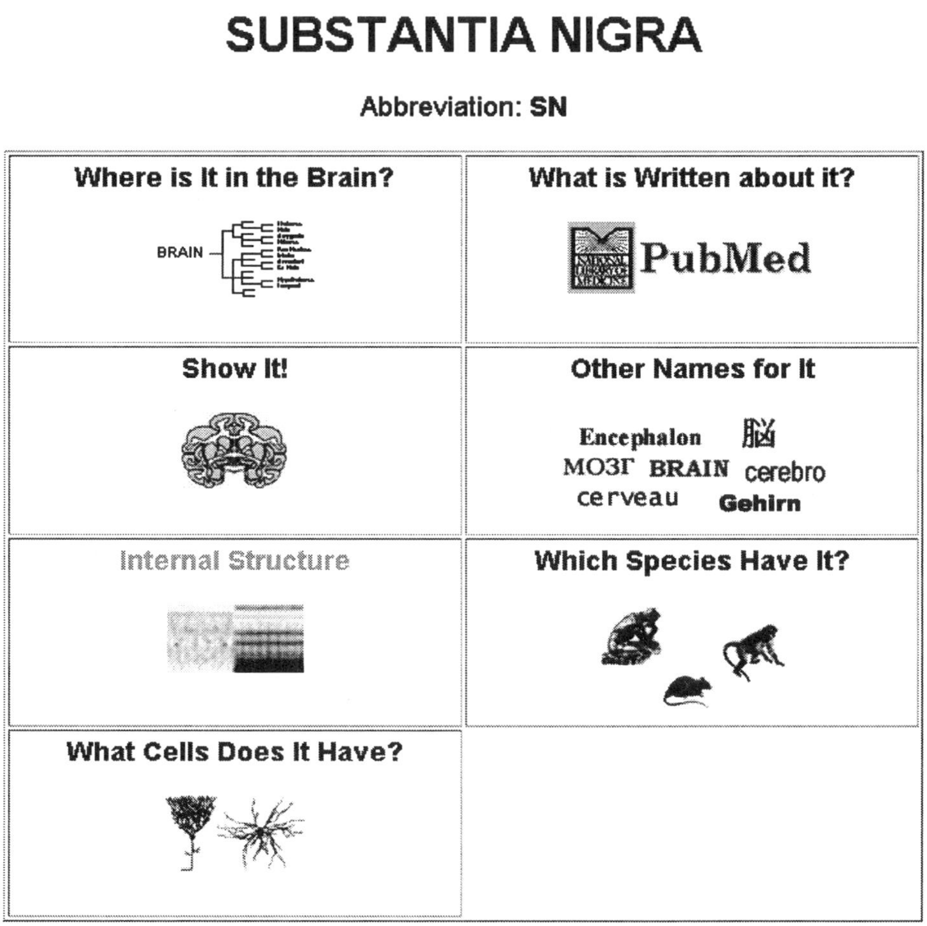

Figure 2. The Central Directory for substantia nigra.

- In either case, Brain Info will take you to the Central Directory for the structure where you can indicate the kind of information that interests you (Figure 2).

1.2 To Search for a Structure by Browsing the Atlas

- Click *Search by Browsing the Brain Atlas.* BrainInfo will display the entrance to the Template Atlas (Figure 3).
- If the structure of interest is a cortical gyrus or sulcus and you can identify it on the brain surface, click one of the four surface views.

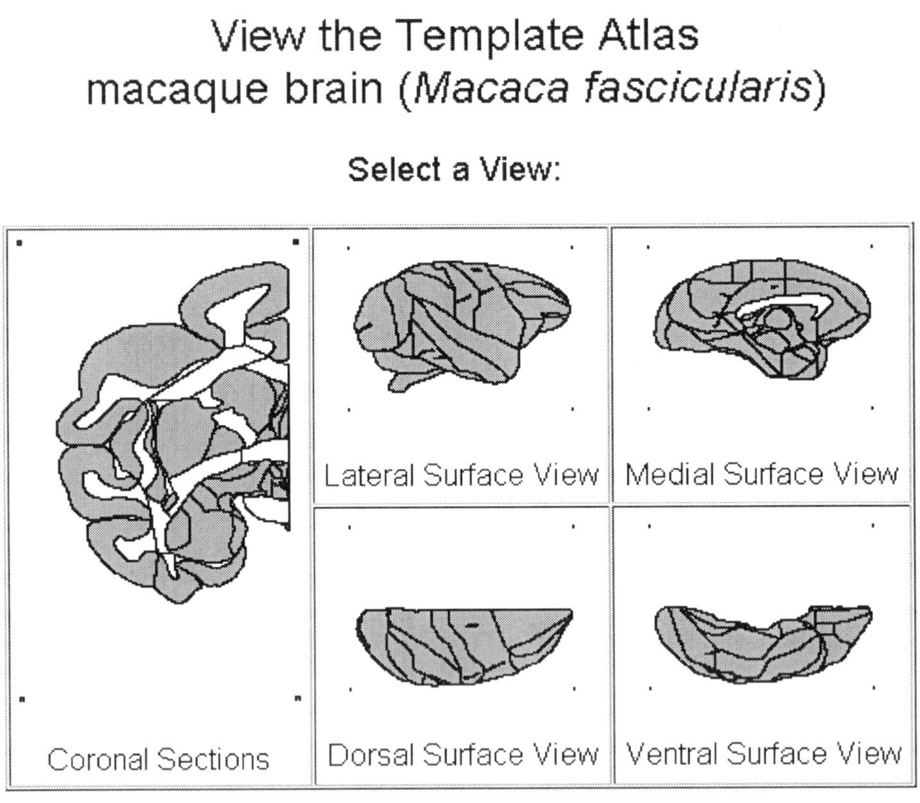

Figure 3. Entrance to the Template Atlas of the Macaque Brain.

- If you can better identify the structure in a cross-section, click the coronal view and click the thumbnail (miniature cross-section) that appears most likely to contain the structure of interest.
- In either case, the selected view appears on the screen.

- If you see the structure of interest, click the Labels box in the gray sidebar; place the cursor over the label that appears on the structure of interest and the full name of the structure will pop up; click the label and Brain Info will take you to the Central Directory for the structure where you can indicate the kind of information that interests you.
- If you do not see the structure of interest in the first view that comes up you can navigate to the appropriate view by several routes: you can go to any other view by clicking the down-arrowhead at the top of the gray sidebar or by clicking Back in your browser's tool bar; from a cross-sectional view you can advance to adjacent sections by clicking *Go Anterior* or *Go Posterior*; once you have found the structure of interest continue according to the preceding instruction.

2. TO DETERMINE THE LOCATION OF A STRUCTURE IN THE BRAIN

The brain is ordinarily divided into three parts: forebrain, midbrain, and hindbrain. These are in turn subdivided into smaller and smaller divisions down to about 600 primary structures. In BrainInfo, the names for all of the parts are organized into a hierarchy with "BRAIN" at the top, the three main subdivisions at the second level, and so on down to the primary structures which may be as many as nine levels deep. To find where in the hierarchy a particular structure name applies:

- Go to the Central Directory for a structure, e.g., pars compacta, using the instructions in Section 1.
- Click *Where is It in the Brain*?
- If the structure is part of the brain as segmented in the NeuroNames Hierarchy, BrainInfo will display the part of the hierarchy where the structure appears. If it is a superstructure, e.g., SUBSTANTIA NIGRA, its name will appear in uppercase at the top of the page; if it is a primary structure, will appear in lower case in the list of substuctures below, in this case, pars compacta and pars reticulata.
- To view higher order structures in the hierarchy, click the left-pointing arrow next to the superstructure; if some of the substructures are themselves further divided, click the right-pointing arrows next to them to view their substructures.
- Click the name of any of the structures listed to go to its Central Directory.
- To view the same part of the hierarchy in Latin, click *Latin* beneath the list of structures. Click *English* to come back.

- If a structure is not part of the brain hierarchy as segmented in the NeuroNames Hierarchy clicking *Where is It in the Brain* takes you to a verbal definition. The structure is described in terms of relationships to structures in the Hierarchy. For example, if one seeks the location of the ventral striatum, one finds the following definition:

  "The term ventral striatum refers to a composite structure of the telencephalon that is defined in the striatopallidal system by connectivity and neurochemical staining. It includes the *nucleus accumbens*, the most ventral portions of the *caudate nucleus* and the *putamen*, the rostrolateral portion of the *anterior perforated substance*, the *islands of Calleja* and a rostral subcommissural portion of *substantia innominata* (Heimer-95)."

- Italicized terms in such definitions are the BrainInfo default names for structures defined in the classical Hierarchy. Click them to go to their Central Directories.

3. TO VIEW ILLUSTRATIONS OF A STRUCTURE

BrainInfo offers the opportunity to view several hundred structures in the human brain and the macaque brain. Many of the illustrations, particularly of the human brain, are located at other websites. They include photographs of the structure in dissected specimens and in cross-section, photomicrographs of sections stained for gray matter and white matter, atlas images showing the boundaries of structures in context, magnetic resonance images (MRIs) and movies of three-dimensional models of structures rotating in space. To view illustrations of a particular structure:

- Go to the structure's Central Directory using the instructions in Section 1 above. The globus pallidus is a good example.
- If BrainInfo has illustrations to show, the button *Show It!* will be lit. Click there.
- Click the thumbnail image of the kind of illustration you want to view.
- If you click the macaque brain cross-section at the top of the page BrainInfo will take you to a display of all of the cross-sections of BrainInfo's Template Atlas of the macaque brain where the structure appears colored blue. If you want a closer view, click one of the thumbnail images and an enlargement will appear. If your structure is a superstructure, each of its substructures will be labeled with an abbreviation; to see the full name of a structure place the cursor on the label. To learn more about it, click the label. To identify surrounding structures, click the Labels box in the gray sidebar.

- If you click a thumbnail other than that for the Template Atlas, BrainInfo will most likely take you to the page in another website that displays the structure of interest. If the structure is labeled there with a name or number other than the BrainInfo default name, a message will appear in the green sidebar saying, e.g., "GLOBUS PALLIDUS: Look for 3,"or "Look for yellow in BASAL GANGLIA." You will also see the name and location of the website that you are viewing. Since you are linked directly into that site, you can use all of the buttons, hypertext and other links provided there to navigate to other information of interest. When you have finished, click *Back to BrainInfo* in the green sidebar to return to the Central Directory.

4. CORTICAL AREAS OF THE MACAQUE BRAIN

BrainInfo can display many structures that do not fit the classical Brain Hierarchy. Currently, most such structures are cortical areas defined on the basis of cyto- and myeloarchitecture instead of gross morphology. BrainInfo allows you to overlay cortical areas as defined and illustrated by different authors on the common background of the Template Atlas so that you can compare their shapes and locations.

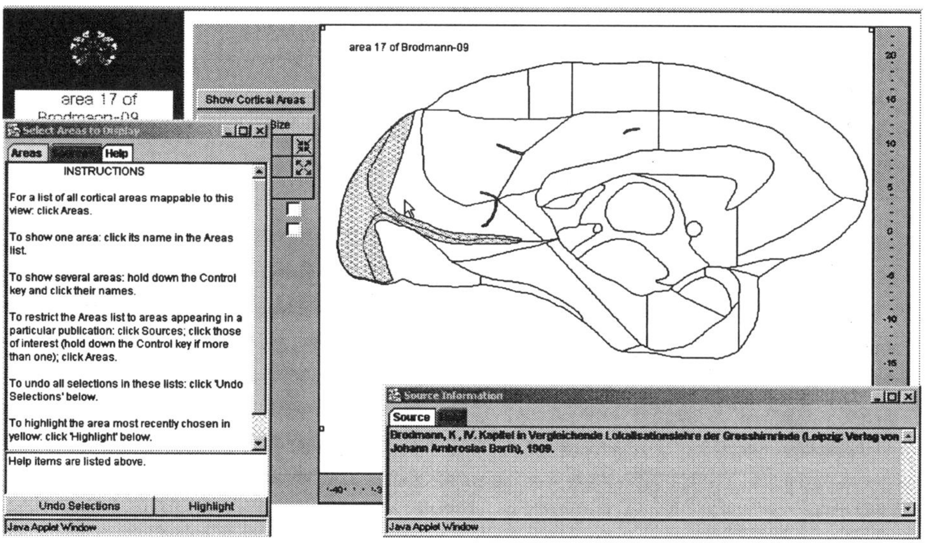

Figure 4. Area 17 of Brodmann-09 on the Template Atlas of the Macaque Brain.

To view a cortical area as defined by Brodmann, Mauss, the Vogts and more recent authors:

- Search for a structure by name (e.g., area 17 of Brodmann-09), using the instructions in Section 1.1 above to arrive at its Central Directory.
- If BrainInfo has illustrations to show, the button *Show It!* will be lit. Click there.
- If BrainInfo has an overlay to show, a macaque brain cross-section with a red stippled square will appear at the top of the page. Click there.
- Small icons showing the location of the structure on the available views will appear, above brief narrative descriptions. Click the one you want.
- After a brief delay, the view appears, without shading except for the selected structure (Figure 4). The outlines of the primary structures are visible through the colored overlay, and the Labels, Grid Lines, Zoom, and Pan buttons are functional. The name of the overlay appears to the upper left.
- Two other windows open, both of which can be moved, resized, or closed as you wish. The window on the left side contains tools for selecting overlays. Although in our example area 17 is already on display, other areas can be selected for display. Click the yellow Areas tab and while pressing the control key, click other areas. The cortical areas are color coded to match the colored icons in the Areas list (Figure 5).

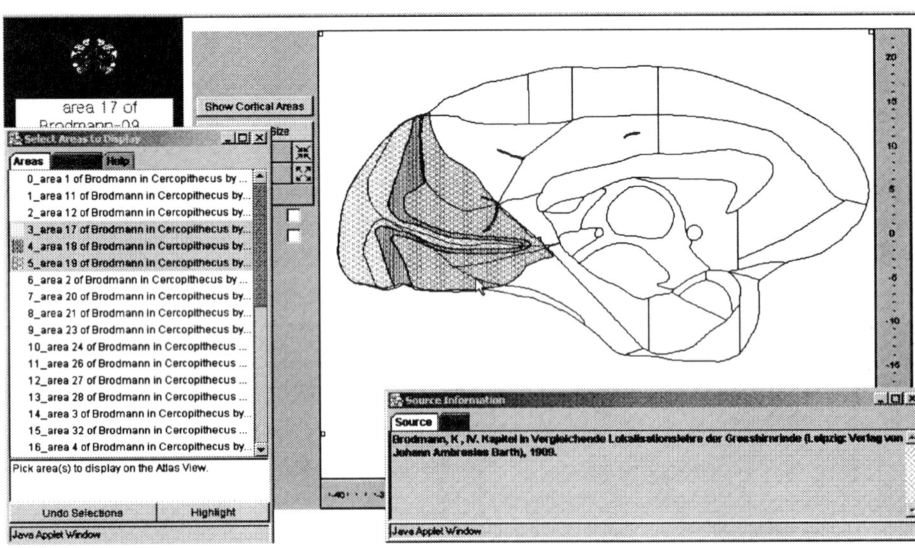

Figure 5. Areas 17-19 of Brodmann-09 on the Template Atlas of the Macaque Brain.

- The Areas list contains all the overlays currently in BrainInfo for a given view. You can restrict it to show only overlays from one source in the literature. Click the Sources tab and then click one of the sources. Click the Areas tab again to find the shorter list of areas from this source.
- The window, at the bottom of the screen, provides information about the source of the most recently activated overlay.
- The tools for selecting overlays are also available if you have searched by browsing the atlas using the instructions in Section 1.2 If overlays are available for a given view the *Show Cortical Areas* button in the gray sidebar will be lit. Click it and make your selections.

5. TO VIEW THE INTERNAL ORGANIZATION OF A STRUCTURE

BrainInfo will soon offer the opportunity to view photomicrographs that illustrate the internal organization of some structures, particularly the cerebral cortical areas defined on the basis of cyto- and myeloarchitecture. To view the internal organization of a structure:

- Go to the Central Directory of a structure, e.g., area 4 of Brodmann, using the instructions in Section 1 above.
- If BrainInfo has information on the subject, the button *Internal Structure* will be lit. Click it and BrainInfo will present thumbnail images of one or more photomicrographs, each representing the structure as reported in a particular species and illustrated in a particular source publication.
- Click a thumbnail to view the full-size photomicrograph.
- To learn more about the source publication, click the author's name to the right of the thumbnail.
- The photomicrograph will appear in a separate window on your screen. If you want to view sections from several authors at once, you can click each of their thumbnail images. Resize each window by click-and-dragging the lower right corner; and click-and-drag the blue Internal Structure bars at the top of the windows to align the images next to each other for comparison.
- If a window disappears behind BrainInfo click its button (*Internal Structure*) at the base of your screen to bring it forward.
- When you have finished, close open windows by clicking the x-box in the upper right corner of each.

6. TO VIEW THE KINDS OF CELL FOUND IN A STRUCTURE

If you are interested in the kinds of neuron found in a structure, BrainInfo may send you to CellPropDB, a database of structural and functional characteristics of neurons in some brain structures. CellPropDB is located at the SenseLab website of Yale University (New Haven, Connecticut, USA; http://senselab.med.yale.edu/senselab). To view the characteristics of neurons in a particular structure:

- Go to the structure's Central Directory using the instructions in Section 1 above. The hippocampus is a good example.
- Click the button *What Cells Does It Have?*
- Find the name of the structure of interest in the three-column hierarchy to the left, and if CellPropDB contains information on the cell-type of interest click the corresponding name in the right-most column.
- If CellPropDB uses a structure name other than BrainInfo's default name, a message will appear in the green sidebar saying, e.g., "STRIATUM: Look for Neostriatum."
- Since you are linked directly into the SenseLab website, you can use all of the buttons, hypertext and other links provided there to navigate to other information of interest. When you have finished, click *Back to BrainInfo* in the green sidebar.

7. TO LEARN WHAT IS WRITTEN ABOUT A STRUCTURE

The most comprehensive web-based guide to the neuroscientific literature is PubMed, a bibliographic database maintained by the US National Library of Medicine (NLM). Since the NLM indexes articles to the level of individual brain structures, PubMed is a rich source of references to studies about specific structures. Unfortunately, however, authors use many different names for the same structure. So, to obtain a comprehensive bibliography of publications on a specific structure, one may have to compose a complex query or run the same query several times using different names, e.g., "Ammon's horn" and "Cornu ammonis" in addition to "hippocampus".

BrainInfo provides assistance with this problem, e.g., when you submit a query about the hippocampus to PubMed through the website, BrainInfo composes a single query that includes all names for the hippocampus. Thus, in a single search one obtains a complete listing of publications that fit the query. To submit a query to PubMed through BrainInfo:

- Go to the structure's Central Directory using the instructions in Section 1. A good example is lateral amygdaloid nucleus.
- Click *PubMed.*
- If you wish to limit your search, enter additional terms (e.g., epilepsy) in the box in the green sidebar.
- Click Submit.
- Once PubMed displays its first bibliography, you can modify the query by clicking PubMed's *Details* and editing it there. For further information about efficient search strategies, click PubMed's *Help.*
- When you have finished, click *Back to BrainInfo.*

8. IS STRUCTURE X THE SAME AS Y?

A common difficulty one confronts in reading the neuroscientific literature arises because most structures have several names (an average of six for structures in the Hierarchy). One can spend an inordinate amount of time perusing textbooks and atlases to determine, for example, whether the prerubral field of one author is the same as Area tegmentalis of another, or field H of another. In fact they are the same, but it may take 20 minutes, even in a well-stocked library, to establish the fact. BrainInfo can answer most such questions regarding neuroanatomic terminology in seconds. To view most of the names that have been applied to a given structure:

- Go to the structure's Central Directory using any name for it and the instructions in Section 1 above. A good example is the nucleus of Darkschewitsch. If you are uncertain how to spell it, you can search using only the first few letters of key words, e.g., "nuc dark".
- Click *Other Names for It*; BrainInfo will provide a table of the most commonly used English and Latin spellings and synonyms, i.e., synonyms that retrieved more than 1% of citations in PubMed in a search of the years 1976 to 2000. For information about an authoritative source that links one synonym to another, click the item in the Source column. To determine the species in which the source publication applied the term, see the Species column; to learn more about relations of that species to others click *Show Taxonomic Relations.*
- If you are interested in seeing all of the English and Latin synonyms, regardless of their frequency of use, click *Show All English & Latin.*
- To see synonyms in other languages, click *Show All Languages.*

9. IS STRUCTURE X SEEN IN PARTICULAR SPECIES?

The answer to the question of whether different species share homologous structures depends greatly on one's criteria for homology. Structures can be compared in terms of location in the brain, morphology, connectivity, internal structure, staining characteristics, gene expression, or a variety of functional characteristics. BrainInfo provides information on homology as judged on the basis of location, morphology and to a certain extent, internal structure. By these criteria a very large proportion of subcortical structures seen in the human brain are also found in other mammalian species, specifically the macaque and the rat. Only in the cerebral cortex does the pattern of sulci and gyri differ so greatly from order to order that different terminologies are needed. Currently BrainInfo presents homologic information in two ways:

9.1 To Determine Whether a Structure is Found only in the Human or only in the Macaque

- Go to the structure's Central Directory using the instructions in Section 1. Try the marginal sulcus or the simian fossa, for example.
- If the structure is found only in the human, its name at the top of the page will be followed by (H); if only in the macaque, its name will be followed by (M)

9.2 To Determine Whether a Structure Found in the Macaque has a Homologue in the Rat

- Go to the structure's Central Directory using the instructions in Section 1. Try the cuneate nucleus, for example.
- Click *Which Species Have It*?
- Look in the first column of the table for the species (human, macaque, rat) in which homologous structures have been reported. The second column tells the name for the structure used by the source author(s).
- For more information about the source, click in the right-most column.
- For more information about the species, click in the first column.
- For an overview of the phylogenetic tree and the locations of the species most used in neuroscientific research, click *Show Taxonomic Relations* at the bottom of the page.

10. STEREOTAXIC MACAQUE BRAIN ATLASES

BrainInfo provides access to three stereotaxic macaque brain atlases: 1) a fully segmented template atlas of the brain of the longtailed macaque (*Macaca fascicularis*) with stereotaxis referenced to the bicommissural line[2]; 2) a photographic atlas of the brainstem of the rhesus macaque (*M. mulatta*) in sitting posture for stereotaxic direction of electrodes in the waking monkey that is referenced to the conventional orbitomeatal plane; and 3) an MRI atlas of the pigtailed macaque (*M. nemestrina*) in three planes of section referenced to the orbitomeatal plane. The latter is located at the Laboratory of Neuro Imaging (LONI) website at the University of California, Los Angeles, USA (http://www.loni.ucla.edu).

10.1 Macaque Atlas Referenced to Bicommissural Line

- In the green sidebar click *Search Atlas*.
- Click *Coronal Sections*
- Click the thumbnail for a cross-section that you want to view full-size, e.g., the section at anterior commissure minus 9 mm (ac –9 mm).
- Click the Grid Lines box in the gray sidebar to apply a 5 mm stereotaxic grid to the image. The dorsoventral and mediolateral coordinates (with the center of the anterior commissure as the origin) appear on gray scales at the right and lower margins of the image.
- To increase the resolution of the grid to 1mm, click either *Zoom In* or the icon to its right. If you click *Zoom In* and then click the part of the image that is your focus of interest, the image will enlarge with the focus of interest at its center. If you click the icon, the image will enlarge in place.
- To return the image and grid to original dimensions, Click *Standard Size.*
- To identify a particular structure, click the Labels box in the gray sidebar, and place the cursor over the label.
- For more detailed information on the use the Template Atlas with stereotaxic procedures click *Home*; click *About BrainInfo*; and scroll to "Stereotaxis". Also, consult *Primate Brain Maps: Structure of the Macaque Brain* (Martin and Bowden, 2000).

[2] Templates for publication and display of data mapped to the macaque brain (coronal sections and surface views) can be obtained from Elsevier Science: http://www.elsevier.nl/inca/publications/store/6/2/0/8/6/0/index.htt. A demonstration of the compact disk on which Elsevier distributes the templates can be viewed at Primate Brain Maps: http://www.elsevier.nl/homepage/sah/pbm/menu.htm.

10.2 Macaque Brainstem Atlas in Sitting Posture

- Click *Other Atlases* in the green sidebar of the Home Page.
- Click *Rhesus Brainstem: Sitting Posture.*
- In the green sidebar, click the thumbnail of the coronal section nearest your focus of interest.
- To explore adjacent sections, click the buttons in the green sidebar labeled *posterior* and *anterior.*
- To examine the section at 2.5x magnification or 5.0x magnification, click the buttons labeled *Zoom 2.5X* or *Zoom 5.0X* in the green sidebar.
- Structures are labeled in every fourth or fifth section. To identify a structure, click the thumbnail image of the section nearest to that in which the structure appears; if the structure is labeled there, click the label and BrainInfo will display the Central Directory of the structure. If the structure is not labeled, click *Nearest Atlas Section* and BrainInfo will display the page of the fully-segmented macaque atlas most likely to show the name.
- For a legend explaining the vertical and horizontal scales and other details click *About This Atlas.*
- For a guide to printing the brainstem atlas, click *About This Atlas* and scroll to "Make Your Own Printed Atlas"

10.3 Macaque Brain in Three Planes of Section

- Click *Other Atlases* in the green sidebar of the Home Page.
- Click the thumbnail for LONI Brain Atlases of *Macaca nemestrina.*
- You will have a choice of viewing the brain in coronal, sagittal, or horizontal sections, with or without extra-cerebral cranial structures and with or without labels. Click the thumbnail of the combination you select. The LONI website provides a number of navigational features that will help you find the structures and stereotaxic areas of interest.
- Note that the abbreviations for structure names are different from those of BrainInfo. To obtain more information about a structure labeled in LONI, click the *Table of Abbreviations* at the top of the screen. If you do not recognize the structure by its name there, you may copy and paste the name into the Dictionary in the green sidebar. BrainInfo will tell you its default name in NeuroNames and, if you click the default name, will tell you all of its synonyms.

11. FURTHER INFORMATION ABOUT BRAININFO

BrainInfo is an evolving neuroscientific resource (see Home Page, click *About BrainInfo* and scroll to "History of BrainInfo"). The nomenclature has grown from 4,000 terms in 1991 to more than 11,000 in 2002 (Bowden and Dubach, 2002). It now includes almost all of the concepts of classical human neuroanatomy down to the level of primary structures, all of their English and Latin names and illustrations of most structures found in the nonhuman primate brain. Information that is currently incomplete but that is steadily growing includes the most common 1000 names of structures in the world's major languages, structures in the rodent brain that are topographically homologous to primate structures, the location of architectonically defined cortical areas, and photomicrographs of the internal structure of the cortical areas. Features to be added include an input method for users to attach comments and to read comments attached by others, and input tools that will allow users to map their own data into the macaque brain atlas and to calculate the degree of overlap between their data and data already mapped to the atlas (Bowden and Martin, 1997).

ACKNOWLEDGMENTS

We thank Evan Song and John Wu for conversion of NeuroNames from HTML to Microsoft ACCESS and for the programming that enables its presentation in BrainInfo on the Web, to Tri Wahyu (Esti) Pangestiningsih, Valery Siderova and Erika Seki for assistance in incorporating Indonesian, Russian and Japanese terms into the database, and to Erik McArthur for preparation of the figures for this report. This work was supported by NIH grants LM-06243 and RR-00166.

REFERENCES

Bowden D. M., Dubach M. (2002) NeuroNames 2002. *Neuroinformatics* (in press, 2002).

Bowden D. M., Martin R. F. (1997) A digital rosetta stone for primate brain terminology. *Handbook of Chemical Neuroanatomy, Vol. 13: The Primate Nervous System, Part 1,* F. E. Bloom, A. Björklund and T. Hökfelt editors, Elsevier Science B.V., Amsterdam, pp. 1-37.

Martin R. F., Bowden D. M. (2000) *Primate Brain Maps: Structure of the Macaque Brain,* Elsevier Science B.V., Amsterdam.

Chapter 19

Federation of Brain Data through Knowledge-guided Mediation

Maryann E. Martone[1], Amarnath Gupta[2], Bertram Ludäscher[2], Ilya Zaslavsky[2] and Mark H. Ellisman[1, 2]
[1]*National Center for Microscopy and Imaging Research and Department of Neurosciences, University of California, San Diego, San Diego CA,*
[2]*San Diego Supercomputer Center, University of California, San Diego, San Diego CA*
Correspondence to: mmartone@ucsd.edu

Abstract: In this chapter, we describe a data integration framework for multi-resolution data and some of the tools we have created for achieving integration of distributed data sources. The chapter is divided into two parts. In the first part, we describe a recently developed database for high resolution 3D data on cell structure and protein localization, the Cell Centered Database or CCDB. In the second part, we describe how we can link data sources such as the CCDB to other databases using a mediation-based approach. By incorporating additional knowledge into the system, we are able to link together sources that are related conceptually, even if they do not contain similar elements in their schemata. For each of the tools and resources described, we indicate its current state of development and availability.

Key words: database, electron tomography, protein localization, confocal microscopy, electron microscopy

1. INTRODUCTION

The grand goal in neuroscience research is to understand how the interplay of structural, chemical and electrical signals in nervous tissue gives rise to behavior. Experimental advances of the past decades have given the individual neuroscientist an increasingly powerful arsenal for obtaining data, from the level of molecules to entire nervous systems. Scientists have begun the arduous process of adapting and assembling neuroscience data at all scales of resolution and across disciplines into computerized databases and other easily accessed sources (Wong and Koslow, 2001). These databases will complement the vast structural and sequence databases created to catalogue, organize and analyze gene sequences and protein products. The utility of such databases has been amply demonstrated, as an entire new field, bioinformatics, has been inaugurated to organize and mine the data (Persidis,

1999, 2000). The question of whether equal rewards will accrue from similar efforts for histological data, whole brain imaging, physiological data, etc., remains to be answered. Nevertheless, at the very least, neuroinformatics efforts will provide organizing frameworks and computational tools to manage neuroscience data, from the lab notebook to published data (Gardner et al., 2001; Miller et al., 2001). In the best case, it will provide new opportunities to integrate large amounts of biological data into unified theories of function and aid in the discovery process (Chicurel, 2000; Fox and Lancaster, 2002).

In this chapter, we describe a data integration framework for multi-resolution and multi-disciplinary data and some of the tools we have created for achieving integration of distributed data sources. The chapter is divided into two parts. In the first part, we describe a recently developed database for high resolution 3D data on cell structure and protein localization, the Cell Centered Database or CCDB (Martone et al., in press). In the second part, we describe how we can link data sources such as the CCDB to other databases using a mediation-based approach. In a mediated system, the user does not interact directly with an individual data source but poses a query to a mediator, which then breaks down the query and retrieves data from the individual sources. By incorporating additional knowledge into the system, we are able to link together sources conceptually so that information can be linked and assembled from disparate sources to achieve integration of biological data across scales and disciplines. For each of the tools and resources described, we indicate its current state of development and availability.

2. THE CELL CENTERED DATABASE

2.1 Data Modeling and Structure of the CCDB

The CCDB was created for high resolution 3D light and electron microscopic reconstructions of cells and subcellular structures. It contains 3D structural and protein distribution information from confocal, multiphoton and electron microscopy, including correlated microscopy. Many of the data sets are derived from electron tomography, a powerful 3D reconstruction technique for electron microscopic specimens. Electron tomography is similar in concept to medical imaging techniques like CAT scans and MRI in that it derives a 3D volume from a series of 2D projections through a structure. In this case, the structures are contained in sections prepared for electron microscopy, which are tilted through a limited angular range. Projections through the section are captured at regular tilt increments, aligned to a common origin and then back projected into a 3D volume (McEwen and Frank, 2001). When combined with high voltage electron microscopy, detailed reconstructions can be made of relatively large expanses of tissue at EM resolution. For example, the CCDB contains reconstructions of extended structures such as spiny dendrites from selectively stained neurons contained within 2-5 μm thick sections (Fig. 1). Single reconstructions can contain up to a 20 μm stretch of dendrite with upwards of 60 spines. Electron tomography is also providing new insights in the 5–40 nm range. Structural features in this range have traditionally been difficult to study because they are smaller than the thickness of a thin section produced for electron microscopy (60-100 nm thick) and so tend to be obscured within the depths of the section. In electron tomography, computed sections through a volume can be generated with an effective resolution of 5-10 nm, thereby uncovering structural detail within this range. Tomography has led to a complete revision of our understanding of even well-studied structures such as mitochondria (Perkins et al., 1997; Mannella et al., 1997) and the neuromuscular junction (Harlow et al., 2001).

The CCDB is built on an object-relational framework using Oracle 8i. The current CCDB has over 80 tables containing a large amount of descriptive data. It models the entire process of 3D reconstruction, from specimen preparation to segmentation and analysis. A volume reconstruction is stored along with all of the raw images and the processing details required to reconstruct the volume from the raw data. Each object segmented from the 3D volume is stored as a separate object indexed to the parent reconstruction. Four types of segmented objects are currently modeled in the CCDB: 1) surface objects: polygonal surface meshes representing 3D objects in the reconstruction, extracted using either isosurfacing methods or manual contouring; 2) contour objects: a series of 2D contours defining an

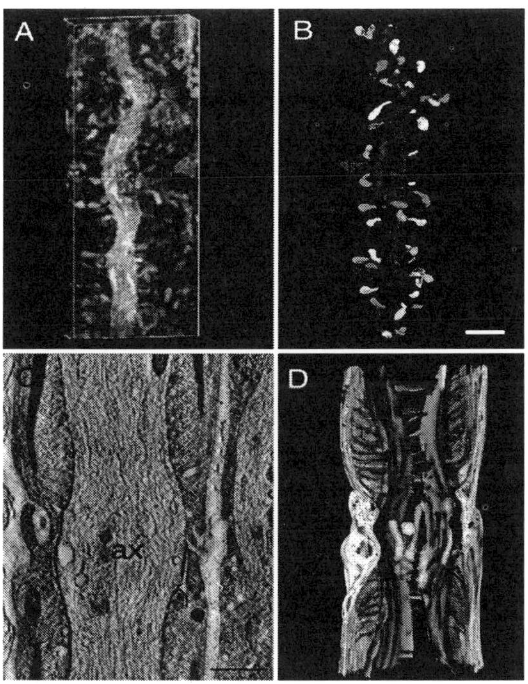

Figure 1. Examples of tomographic data sets in the CCDB. A and B show a selectively stained spiny dendrite from a Purkinje cell. A is a projection of the volume reconstruction (dendrite appears as white against dark background). B is the segmented dendrite. C and D show a tomographic reconstruction of the node of Ranvier. C is a single computed slice through the volume. D is a surface reconstruction of the various components comprising the node. Scale bar in B = 1 μm; in C = 0.5 μm

object; 3) volume objects: subvolumes containing an object of interest; and 4) tree objects: skeletons of branching objects like dendrites and axons, derived from Neurolucida (Microbrightfield, Colchester, VT, USA). Each object is stored along with any measurements like surface area, volume, length, number and labeling intensity. Whenever possible, parsers are written for the output of analysis programs so that results can be uploaded directly into the CCDB. For example, measurement summaries for tree objects are uploaded directly from the output of NeuroExplorer, an analysis program for Neurolucida-derived data.

The data contained in the CCDB will be derived from both published and unpublished data. A constant concern in creating and maintaining databases of experimental information is the quality of data retrieved from a query. In the CCDB, the evaluation of the quality and accuracy of morphometric or protein distribution modeling is up to the user. Thus, the CCDB contains several features to aid in this process including: 1) the inclusion of morphometric data stored with the objects to allow the user to compare the statistics of a given data set to other stored and published data to determine whether they fall within expected ranges; 2) access to the raw data along with all the imaging and processing steps to allow the accuracy and quality of the final reconstruction to be assessed by an experienced user; 3) field to indicate whether the data come from published studies; 4) evaluation tables for users to store estimations of the quality of experimental, imaging, protein labeling and reconstruction results; and 5) the ability for users to supply additional or alternative interpretations of a given data set, which will be retrieved along with the data. In this way, the CCDB can serve as an interactive forum for data interpretation.

2.2 Accessing the CCDB

The CCDB can be accessed at http://www.ncmir.ucsd.edu/CCDB. The home page is shown in Fig. 2. The CCDB is still under development but is available on-line for limited use. We welcome feedback from interested users, although some of the functions described in this chapter may not be fully implemented by the time of publication. Information on the current functionality and updates to the site are available from the "Status/Update" link off of the main page. The main page also contains links to a description of the CCDB and to visualization and analysis tools available for data contained in the CCDB. The CCDB itself is accessed by the "Enter the database" link which takes the user to the log in page. The log in page allows users to create new accounts, log in to the database or view the CCDB schema by clicking on the appropriate links. Some sample query results are also available. The CCDB is freely accessible, although users must register for an account. Accounts are created by clicking on "Create Account."

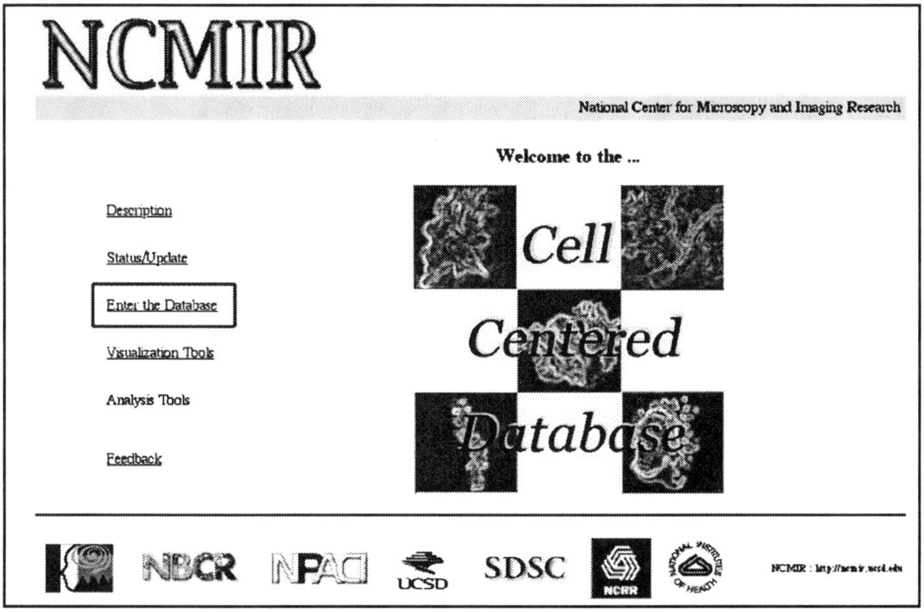

Figure 2. Home page of the CCDB. Clicking on "Enter the database" takes the user to the log in page.

Upon logging in, the user is presented with a simple query interface shown in Fig. 3. Users may query on a limited number of descriptive fields such as cell type or structure, by typing in values or selecting from the pull down menus. More advanced query interfaces are currently being developed for the CCDB which will take advantage of the rich data model of the CCDB to retrieve data based on morphometric features contained in the data. An example of such a query form is shown in Fig. 3 (right panel) for neuronal branching data contained in Neurolucida files. Users can retrieve neuron data based on branching patterns, tortuosity of dendrites and other metrics. This form can be viewed, along with sample queries, by clicking on the "Advanced Query Form" from the "Enter the database" page, although at the time of writing, this form is not linked to the CCDB. From these types of feature based queries, one can begin to look at relationships among various data sets. We might ask, for example, whether variability in neuronal branching patterns correlates with any subject characteristic.

In response to a query, the CCDB informs the user of the total number of data sets matching the criteria and presents a short set of descriptors. For each data set returned, the CCDB presents thumbnails of the types of image data available for that particular result (Fig. 4). The user may view a 2D image by clicking on the thumbnail or may download the data. In the current version of the CCDB, the image data itself is not stored directly in the database; that is the actual voxel values and spatial arrays are not stored in the

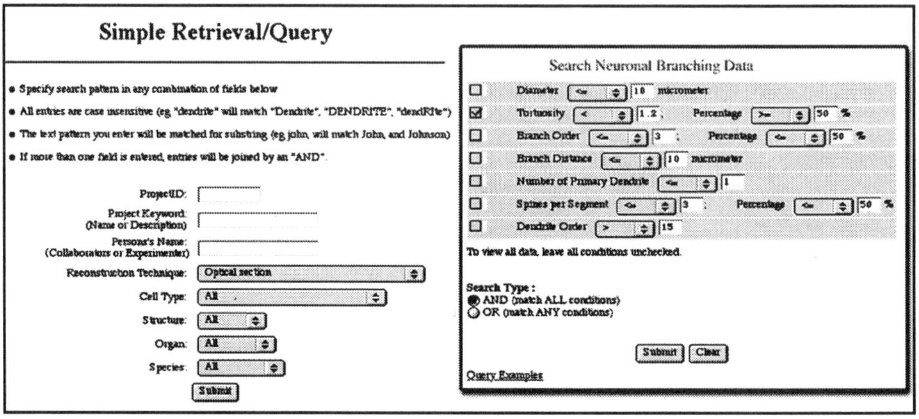

Figure 3. Simple query form for the CCDB (left). An example of the type of advanced query forms being developed for attribute queries is shown on the right.

database.

Instead, the descriptive and analysis data are stored along with pointers to the image file. To store the images for the CCDB, we have opted to use the Storage Resource Broker (SRB), a data management system for storing and accessing distributed data (http://www.npaci.edu/DICE/SRB/). The SRB is sophisticated client-server middleware that provides a uniform interface for connecting to data resources over a network. Unlike conventional access methods, e.g., file servers, ftp, or http, SRB is grid-based software providing transparent access to data, relieving the user from dealing with aspects such as physical location of imaging data, concrete storage devices, and device-dependent access protocols. Thus, regardless of where the data lives, whether in a single location or distributed across

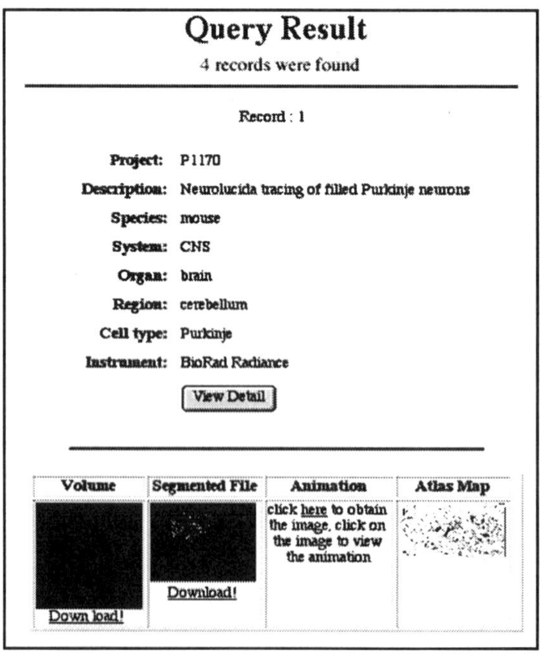

Figure 4. Return of query results from the CCDB. All of the types of image files associated with a given data set are listed at the bottom. "Atlas Map" shows the location of a data set with respect to a brain atlas (see below).

several databases, file systems, and high-performance storage systems, SRB provides access to the data via an SRB identifier. When a user requests a file from the CCDB, the file is retrieved via its SRB ID. Eventually, the SRB will handle additional functions like file format conversions so that the user can request the data in most common formats. The CCDB acts as a single client of the SRB, so that separate SRB authentications and accounts do not have to be obtained for each user.

Data in the CCDB are stored in a variety of formats currently used at NCMIR and may be viewed with tools available for download. A list of these tools is available under the "Visualization Tools" link on the CCDB main page (visible in Fig. 2), which provides the platform dependencies of each tool along its utility for the different types of data available (volumes, contours, tree structures and surfaces). A general purpose java-based viewer for CCDB data, Jviewer, is being developed, which can be used to view images in the CCDB directly from the web interface. Also available at the bottom of the Visualization Tools web page is a link to file format converters available through NCMIR.

2.3 Spatial and Semantic Indexing of Data in the CCDB

One of the goals of our work, and informatics in general, is to develop systems for integration of data across scales and disciplines. In our case, we wish to relate the cellular level data in the CCDB to data acquired at the tissue and molecular levels. In order to achieve this goal, we must develop shared frameworks in which to place individual data sets so that they can be related to data taken in other experiments. One obvious framework is spatial, using standard brain atlases to provide registration to a common coordinate system. A second is semantic, taking advantage of efforts in various communities to develop shared ontologies. An ontology can be thought of as a network of terms, concepts and the relationships between them.

To aid in the registration of data in databases like the CCDB, we are developing a web-based tool for situating brain data in both a spatial and semantic context. The Spatial Mark Up and Rendering Atlas Tool ("Smart Atlas"), based on a geographical mapping tool, is a web-based tool that allows users to define polygons on a series of 2D vector images and annotate them with names, relationships and ontology concept IDs. Using this tool, we are creating a "Smart Atlas" and spatial database for neuroscience data using a brain atlas as the interface. For the purposes of demonstrating the utility of this tool, we are using a commercially available brain atlas, Paxinos and Watson (1998), as the interface. Because the atlas is a commercial product, the web page is password protected. The Smart Atlas tool, itself, though will be made available to other scientists for use with their own atlases.

Therefore, we describe its utility in some detail below. A demonstration page is being set up and will be available from the log in page of the CCDB.

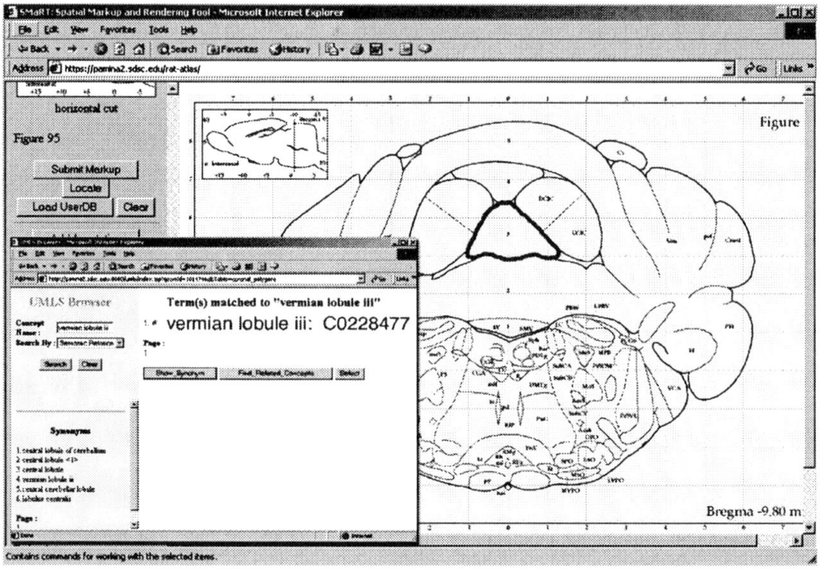

Figure 5. Spatial Mark Up and Rendering Tool for spatial and semantic registration of brain data. The atlas slice shown is from the Paxinos and Watson rat brain atlas. The user has defined cerebellar lobule iii and attached the concept ID from the UMLS.

The Smart Atlas interface is illustrated in Fig. 5. The Adobe Illustrator files supplied with the atlas were converted into SVG format using a plug-in available at http://www.adobe.com/support/downloads/. The spatial coordinates of the line segments in each image were stored in an Oracle database, which serves as the back end of the Smart Atlas. The user then "marks up" the atlas using the Smart Atlas tool. First, the user selects those line segments comprising a brain region to define a polygon (Fig. 5). The user then annotates the polygon with the name of the brain region. In addition, the user links the brain region to one or more ontologies through a browsing tool accessed through the Smart Atlas. In the case of the CCDB, we are using the Unified Medical Language System (UMLS), developed for biomedical sciences by the National Library of Medicine (Ingenerf et al., 2001). Strictly speaking, the UMLS is a metathesaurus, but for many purposes it can be used like an ontology. The user browses the UMLS and selects the relevant concept. The polygon is now tagged with both the name of the region and the UMLS concept.

Once the atlas has been marked up, it can be queried from the same browser. Clicking on any point in the atlas will return the stereotaxic coordinates. Clicking on a brain region will return the name of that region, along with any synonyms, and highlight all planes containing that structure (Fig. 6).

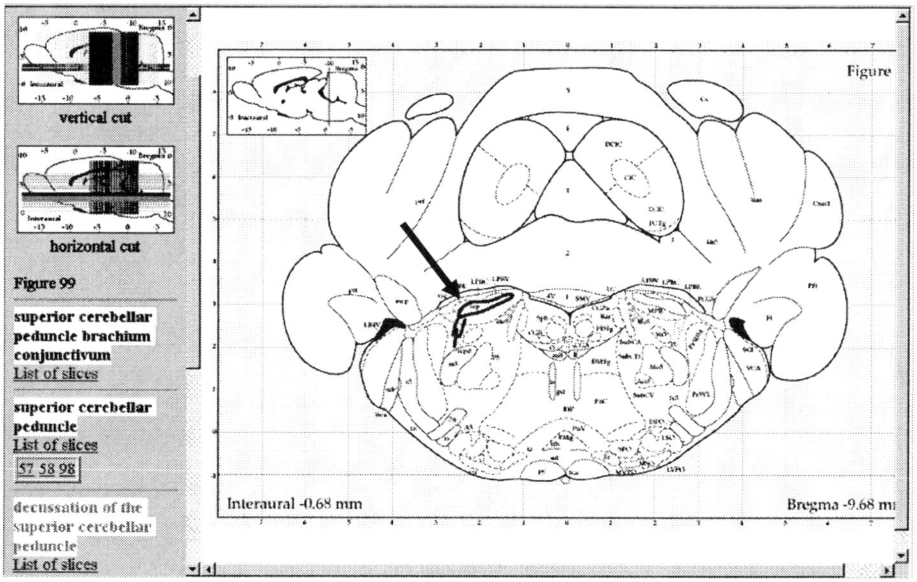

Figure 6. Querying the marked up brain atlas. The user has highlighted the superior cerebellar peduncle (arrow). The list of slices containing the structure are highlighted in the image maps in the upper left corner.

The Smart Atlas can now be used to register a researcher's data to a specific spatial location. In this case, the user draws an arbitrary polygon representing the approximate data location on one of the atlas planes (Fig. 7). The user is then presented with a form which can be used to add annotations or to link to an ontology concept. Although the UMLS is used in the examples shown here, the user will eventually be able to use multiple ontologies, including those of their own creation, for semantically indexing data. Tools are also being developed to define new terms and relationships in existing ontologies.

The utility of indexing data in a database using the Smart Atlas tool is several fold. By using an ontology concept ID for indexing CCDB data, the user is relieved from having to specify multiple terms to characterize the data. Once the concept is chosen, the data set inherits the parents and children of this node and is thereby placed in a larger semantic context. The full utility of this system will be illustrated in the next section, which describes our approach to linking brain databases into a federated, cross-queriable system.

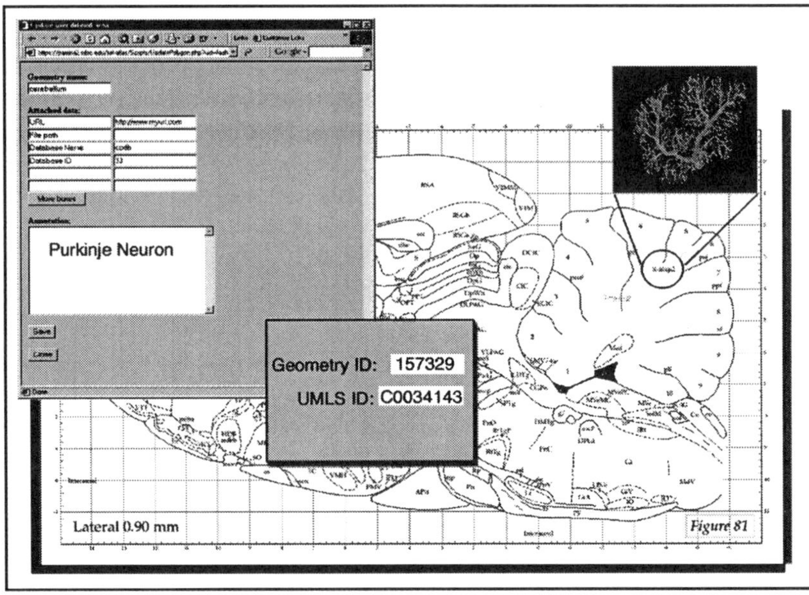

Figure. 7. Registration of a data set with the Smart Atlas. The user draws a polygon representing the location of a data set, in this case a filled Purkinje neuron. The user then specifies the database containing this data, enters an annotation and selects a concept from the UMLS or some other ontology. The concept ID is stored back in the data base.

Spatial indexing of data to a coordinate system will allow the user to pose spatial queries, e.g., find data associated with this region; find all structures within 1 mm of this area. It also allows the user to take advantage of efforts to develop spatial transformations and alternative representations of brain topology, e.g., flat maps of cortex and cerebellar cortex (Dickson et al., 2001; Van Essen et al., 2001). The locations of filled or immunolabeled neurons from the CCDB can be mapped onto these flat maps to give additional information on the distribution of neuronal classes or compare features such as dendritic arbor size as a function of location. Thus, by indexing to a common coordinate system, we can facilitate interoperability of informatics tools.

3. FEDERATION OF BRAIN DATA

No matter how sophisticated and complete the CCDB is in modeling and querying the data, it only covers a very small portion of the biological spectrum. The oft stated goal of informatics research is to create computer-based approaches to allow scientists to integrate and relate data obtained at different scales, experimental systems and sub-disciplines. To tackle such an ambitious goal, various approaches have been taken. These approaches range from data warehousing, where data is gathered from multiple souces and deposited into a larger database (Rachedi et al., 2000), to the creation of "virtual federations" where individual data sources remain separate but are "wrapped" in a standard language such as XML (extensible markup language) that allows their content to be understood by a federation engine known as a mediator. The mediator is responsible for selecting, restructuring, and merging information from autonomous sources and for providing an integrated view of the information.

Although mediator-based systems have been around for some time, their use to integrate biological databases presents a challenge. When one is linking databases containing similar information, e.g., gene sequence databases, the linkage between data sources are straightforward and can be expressed as relatively simple structural statements which identify common elements in their respective schemata. The task becomes more difficult, however, when attempting to relate data sources which cannot be joined on a purely structural level but which nevertheless contain relevant data. For example, a database at Montana State University on sensory neurons encoding wind direction in the cricket may not share any attributes in common with a database of anatomical structures in the monkey visual cortex. Despite the lack of common semantic links, most neuroscientists can easily relate these two data sources at the conceptual level, by recognizing that they both are sensory systems and may share certain aspects of population coding of sensory stimuli. In fact, neuroscientists usually can navigate with relative ease from the level of individual molecules to brains and across experimental disciplines, because they possess the requisite knowledge to conceptually relate data at each level. We have developed a novel mediator-based integration paradigm which exploits such expert knowledge to begin to address the problem of data integration of heterogeneous neuroscience data, a system which we call "knowledge-guided or model-based mediation".

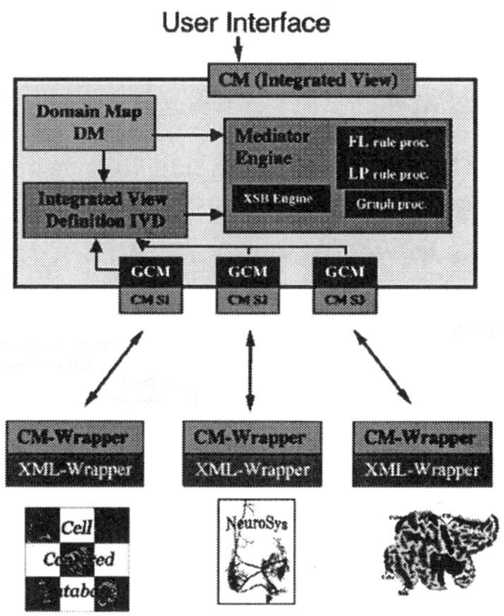

Figure 8. Overview of the prototype KIND mediator. The lower icons represent 3 different sources linked to the mediator. The box in the upper panel represents the mediator layer.

An overview of the prototype mediator, called the KIND (Knowledge Integration of Neuroscience Data) mediator, is shown in Fig. 8 and is described in more detail in Gupta et al., (2000) and Ludaescher et al., (2001). Additional references and demonstrations are provided at http://www.npaci.edu/DICE/Neuro/. There are two places in the system where additional knowledge is incorporated. At the source level, the wrapper language exports a conceptual model (CM) of each data source containing information about relationships, classes and values using an object-oriented language, F-logic, as the deductive engine. At the mediator level, conceptual knowledge is encoded in the form of a semantic network of terms and relationships which we call a "domain map". Domain maps can be thought of as ontologies with more formal logic-based semantics. The purpose of the domain map is to provide a declarative means for specifiying additional knowledge that is not present in the source but is required to bridge two information sources. When a standard SQL query is launched, the mediator breaks it down into its component parts, accesses the appropriate data sources using knowledge sources contained in the wrapper or the domain map, and reassembles the results of the query into an integrated view.

In the following examples, we illustrate the utility of the KIND mediator for answering queries which require bridging of multiple data sources, ranging from retrieval of multi-resolution data to true data integration where a new piece of information is derived by combining data from two sources. The current system consists of the KIND mediator with the following sources registered: 1) a prototype of the CCDB; 2) the EF-Hand Calcium binding protein database, a database of sequence data available from Vanderbilt University (http://structbio.vanderbilt.edu/cabp_database/); 3) the NeuronDB database from the SenseLab project (Shepherd et al., 1998; Miller et al., 2001); 4) TAXON, a database created from taxonomy information available from the National Center for Biotechnology Informatics (NCBI).

The queries below are written in natural language for the sake of clarity, but in all cases were executed in standard SQL. Similar demo queries can be found at: http://www.npaci.edu/DICE/Neuro/ (go to "Demonstrations: Knowledge based mediation" and click on "Protein Localization Query") and must be executed with an XML-capable browser such as Internet Explorer.

3.1 Examples of data federation using the KIND mediator

3.1.1 Retrieval and navigation through multi-scale data

An example of how the mediator is used in conjunction with the Smart Atlas is shown in Fig. 9. In this case, the user launches a simple query from the Smart Atlas Tool by clicking on the cerebellum in order to find all data associated with this brain region. It turns out that there are no sources hooked to the mediator which have data indexed under the term "cerebellum", but the

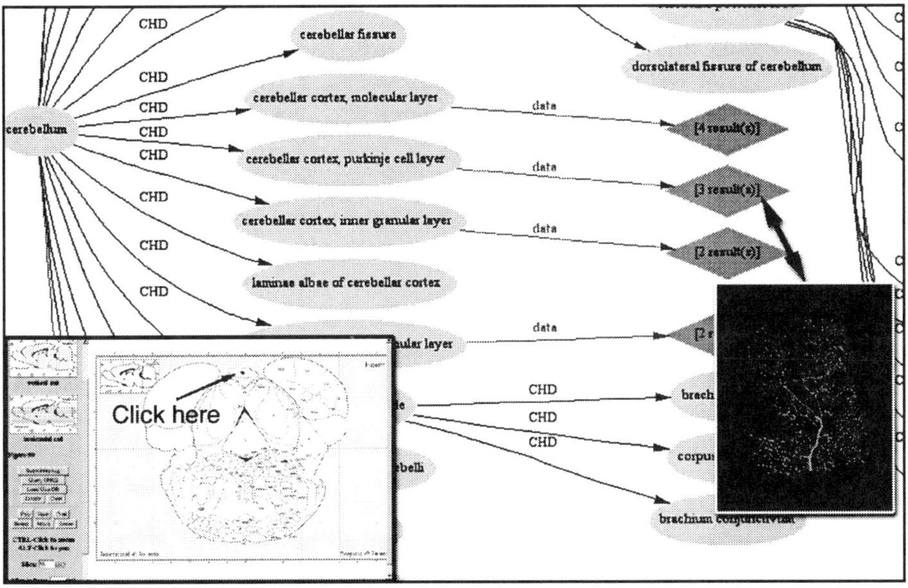

Figure 9. Results of query launched from Smart Atlas. The ovals represent concepts from the UMLS and the diamonds where data is found. Clicking on the diamond returns the data.

CCDB has data on Purkinje neurons and other structures found within the cerebellum. The mediator locates the terms under "cerebellum" in the domain map, queries the sources and returns the relevant portion of the domain map to the user with the available data placed in context. In this way, the user can immediately see how the available data relates to the concept they chose. The user then clicks on the data node to retrieve the data.

3.1.2 Making use of ancillary data sources to increase efficiency of data retrieval

"What is the cerebellar distribution of proteins with 90% homology to human NCS-1 in the rat? In all rodents?" To answer this query with single sources, the researcher would first go to the calcium binding protein database and retrieve the sequences with 90% homology to human NCS-1. Each retrieved protein would then have to be used to query the available histological data sources. Using the KIND mediator, however, the researcher specifies the protein, degree of homology and anatomical region in the mediator and launches the query. First, the mediator retrieves the list of proteins with homology to NCS-1 from the Calcium Binding Protein database, probes the available sources and then returns the results. Because the user specified "cerebellum", all structures defined as belonging to cerebellum are retrieved. To retrieve results for all rodents, the mediator goes to the taxonomy database to retrieve species under "rodents" and re-probes the sources.

3.1.3 Integrating information from multiple sources

"What is the overlap between the parallel fiber input into the Purkinje neuron and the distribution of ryanodine receptor?" In the third example, we show how knowledge-based mediation can answer a query by integrating information across sources to derive information that is not present in any single source. Two registered sources have information about Purkinje neurons, the CCDB and the Senselab Neuron database. The CCDB has information on the distribution of the ryanodine receptor, but no information on connectivity. The Senselab database has information on connectivity, but no information on the ryanodine receptor. To address this query, an integrated view definition must be created which structurally and conceptually links the two data sources at the level of the domain map. First, the mediator retrieves data from the SenseLab database on the distribution of the parallel fibers. It then locates the relevant portion of the domain map and queries the CCDB via the domain map to find the intersection. The full demonstration of this query can be viewed at: http://www.npaci.edu/DICE/Neuro/, under "Demonstrations: Knowledge-based mediation" by clicking on "Protein Localization Query (with domain map)" and viewing query #3.

3.2 Extending the Domain Map

We have been extending the concept of the domain map to include more dynamic processes such as signal transduction. In a process map, the nodes represent states and the edges represent transitions. For example, we have taken some of the pathways involved in the production of long term potentiation and represented them in a graphical form. As ontologies and process maps are developed, they become not only a useful adjunct to database indexing and data management, but an important source of knowledge in their own right. To take advantage of these graphical bodies of knowledge, we have developed a tool for browsing, querying and tying data to these process maps. The Knowledge Map Explorer Tool (Know-Me) is shown in Fig. 10 and can be demonstrated at http://pamina2.sdsc.edu:18080/knowme/pm.jsp. Users can select nodes to elaborate processes, execute queries like "compute the shortest pathway between two nodes" and query for experimental evidence supporting a given relationship.

4. CONCLUSIONS

The KIND mediator is a prototype system, but will form the information integration strategy employed by the newly created the Biomedical

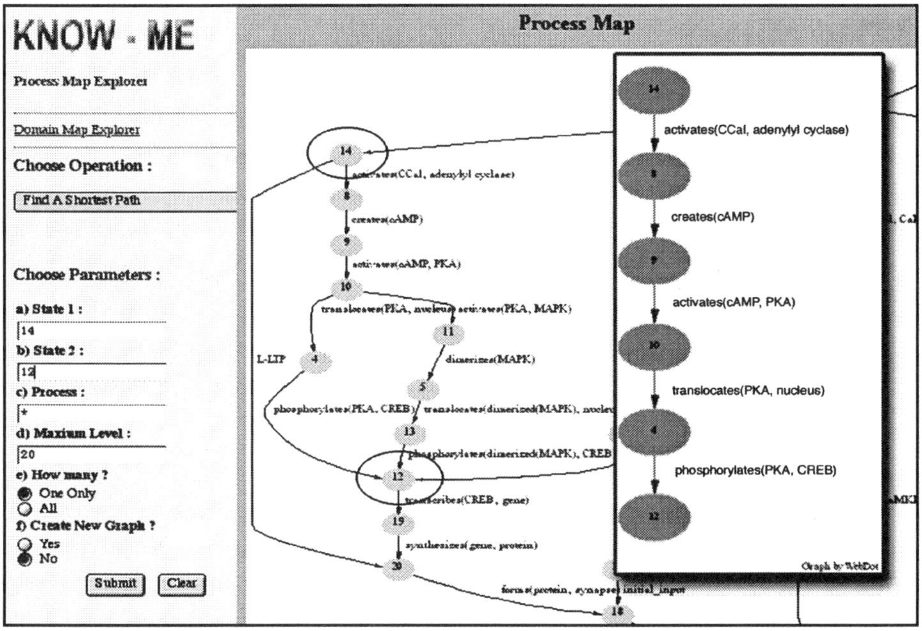

Fig. 10: The Know Me Tool: Part of the process map for long term potentiation is shown in the main window. Each step in the process is identified by a number while the edges are labeled with the process name. In the example shown , the "compute shortest pathway" function was used to determine the shortest pathway between step 14 and12.

Informatics Research Network (http://www.birn.ncrr.gov/) or BIRN. The BIRN project will ultimately link together technology development resources supported by the National Center for Research Resources of the NIH into a virtual data grid, where data can be shared and accessed through a mediator-based system over Internet2. Initiatives like the BIRN will serve as test beds for creating the infrastructure necessary for scientists to adapt the scientific process to the new electronic forum enabled by network connectivity and high performance computing. The challenges involved in adapting the scientific enterprise for this global forum are many. They range from sociological issues involved in data sharing to issues of curation and validation to technical issues of interoperability and moving large data sets (Kotter, 2001; Toga, 2002). However, we should not let these challenges blind us to the advantages afforded by immediate access to data. New methods for visualizing, indexing and exploring data will likely help to accelerate the discovery process by identifying inconsistencies, controversies and knowledge gaps much more rapidly than is possible through the literature. As more data sources become linked through technologies such as database federation, we will be able to navigate through many levels of biological complexity and come closer to the goal of understanding biological systems in their entirety.

ACKNOWLEDGEMENTS

This work supported by NIH grants from NCRR RR04050, RR RR08605 and the Human Brain Project DC03192, an NSF grants supporting the National Partnership for Advanced Computational Infrastructure NSF-ASC 97-5249 and MCB-9728338. The authors would like to acknowledge the contributions of Ms. Xufei Qian, Shenglan Zhang, Mona Wong and Mr. Joshua Tran for their work on the tools presented here.

REFERENCES

Chicurel, M. (2000) Databasing the brain. *Nature,* 406, 822-825.

Dickson, J., Drury, H. and Van Essen, D. C. (2001) 'the surface management system' (sums) database: A surface-based database to aid cortical surface reconstruction, visualization and analysis. *Philos Trans R Soc Lond B Biol Sci,* 356, 1277-1292.

Fox, P. T. and Lancaster, J. L. (2002) Opinion: Mapping context and content: The brainmap model. *Nat Rev Neurosci,* 3, 319-321.

Gardner, D., Abato, M., Knuth, K. H., DeBellis, R. and Erde, S. M. (2001) Dynamic publication model for neurophysiology databases. *Philos Trans R Soc Lond B Biol Sci,* 356, 1229-1247.

Gupta, A., Ludaescher, B. and Martone, M. E. (2000) Knowledge-based integration of neuroscience data sources. *Proceedings of the 12th International Conference on Scientific and Statistical Database Management (SSDBM'00) IEEE Computer Society,*

Harlow, M. L., Ress, D., Stoschek, A., Marshall, R. M. and McMahan, U. J. (2001) The architecture of active zone material at the frog's neuromuscular junction. *Nature,* 409, 479-484.

Ingenerf, J., Reiner, J. and Seik, B. (2001) Standardized terminological services enabling semantic interoperability between distributed and heterogeneous systems. *Int J Med Inf,* 64, 223-240.

Ludaescher, B., Gupta, A. and Martone, M. E. (2001) Model-based mediation with domain maps. *Proceedings of the 17th International Conference on Data Engineering, IEEE Computer Society 2001,* Heidelberg, Germany, April 2-6, 2001.

Mannella, C. A., Marko, M. and Buttle, K. (1997) Reconsidering mitochondrial structure: New views of an old organelle. *Trends Biochem Sci,* 22, 37-38.

Martone, M. E., Gupta, A., Wong, M., Qian, X., Sosinsky, G., Ludäscher, B. and Ellisman, M. H. (in press) A cell-centered database for electron tomographic data. *J. Struct. Biol.,*

McEwen, B. F. and Frank, J. (2001) Electron tomographic and other approaches for imaging molecular machines. *Curr Opin Neurobiol,* 11, 594-600.

Miller, P. L., Nadkarni, P., Singer, M., Marenco, L., Hines, M. and Shepherd, G. (2001) Integration of multidisciplinary sensory data: A pilot model of the human brain project approach. *J Am Med Inform Assoc,* 8, 34-48.

Paxinos, G. and Watson, C. (1998) *The rat brain in stereotaxic coordinates,* Academic Press, San Diego.

Perkins, G., Renken, C., Martone, M. E., Young, S. J., Ellisman, M. and Frey, T. (1997) Electron tomography of neuronal mitochondria: Three-dimensional structure and organization of cristae and membrane contacts. *J Struct Biol,* 119, 260-272.

Persidis, A. (1999) Bioinformatics. *Nat Biotechnol,* 17, 828-830.

Persidis, A. (2000) Data mining in biotechnology. *Nat Biotechnol,* 18, 237-238.

Rachedi, A., Rebhan, M. and Xue, H. (2000) Gabaagent: A system for integrating data on gaba receptors. *Bioinformatics,* 16, 301-312.

Shepherd, G. M., Mirsky, J. S., Healy, M. D., Singer, M. S., Skoufos, E., Hines, M. S., Nadkarni, P. M. and Miller, P. L. (1998) The human brain project: Neuroinformatics tools for integrating, searching and modeling multidisciplinary neuroscience data. *Trends Neurosci,* 21, 460-468.

Toga, A. (2002) Neuroimage databases: The good, the bad and the ugly. *Nat Rev Neurosci,* 3, 302-308.

Van Essen, D. C., Drury, H. A., Dickson, J., Harwell, J., Hanlon, D. and Anderson, C. H. (2001) An integrated software suite for surface-based analyses of cerebral cortex. *J Am Med Inform Assoc,* 8, 443-459.

Wong, S. T. and Koslow, S. H. (2001) Human brain program research progress in biomedical imaging/neuroscience, 2001. *J Am Med Inform Assoc,* 8, 510-511.

Chapter 20

Facilitating Data and Software Sharing in the Neurosciences – a Neuroinformatics Portal

Raphael Ritz, Rainer Förster, and Andreas V.M. Herz
Institute for Theoretical Biology, Humboldt–Universität zu Berlin
Invalidenstraße 43, D–10115 Berlin, Germany
Correspondence to: ritz,rfoerster,herz@itb.biologie.hu-berlin.de

Abstract: Following the open source philosophy, more and more neuroscientists are willing to share their primary data as well as custom software with the scientific community. To facilitate this interaction, we are establishing websites that link publicly available resources (such as experimental data, data analysis tools, computer models) and contain extensive annotation. Special emphasis is given to high quality standards of the linked resources and to integrate both users and providers in this process. Building on open source software tools, we therefore implemented a web–based content management system including a reviewing process to enable neuroscientists to contribute to the underlying metadata database. We also developed a neuroscientific classification system. Currently we are exploring the potential benefits of standardized usage of metadata and of the evolving field of web services. This project is an integral part of the international efforts coordinated by the European Union and the OECD to foster the developing field of computational neuroscience and neuroinformatics.

Key words: metadata database, dynamic website, interoperability, classification, neuroscientific ontology, web–publishing, Zope

1. BACKGROUND AND MOTIVATION

Novel experimental and computational techniques have led to major transitions in the neurosciences, all the way from the molecular to the system level. At the same time, more and more scientists now share their experimental data, analysis tools and computer models and have thus started a new research culture reflecting the open source philosophy. However, many of the data and computer programs already publicly available are not known to the general neuroscience community and information is often difficult to locate.

The Working Group on Neuroinformatics of the OECD Megascience Forum has identified this major deficit in a recent progress report (OECD, 1999). To overcome the problem, the Working Group has issued a proposal to create *Neuroinformatics Portals*, i.e., internet based global knowledge management systems for all data relating to nervous system structure and function. In November 2000 the German Federal Ministry of Education and Research started a three year pilot project to help in jump–starting portals. This project is hosted by the Institute for Theoretical Biology at the Humboldt–Universität zu Berlin.

Already in 1999, the European Union established the *Thematic Network Computational Neuroscience and Neuroinformatics*. The primary objective of the thematic network is to promote the fields of computational neuroscience and neuroinformatics and to start building a European neuroinformatics infrastructure. One of the initiatives within the thematic network was to establish a website that contains a database of ongoing activities and research groups in the fields of neuroinformatics and computational neuroscience. This project is hosted by the Laboratory of Theoretical Neurobiology at the University of Antwerp and started the website `http://www.neuroinf.org`. It provides the international neuroinformatics community with information and access to courses and workshops and manages the `comp-neuro` mailing list.

2. TASK

Our long–term goal is to contribute to a global internet portal for the entire field of neuroscience with a particular emphasis on facilitating the exchange of data and software but also providing various other kinds of information, such as *who is doing what and where* or services like news, bulletin boards and threaded discussions. The current state of our project can be checked at `http://www.neuroinf.de`. A screen shot of the homepage and a more detailed site map are shown in Fig. 1 and 2, respectively.

We do not plan, however, to serve as a primary data repository where people can upload and publish raw data sets for instance. All we are interested in is the collection and dissemination of so–called metadata, meaning structured data about data. People can use the portal to advertize their resources which are already available somewhere else on the internet. The potential benefit of doing so is to become more visible to the community.

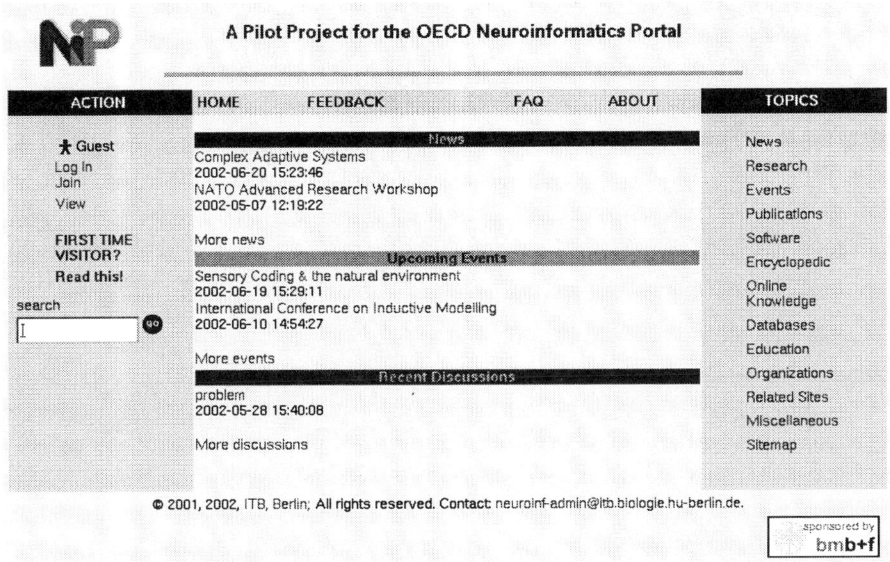

Figure 1. Screenshot of our homepage at `http://www.neuroinf.de` illustrating the general layout of the site. The menu on the left–hand side displays all possible actions currently available to the user. These actions depend on the role of the user (guest, member, reviewer, manager, etc.) as well as on the content type displayed (overview or any particular item) and the current state of the displayed object (private, pending review, or published).

3. CHALLENGES

All major community sites on the internet are facing the following challenges:

1. How to get all the relevant information into the underlying database and how to get it updated?

2. How to make the website reflect this information in a consistent manner?

3. How to assure a certain quality?

4. How to structure the site and allow for rapid and goal–directed searches?

In the long run we also need to address additional issues:

5. How to establish interoperability with other websites?

6. How to provide computational services through the web?

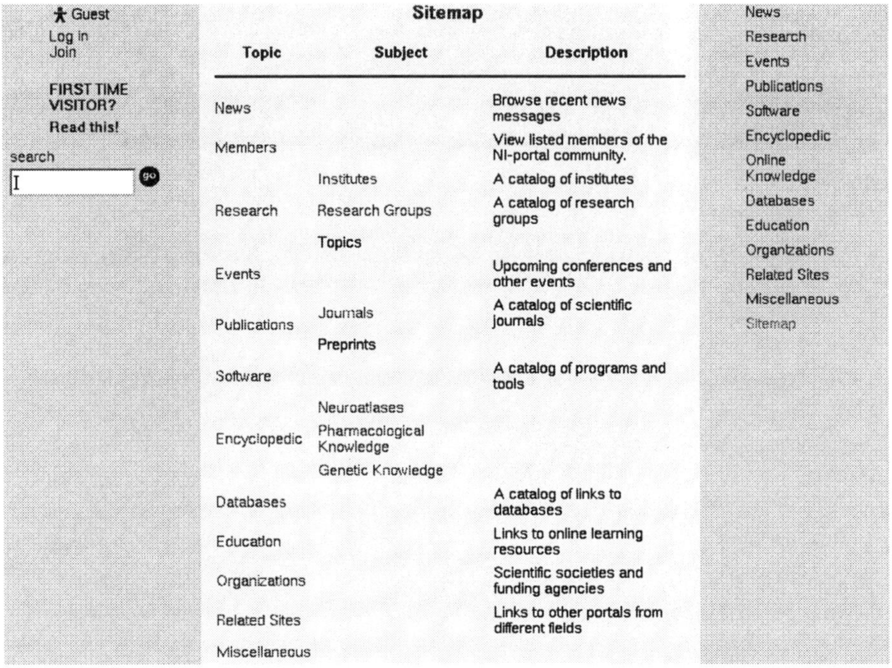

Figure 2. Sitemap illustrating some of the contents currently covered by `neuroinf.de`.

None of these problems can be solved by the traditional static HTML–programming approach as no one person or team is able to keep up with all the additions and changes necessary once a site has grown sufficiently. Solving these problems requires a well–designed structure of the portal and efficient organization of data manipulation. The following section describes some of the approaches that we have taken or are currently considering.

4. SOLUTIONS

4.1 Getting the Data: Community Involvement

The best way to get at all the information necessary for making the site useful is to get the community involved. Ideally, everybody should be able to provide whatever information he or she wants to share, including individual information (who is doing what where) as well as information about software tools, data sets or any other topics of potential interest to the community. If the person who provides a certain piece of information can also be made responsible for keeping it up–to–date, chances are high

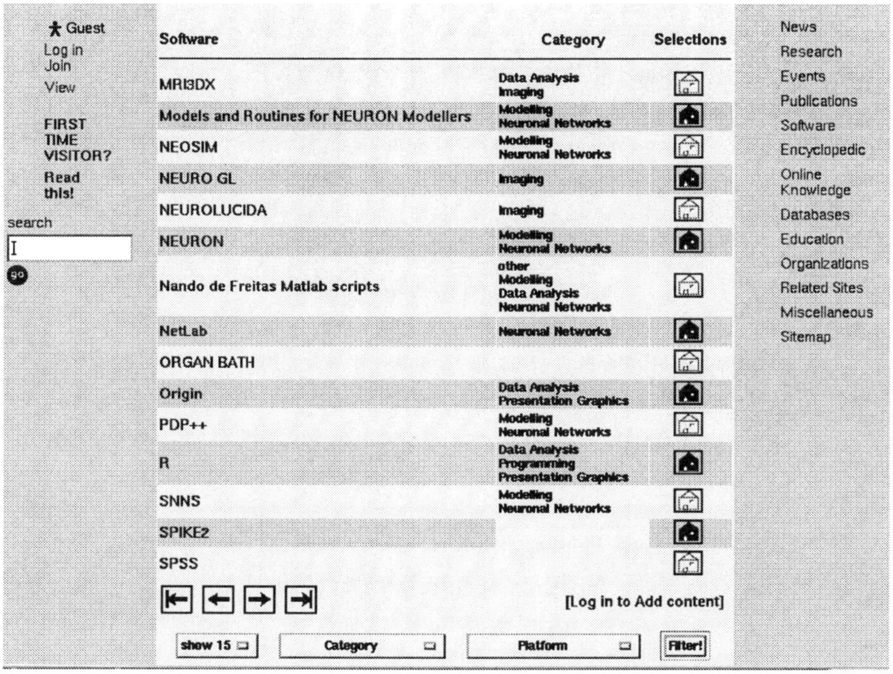

Figure 3. Extract from the listing of software tools. If needed, the user can restrict the items shown to particular categories or platforms. Clicking on an entry provides more detailed information as shown in the next figure.

that the site will not be outdated soon. In addition, we are taking efforts to deploy automatic techniques to detect if there are updates on the original sites.

4.2 Up–to–date Pages: a Dynamic Site

Having the users provide information is one important issue, having the actual display of the site reflecting the *current knowledge* is another. To achieve this, the site has to be dynamic, which means that all the different pieces of information need to be stored and updated in a database. Most web pages of our site are therefore not hard–coded using HTML but – on request – they are dynamically generated from the database. Figure 3 shows as a particular example of such a dynamic page a listing of software tools. Clicking on an item then reveals more detailed information about the particular resource (Fig. 4).

With respect to time–critical issues such as the announcement of events this approach allows for an automatic removal of entries from an

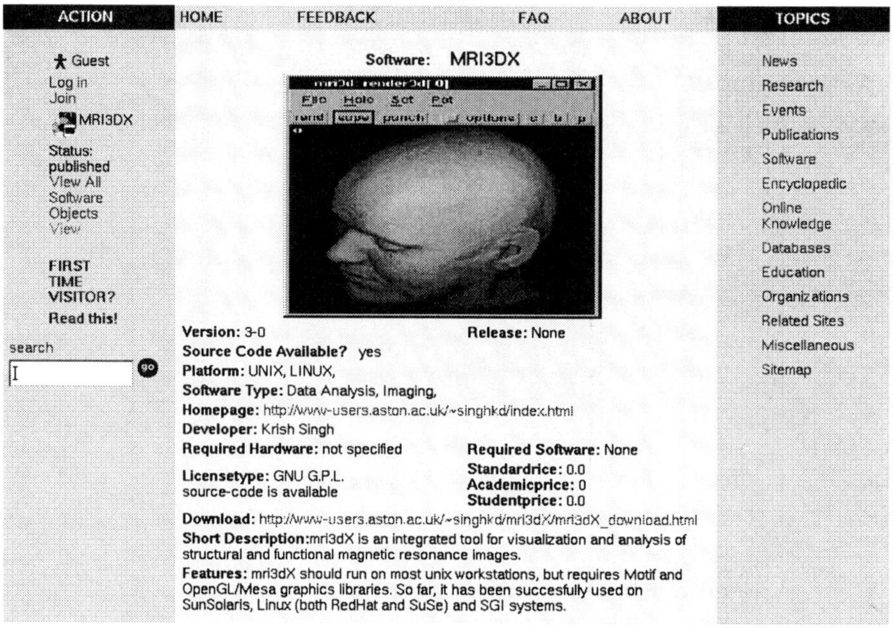

Figure 4. Detailed view of a particular software item.

Upcoming Events list, as soon as they have passed. Furthermore, the user can define an *Effective Date* as well as an *Expiration Date* for each individual object (through following the link EDIT ALL METADATA) that way controlling himself the period of time during with the item will be displayed via the portal.

4.3 Quality Control: Define Workflow

If everyone were able to provide content without any editorial control, the site might easily be overwhelmed by irrelevant or inflammatory items. There needs to be a way of assuring a certain quality of the information published. We therefore implemented a web–based review process very similar in spirit to the traditional scientific review process – but it is not anonymous.

In practice this means that we first want to know the identity of a user before he is allowed to enter anything into our site. Therefore one needs to register before submitting data. The various content types currently available to our members are shown in Fig. 5 and the particular interfaces for software tools can be seen in Fig. 6. Any data submitted for view will be checked by one of the site reviewers or managers for consistency and

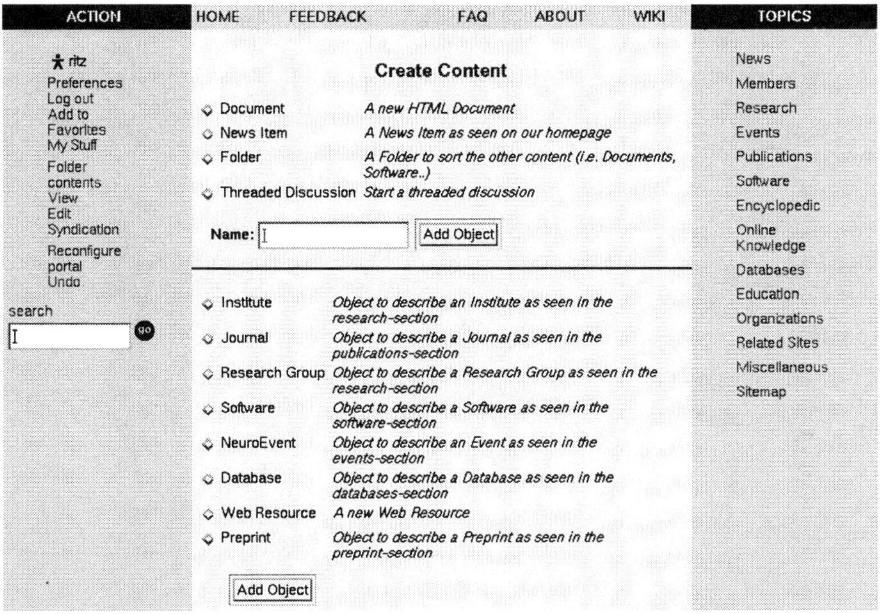

Figure 5. Content types that our members currently can create. To reach this interface one needs to first log in to our site. Then within the members home directory (accessible via the MY STUFF action) clicking on NEW prompts the options shown.

completeness before it is made public. In addition, every published item can be commented on just by clicking REPLY on the left–hand side. This means that we are not peer–reviewing the proposed resource according to scientific standards. This is left to the community by commenting on it. We are just checking the (meta-)data provided through the user interfaces of the portal.

4.4 Structuring the Site: Classification

The fourth challenge listed above shall be overcome by appropriate classification of each item at the time of entry (i.e., by the member who submits it). The most obvious way of classifying, namely through a number of freely chosen keywords, would not work because it would not allow to automatically group and relate different entries. We therefore implemented a predefined classification scheme based on a controlled vocabulary. The scheme consists of six hierarchically organized categories to select the species, the anatomical structure, the neural system, the phenomenon or behavior under study, the experimental condition, and the method used. Through selection from each menu a string of six key-

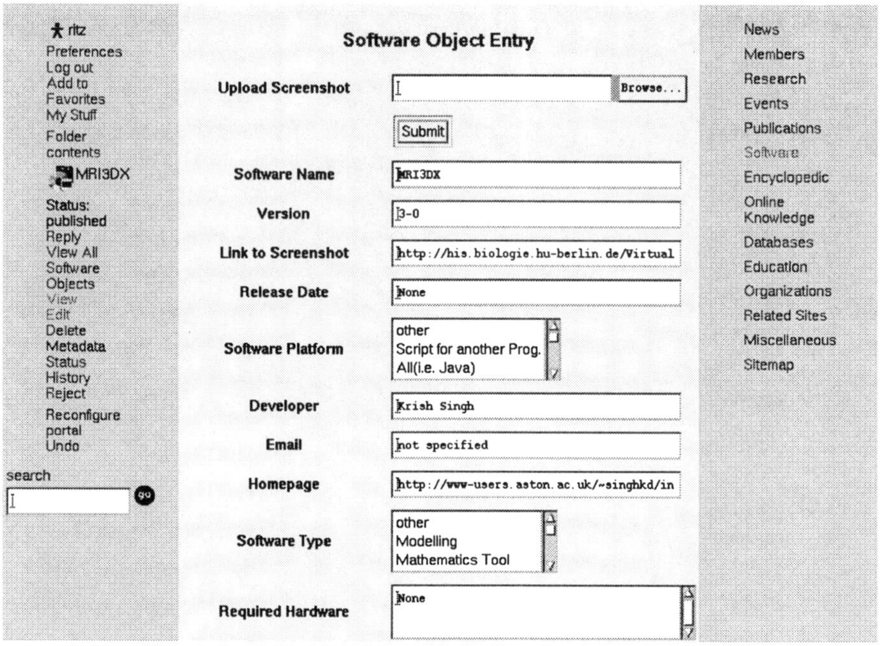

Figure 6. All content items a user has generated can be edited at any time. Here, the interface for editing the software item displayed previously is shown.

words gets constructed. This is the basic classification step. Each item can be classified using multiple such strings. Based on this scheme it is now possible to selectively group the content either for specific listings within the portal directly or in response to a (guided) search of the site.

4.5 Interoperability I: Expose Metadata

At the heart of automated data sharing between different web sites is the consequent and standardized usage of structured data about data. Such data about data are also called metadata. In 1994 a group of experts on this issue met in Dublin, Ohio and developed what is now referred to as the *Dublin Core*: a simple yet effective element set for describing a wide range of networked resources. The Dublin Core standard comprises fifteen elements, the semantics of which have been established through consensus by an international, cross–disciplinary group of professionals from librarianship, computer science, text encoding, the museum community, and other related fields of scholarship.

The Dublin Core element set is outlined in Table 1. Each element is optional and may be repeated. Each element also has a limited set of

Table 1. Dublin Core Metadata Element Set

Name	*Definition*
Title	A name given to the resource.
Creator	An entity primarily responsible for making the content of the resource
Subject	The topic of content of the resource.
Description	An account of the content of the resource.
Publisher	An entity for making the resource available.
Contributor	An entity responsible for making contributions to the content of the resource.
Date	A date associated with an event in the life cycle of the resource.
Type	The nature or genre of the content of the resource.
Format	The physical or digital manifestation of the resource.
Identifier	An unambiguous reference to the resource within a given context.
Source	A reference to a resource from which the present resource is derived.
Language	A language of the intellectual content of the resource.
Relation	A reference to a related resource.
Coverage	The extent or scope of the content of the resource.
Rights	Information about rights held in and over the resource.

qualifiers, attributes that may be used to further refine (not extend) the meaning of the element. The Dublin Core Metadata Initiative (DCMI) has defined standard ways to *qualify* elements with various types of qualifiers. A set of recommended qualifiers conforming to DCMI *best practice* is available, with a formal registry in process.

One crucial step in enabling interoperability between different web sites is a standardized way to expose and access the metadata. The *Open Archives Initiative* (OAI) developed the *Metadata Harvesting Protocol* (MHP) for this purpose. Generally one distinguishes between *data providers* who administer systems that support the OAI protocol as a means of exposing metadata about the content of their systems and *service providers* who issue OAI protocol requests to the systems of data providers and use the returned metadata as a basis for building value–added services.

From a more general perspective this is just one particular example of a *web service* which will be introduced in the following section.

4.6 Interoperability II: Web Services

An important aspect of the development of the World Wide Web is the integration of computational capabilities (software and CPU time) into the web. Such capabilities are usually referred to as *Web Services*. Web services provide a standard for interfaces between applications and content services on the internet. The key design principle is that one web service can be called from another web service to use the methods provided by the first as if it was a locally integrated module even though the services may reside at a different site and be implemented on a different software platform (operating system, programming language, etc.). This scheme works because the interface to each and every service is designed according to a well–defined standard enabling the service to be called from a third party without knowing the underlying software infrastructure.

This implies a general solution to the problem of integrating different (even proprietary) systems. The potential impact of this new distributed computing paradigm is huge and might drastically change the way we will do computations in the future.

5. IMPLEMENTATION

5.1 Dynamic Site

The site is implemented using open source and original, self–developed software. We use the web–application server Zope together with its content management framework CMF behind an Apache web server. Zope and all its components are programmed in Python except for a few performance critical parts which are implemented using C.

The databases used are Zope's own Zope Object DataBase (ZODB), a persistent and transactional object database that can also be used on its own and a relational database where appropriate.

To improve the general accessibility we plan to globally mirror the site. The mirror servers will be synchronized using Zope's Enterprise Objects (ZEO).

5.2 Exposing Metadata

Metadata encoding based on the Dublin Core can be done in several different syntaxes, including HTML and RDF/XML. The Resource Description Framework (RDF) allows multiple metadata schemes to be read by humans as well as parsed by machines. It uses the eXtensible Markup Language (XML) to express structure thereby allowing metadata communities to define the actual semantics. This decentralized approach reflects the fact that no one scheme is appropriate for all situations, and further that schemes need a linking mechanism independent of a central authority to aid description, identification, understanding, usability, and/or exchange.

RDF allows multiple objects to be described without specifying the detail required. The underlying glue, XML, simply requires that all name spaces be defined. Once defined, they can be used to the extent needed by the provider of the metadata.

For example, the following code fragment could describe a users guide to our portal.

```
<rdf:RDF xmlns:rdf="http://www.w3.org/1999/02/22-rdf-syntax-ns#"
         xmlns:dc="http://purl.org/dc/elements/1.1/">

  <rdf:Description rdf:about="http://www.somewhere.de">
      <dc:creator>Raphael Ritz</dc:creator>
      <dc:title>Internetplattform Neuroinformatik</dc:title>
      <dc:description>Users guide for the neuroinformatics portal.
      </dc:description>
      <dc:date>2002-01-24</dc:date>
  </rdf:Description>
</rdf:RDF>
```

Using XML and RDF, the Dublin Core can now be mixed with other metadata vocabularies. For example, the simple Dublin Core description above might be used together with other vocabularies such as vCard that can describe the author's affiliation and contact information, or a more specialized *neuroinformatics* vocabulary that describes the scientific scope in greater detail. Specifically, we are considering to use NeuroNames (this volume, chapter 18) for encoding anatomical information, NeuroML (Goddard et al., 2001) for specifying models, or BrainMetaL (Gardner et al., 2001) for more general descriptions.

The key point for the user of our site now is that she or he does not need to know anything about the technical issues of encoding the resource appropriately but that this is generated automatically through the registration at the portal.

5.3 Web Service Standards

The success of the internet is mainly based on standardized protocols (rather than any particular implementations) enabling different technologies to communicate with each other.

To enable the integration of different applications, web service standards have been developed (by Microsoft, IBM, Sun and others in the XML standard protocol commission) and have been submitted to the W3C for recommendation.

The base protocol is called Simple Object Access Protocol (SOAP). It enables the core transmission of objects between different sites over HTTP or SMTP (i.e., the web or e–mail). It is supported by all major software vendors and the open source community, and there are implementations for Java, Python, Apache, WebSphere, Visual Basic, etc. already available.

To standardize the formal description of interfaces the Web Services Description Language (WSDL) has been developed. The description has to include all the information necessary (supported protocols, address and port number, procedures and functions defined including their supported input and output formats).

To combine different web services in one work flow, the Web Services Flow Language (WSFL) has been developed. The WSFL enables developers to easily create, execute, and combine web services into complex applications.

To find a particular web service, a registration system called Universal Discovery, Description, and Integration (UDDI) has been established. The UDDI directory can be accessed manually (by software developers) as well as by applications at runtime.

We hope to be able to use these developing standards in the future to further enhance the usefulness of the portal. Direct applications will be the automated collection of (meta-)data from other online repositories and an improved accessibility of the portal's data from third parties.

6. OUTLOOK

Given the motivation described at the beginning of this article the portal introduced here can only be a first step toward the more ambitious goal of inducing a paradigm shift within the neurosciences toward data and software sharing.

While changing the attitude many scientists have with respect to *their* data is partly a cultural challenge (Koslow, 2000) there are also technical challenges (Toga, 2002): To realize the potential of neuroinformatics to re–connect and integrate neuroscience researchers and their work, we need to implement a connected network of databases and tools that offers a systematic coverage of neuroscience, in all its genetic, molecular, cellular, local circuit, systems, and behavioral aspects, and in all the species of central interest for neuroscientists.

The portal currently helps in disseminating information about these data, databases and available software. But the real task on top of this is to make these databases and software tools interoperable. Ideally, it will be possible to seamlessly search through a large set of databases from the molecular to the organismic level and to apply the rich repertoire of analysis and modeling tools – but obviously there is still a long way to reach this goal.

Web Resources

Apache A free, open source web server
http://www.apache.org

CMF Zope's Content Management Framework
http://cmf.zope.org

DCMI Dublin Core Metadata Initiatitive
http://www.dublincore.org

MHP Metadata Harvesting Protocol; developed by the Open Archives Initiative
http://www.openarchives.org/OAI_protocol/openarchivesprotocol.html

OAI Open Archives Initiative
http://www.openarchives.org

Python A free, open source, object–oriented scripting language
http://www.python.org

RDF Resource Description Framework
http://www.w3.org/RDF

SOAP Simple Object Access Protocol
http://www.w3.org/TR/SOAP

UDDI Universal Discovery, Description, and Integration
http://www.uddi.org

vCard an electronic business card by the *Internet Mail Consortium*
http://www.imc.org

W3C the World Wide Web Consortium
http://www.w3.org

WSDL Web Services Description Language
http://www.w3.org/TR/wsdl

WSFL Web Services Flow Language
http://www-106.ibm.com/developerworks/webservices/library/ws-ref7

XML eXtensible Markup Language
http://www.w3.org/XML

ZEO Zope's Enterprise Objects; a replicated storage server http://www.zope.org/Products/ZEO.

ZODB A free, open source object database management system http://www.zope.org//Wikis/ZODB/StandaloneZODB

Zope A free, open source web–application server http://www.zope.org

Acknowledgments

We would like to thank Stefanie Hertel, Ira Mittmann, Stefan Kröger, Kristian Rother and Benjamin Ott for their contributions to set up `www.neuroinf.de`, Maryann Martone for helpful comments on the manuscript and the BMBF (German Federal Ministry for Education and Research) for continuing support.

References

Gardner, D., Knuth, K.H., Abato, M., Erde, S.M., White, T., DeBellis, R. and Gardner, E.P. (2001) Common data model for neuroscience data and data model exchange. *J. Am. Med. Inform. Assoc.* 8, 17–33.

Goddard, N.H., Hucka, M., Howell, F., Cornelis, H., Shankar, K. and Beeman, D. (2001) Towards NeuroML: model description methods for collaborative modelling in neuroscience. *Philos. Trans. R. Soc. Lond. B* 356, 1209–1228.

Koslow, S. H. (2000) Should the neuroscience community make a paradigm shift to sharing primary data? *Nature Neurosci.* 3, 863–865.

Organization for Economic Co–operation and Development (1999) Final report of the OECD Megascience Forum: working group on biological informatics. http://www.oecd.org/pdf/M000014000/M00014759.pdf

Toga, A. W. (2002) Neuroimage databases: The good, the bad and the ugly. *Nature Rev. Neurosci.* 3, 302–309.

Index